Emerging Conceptual, Ethical and Policy Issues in Bionanotechnology

Philosophy and Medicine

VOLUME 101

Founding Co-Editor
Stuart F. Spicker

Senior Editor

H. Tristram Engelhardt, Jr., *Department of Philosophy, Rice University, and Baylor College of Medicine, Houston, Texas*

Associate Editor

Lisa M. Rasmussen, *Department of Philosophy, University of North Carolina at Charlotte, Charlotte, North Carolina*

Editorial Board

George J. Agich, *Department of Philosophy, Bowling Green State University, Bowling Green, Ohio*
Nicholas Capaldi, *College of Business Administration, Loyola University, New Orleans, New Orleans, Louisiana*
Edmund Erde, *University of Medicine and Dentistry of New Jersey, Stratford, New Jersey*
Christopher Tollefsen, *Department of Philosophy, University of South Carolina, Columbia, South Carolina*
Kevin Wm. Wildes, S.J., *President Loyola University, New Orleans, New Orleans, Louisiana*

For other titles published in this series , go to
www.springer.com/series/6414

Fabrice Jotterand
Editor

Emerging Conceptual, Ethical and Policy Issues in Bionanotechnology

 Springer

Editor
Fabrice Jotterand
University of Texas
Southwestern Medical Center
Dallas, USA

ISBN 978-1-4020-8648-9   e-ISBN 978-1-4020-8649-6

Library of Congress Control Number: 2008928519

© 2008 Springer Science+Business Media B.V.
No part of this work may be reproduced, stored in a retrieval system, or transmitted in any form or by any means, electronic, mechanical, photocopying, microfilming, recording or otherwise, without written permission from the Publisher, with the exception of any material supplied specifically for the purpose of being entered and executed on a computer system, for exclusive use by the purchaser of the work.

Printed on acid-free paper

9 8 7 6 5 4 3 2 1

springer.com

Contents

Introduction

Beyond Feasibility: Why Ethics Is Important for Bionanotechnology 3
Fabrice Jotterand

Knowledge Production in Nanotechnoscience

The World View of Nanotechnology – Philosophical Reflections 13
Andreas Woyke

Nanomachine: Technological Concept or Metaphor? 27
Xavier Guchet and Bernadette Bensaude-Vincent

**No Future for Nanotechnology?
Historical Development *vs.* Global Expansion** 43
Alfred Nordmann

Ethics and (Bio)Nanotechnology

Bionanotechnology: A New Challenge for Ethical Reflection? 67
Christoph Baumgartner

Nanoparticles: Risk Management and the Precautionary Principle 85
Armin Grunwald

Anticipating the Unknown: The Ethics of Nanotechnology 103
Joseph C. Pitt

Applications of Nanotechnology in the Biomedical Sciences:
Small Materials, Big Impacts, and Unknown Consequences 117
Audy G. Whitman, Phelps J. Lambert, Ossie F. Dyson,
and Shaw M. Akula

Public Policy and (Bio)Nanotechnology

Nanobiotechnology and Ethics:
Converging Civil Society Discourses 133
Alexandra Plows and Michael Reinsborough

Allotropes of Fieldwork in Nanotechnology 157
Christopher Kelty

Law, Regulation and the Medical Use of Nanotechnology 181
Kenneth A. DeVille

Human Enhancement and (Bio)Nanotechnology

Stage Two Enhancements 203
George Khushf

Nanotechnology, the Body and the Mind 219
M. Ellen Mitchell

Nanotechnology and Human Flourishing:
Toward a Framework for Assessing Radical
Human Enhancements 239
Ronald Sandler

Author Index 255

Subject Index 261

Introduction

Beyond Feasibility: Why Ethics Is Important for Bionanotechnology

Fabrice Jotterand

Since Richard Feynman imagined arranging atoms one by one in 1959 to current research in nanoscience and (bio)nanotechnology, scientific progress continues to feed societal expectations. Increasingly, we read about new scientific developments in (bio)nanotechnology that could bring solutions to vexing problems in human existence. Radical extension of life expectancy, human-machine interface, and nano-devices for targeted drug delivery represent only few examples of the potential applications of nanotechnology. These emerging technologies have tremendous applications and could enhance the quality of life of many people. However, they also raise important social, ethical and legal/regulatory questions. With the pressure to pursue scientific progress in order to find solutions to pressing human problems, the dangers of unreflectively developing (bio)nanotechnology lurks around the corner.

Clearly we cannot stop scientific and technological progress. It is part of human nature to explore and conquer the infinitely big (space) and the infinitely small (nano-world). This quest for knowledge, however, is not limited to the understanding of novel properties of matter at the nanoscale. Ultimately the application of basic scientific knowledge to practical problems is always connected to human ends and goals. Hence, it is the *exploitation* of these novel phenomena at the nanoscale that requires a critical inquiry beyond the "imperative of feasibility". Without clearly enunciating the hidden assumptions and the metaphysics of the (bio)nanotechnology project, we run the risk of killing the project itself and missing the opportunity to harvest its potential benefits (nanoscale biostructures [artificial bones, tissue engineering and cell therapy]; new types of drugs, either based on the human genome or structural genomics or biomimetics; targeted drug delivery; nanobots [sensors to monitor body parameters such as pulse, blood pressure, etc.]; nanomaterials; various types of nano-devices [neuro-digital interfaces])

George Santayana warns us that "those who ignore history are destined to repeat it". Recent history from the GMO (Genetically Modified Organisms) experiment taught us that "feasibility" is not necessarily synonymous with commercial success and public acceptance of new technologies. Despite the potential benefits of GMOs (crops more resistant to diseases and climate; more productive farm animals; more productive lands, production of vaccines and medicines, etc.) the public is concerned

---

University of Texas at Dallas, UT Southwestern Medical Center, Dallas, TX

about GMOs, especially in the European context. The reasons are multiple. There is a concern about the safety and the possible unintended consequences (toxicity, Frankenfood, etc.) of genetically modified organisms once they are introduced in the food chain. Another concern is about the environment, in which genetic mutation could spread and would result in the tampering with nature ("Playing God" argument). Finally there is the question of labelling (label indicating whether the product was genetically modified), which is not required in some countries, including the United States. However, some activists argue that consumers have the right to know the process of fabrication and the ingredients of the products they buy. In addition, they contend that the public should be more involved in debates concerning the development of new products. This is not to say that the public has not been part of the discussion. As Alexandra Plows and Michael Reinsborough remarks in their contribution to this volume, "the necessity for better public engagement, are not new frames at all, but are increasingly louder, and perhaps better heard…." However, they also point out that the current emphasis is on the necessity for "upstream public engagement" which means an early dialogue between the scientific community and relevant parties.

The lack of early public involvement has led to bad PR on the part of the scientific community and created fear in the mind of many citizens. Ultimately, this situation led to an economic and scientific failure (in the sense of the marketability of the products of scientific development). In short, the lack of transparency on the part of the scientific community and the failure to involve the public "upstream" produced fear and mistrust. What seems imperative in order to avoid the same situation in relation to bionanotechnology is to put forward publicly a plan for a responsive development of nanotechnology. As Mihail C. Roco and William Sims Bainbridge from the National Science Foundation (NSF) point out, "proper attention to the ethical issues and societal needs" can result in the beneficial outcomes for the industry, society and human life. If well informed, the public and the politicians will be willing to finance and support new technologies. Therefore, it seems essential to have an open debate that includes scientists, politicians, the public, etc. early on with regards to the ethical, legal and social implications of bionanotechnology so as to avoid past mistakes.

The Structure of the Volume

This volume provides a critical overview of the nature of nanotechnology (and its applications in the biomedical sciences, i.e. bionanotechnology) and the philosophical and ethico-legal issues it raises. This collection of 13 articles represents an exploration by scholars from various disciplines (philosophy, anthropology, law, social sciences, psychology, and natural sciences) in North America and Europe. The book contains four major parts respectively entitled (1) Knowledge Production in Nanotechnoscience; (2) Ethics and (Bio)Nanotechnology; (3) Public Policy and (Bio)Nanotechnology; and (4) Human Enhancement and (Bio)Nanotechnology. In the first section, authors examine the nature of nanotechnology as a scientific project and critically reflect on its philosophical underpinnings. The next section introduces the readers to a new area of investigation that explicitly addresses the ethics of nanotechnology/bionanotechnology. More specifically, it examines the theoretical framework(s) necessary to sustain rich ethical reflections at the core of the development of nanotechnology. The third section expands on the ethics of nanotechnology/bionanotechnology but focuses on legal and public policy issues and how the public perception of nanotechnology could ultimately shape policies and regulations. Ultimately these three perspective (the nature of nanotechnology, ethical approaches and regulatory issues) will shape and frame the discourse on nanobiotechnology. The final section focuses on how scientific progress could affect humans through enhancement technologies and critically assesses whether such progress actually contributes to human flourishing.

It is important to recognize that other areas (military, economics, politics, etc.) deserve careful and critical examination. However, a comprehensive approach that would include such analysis is beyond the scope of this volume. The focus of this work is on the application of nanotechnology in the biomedical sciences.

Knowledge Production in Nanotechnoscience

In the first article, Andreas Woyke shows how an "intensification of a technocratic view of nature" influences our present understandings of science and the world. He argues that nanotechnology is not just a step forward in the development of science and technology. It is a scientific project with deep philosophical and metaphysical assumptions, which lead to a new understanding of science itself. Furthermore, in this new order of things, Woyke shows how human beings are absent, relegated to a subsidiary role. The removal of human agency is problematic because it provides a one-sidedness to the new technological paradigm. Humans are no longer controlling technology but are rather spectators of its development. The result is a new world view in which ethical, social and ecological issues are ignored. In response, Woyke suggests that our reflections should move toward a holistic philosophy of nature and an integrative anthropology. In so doing, he contends, we set the necessary framework for a responsible development of nanotechnology.

The essay by Bernadette Bensaude-Vincent and Xavier Guchet aims at clarifying the concept of technology and the "machine metaphor" in relation to nanotechnology and critically looks at Eric Drexler's concept of molecular assemblers. The authors first examine the differences between classical machines (Cartesian automata) and complex machines (in Von Neumann's sense). One of the main differences is that in classical machines the constructor controls each part and determines their specific aims. On the other hand, complex machines are autonomous because their behaviour is not uniquely determined by the designer. To a certain degree complex machines are unpredictable and escape human control.

Against this background Bensaude-Vincent and Guchet ask whether Drexler's molecular assemblers fall under the Cartesian/classical or complex machine category. In their final analysis, they critique Drexler's concept of molecular assemblers on the basis that he applies the characteristics of macro-machines to nanomachines. In addition they point out that the chemical and physical properties (stickiness, viscosity, and Brownian motion) at the nanoscale are different from the ones at the macro level. Hence, Drexler's model is intrinsically inadequate because inapplicable. More importantly they note that Drexler uses the two notions of machines, that is, nanobots operating like industrial robot (Cartesian/classical machines) and nanobots as self-replicators (complex machines) without clearly outlining why his methods make molecular assemblers complex machines (self-replicator nanobots).

While recognizing the problem related to Drexler's vision of molecular assemblers, Bensaude-Vincent and Guchet do not reject the idea of bottom-up fabrication. Molecular assemblers, they argue, will be possible only in so far as they emulate "the design principles of nature". This means that engineering at the nanoscale can only come out of biology. Bensaude-Vincent and Guchet suggest that a plausible alternative to Drexler's molecular assemblers is Jones' concept of "soft machines" (nanomachines designed according to biological principles). The reason is that the physical and chemical properties (Brownian motion, stickiness, and viscosity) at the nanoscale are the necessary milieu for soft machines to operate but a barrier to Drexler's molecular assemblers.

The final article of this section looks at the future(s) of nanotechnology. Bluntly, Alfred Nordmann argues that we cannot predict its future. All we can say is that it is "a horizon of expectation in which something unheard-of or unspeakable will appear". While other scholars envision various futures for nanotechnology (Armin Grunwald: prudence; Jean-Pierre Dupuy: catastrophisme éclaire; and George Khushf: hope), Nordmann thinks we should simply abandon any speculative reflection on its future. Our analysis (whether at the political, environmental or ethical levels) should not focus on future generations or future developments (what he calls "nanotechnology's expansion in historical time"), but should favour nanotechnology's "territorial conquest or expansion into space." In so doing, Nordmann thinks that a critical stance towards nanotechnology would frame our reflections in terms of the particular "claims made upon our bodies and the environment that structures our actions." This, he recognizes, is an outrageous idea but has already been achieved in studies on technosciences, which are concerned with the acquisition of present capabilities to solve technical problems.

The Ethics and (Bio)Nanotechnology

The second chapter examines the ethical implications of nanotechnology and the type of theoretical framework necessary to sustain a meaningful ethical analysis. Interestingly, in the previous sections Woyke and Nordmann partially refers to moral concerns about nanotechnology. Woyke stresses the necessity to integrate anthropological considerations in the development of nanotechnology. Nordmann's framework for a critical stand towards nanotechnology likewise includes considerations about our bodies and the environment. What this suggests is that the production of knowledge does not occur in a moral vacuum, but inherently assumes particular values. Hence the importance of ethics in science and technology.

In the first essay, Christoph Baumgartner asks whether the emergence of nanotechnology raises novel and unique ethical issues, and whether we need a comprehensive analysis of these issues. In his analysis, Baumgartner clearly answers negatively. The field of nanotechnology does not justify the creation of a new discipline called "nanoethics." He points out that all major issues raised by nanotechnology are characteristic of other areas of biotechnology and, therefore, nanotechnology is only part of a continuum of reflections that began with other types of technological and scientific advancements. However, because of the *enabling* and *transformative* nature of nanotechnology, Baumgartner contends that we need a comprehensive assessment of its ethical implications. In addition, he points out that our critical investigation should go "beyond the methodological and conceptual framework of biomedical ethics". The merging of biotechnology with the environment and human nature requires that our ethical reflections encompass biomedical ethics, social ethics and environmental ethics.

The precautionary principle is often invoked as a tool to assess risks of emerging technologies. In his essay, Armin Grunwald addresses the question of whether we should use the precautionary principle as a tool for risk management analysis strategies in relation to nanotechnology. After providing an overview of the potential risks associated with nanotechnology and an introduction into the precautionary principle, Grunwald critically assesses its applicability with regards to nanotechnology. He contends that classical risk assessment (i.e., the precautionary principle) is not applicable to nanotechnology. The application of the precautionary principle requires strong scientific evidence (as opposed to the *possibility* of hazards) of the hazardous impact of nanoparticles. Such data is still missing because nanoparticles and nanomaterials have different properties at the nanoscale and further risk assessment is required. For Grunwald, unless we gather the necessary data concerning the potential impact of nanoparticles on the environment and human health the precautionary principle is inadequate for risk assessment.

The third essay, by Joseph Pitt, is more ambitious because its author asserts that pragmatism is the only moral theory helpful with regards to the ethics of (bio)nanotechnology. The reason is that the development of any technology is always in relation to the improvement of the human condition. We develop certain technologies with specific practical problems in mind that hinder our ability to fully experience the Good Life. Hence, Pitt argues that the integration of practical concerns within the framework of the Good Life can only take place with reference to pragmatism. It is the only moral theory that allows a techno-ethical vision of human life.

In the final article of this section, Audy G. Whitman, Phelps J. Lambert, Ossie F. Dyson and Shaw M. Akula provide reflections from scientists working on the application of nanotechnology in medicine. Its authors point out that nanotechnology holds many promising applications that could change and improve our quality of life, but warn us that without proper precautions nanotechnology could become destructive to humans and their environment. They likewise support a responsible development of nanotechnology that is open to public scrutiny, honest and does not take shortcuts as to its ethical implications. They point out that without adequate reflections on its ethical ramifications in the biomedical sciences, and a strong commitment to take seriously the responsibilities that come with such scientific endeavour, "nanotechnology may prove to become detrimental."

Public Policy and (Bio)Nanotechnology

Earlier I pointed out that one of the major causes for rejecting GMOs was the lack of transparency and public involvement. The third chapter looks specifically at the public perception and the regulation of (bio)nanotechnology. The first essay, by Alexandra Plows and Michael Reinsborough addresses the issue of public discourse in the face of the emergence of technological uncertainties. The birth of the field of nanotechnology created a specific "epistemic culture" among scientists, which finds its counterpart in another "epistemic culture" oriented toward criticizing it – what Plows and Reinsborough call "a critique of the politics of technology". Their essay traces the emergence of these counter movements, particularly within the European context.

Interestingly, Plows and Reinsborough remark that these movements are not anti-technology per se. Many activists are aware of the potential benefits of converging technologies, but they also recognize the danger of letting the industry and corporations have sole control over these technologies. One might question the validity of the practices or motives behind the actions of activists yet these counter movements point to an important issue: civil society should have a more important role ("upstream public engagement") on the regulation of emerging technologies, which is exactly what the promoters of GMOs failed to do.

In the next article, Christopher Kelty examines the distinctive contribution anthropology (especially socio-cultural anthropology) could make to the study of nanotechnology, and analyzes recent research done by anthropologists on the subject of nanotechnology, the environment and human health. In particular the author contends that anthropology can make an important contribution because scientific and technological development necessarily includes human concerns (human safety, values, etc.)

Finally, Kenneth A. DeVille looks at the regulation of bionanotechnology and the current legal system. In particular, the author questions the ability of the current legal system to regulate bionanotechnology. While DeVille recognizes that existing laws provide enough information to ensure safe utilization of nanotechnology in the biomedical sciences, he also argues that additional regulatory safeguards will be necessary. For instance, in the United States, drugs and devices are approved and regulated by the Food and Drug Administration (FDA). However, "medical nano-products" are neither drugs nor devices but the merging of the two and therefore stand outside the FDA jurisdiction. Other issues that will require further regulations include the adjustment of the period for assessing limitations and risks of "nanoremedies" due to the nature of nanoparticles (size and mobility in the human body) and tort litigation. In the later case, although nanotechnology will raise complex and novel issues, most cases will use the current doctrines of negligence (medical malpractice) and product liability. However, how a plaintiff will demonstrate that a physician did not use the standard of care and caused damages by negligence will be more complicated in relation to nanotechnology. Due to the innovative dimension of nanotechnology, the idea of standard of care will change drastically. In short, innovative technologies will require the incorporation of new bodies of laws into current regulations so as to reflect the transformative dimension of nanotechnology.

Human Enhancement and (Bio)Nanotechnology

Potential changes are not limited to legal/regulatory aspects. Ultimately transformation could occur within our own bodies. Among the most radical applications of nanotechnology, human enhancement represents the most controversial issue. While the idea of enhancing human capacities is not novel per se, current technology allows us to enter the Cyborg age. One can predict that in the next few years we will witness an increasing number of experiments involving human subjects and machines (human-machines interface, brain – computer interface). As such, the essay by George Khushf anticipates these new developments in science and technology. He draws a distinction between early stages of the enhancement debate and the current stage. The latter is different in kind (at the core of human existence and its environment) and poses various challenges unknown so far. Because of the particularity of these challenges, which call into question previous ethical and policy debate on emerging technologies, society at large should foster new "cultural and intellectual resources" for our reflections on these issues.

The next paper, by M. Ellen Mitchell, explores the psychological implications of a spectrum of real and potential nano-biomedical advances, with a particular emphasis on the area of enhancement. She reviews various biomedical applications in the past and present advances in nanotechnology. In her analysis, she notices a profound difference between past and present technological developments. Contrary to past concerns about the use and misuse of technology, current analyses should focus on the "intended purposes, the pervasiveness, and the invasiveness" of technology.

Mitchell recognizes the transformative power of current technology. Hence, she argues that our efforts should primarily be on the impact of technology on humans (identity and functionality) and their environment, rather than potential misuse. Specifically she reviews three specific areas relevant to the discussion on human enhancement: (1) executive functioning (higher order brain activities such as judgment, attention, planning, etc.); (2) self-regulation (bodily functions, mood, behaviour, thought, feeling, or violence); and (3) capacity for delay and overall strength of negative capability (inability to cope with postponement). According to Mitchell, novel nano-devices or drugs will impact these three areas of human psychology. However, she points out that the variations in individuals and the complexity of the human mind make it difficult to establish a normative approach to human enhancement. In addition, whether these advances are "life enhancing" is also a difficult issue that requires thoughtful reflections.

The final essay of the volume addresses the issue of human enhancement in relation to human flourishing. Ronald Sandler argues that focusing on what is constitutive of human nature is misguiding because it does not provide any clue as to what is acceptable or unacceptable in relation to enhancement. Rather the question should be framed in terms of which "deviations from our form of life are desirable" and whether the risks involved are acceptable.

Concluding Remarks

Undeniably, nanotechnology and its applications in the biomedical sciences could provide new opportunities that could change and improve our way of life. Industrial applications, and the development of procedures and devices in medicine, generate great hopes for the resolution of social issues (e.g., pollution, energy) and the improvement of the human condition (e.g., the curing of diseases). As with other technological innovations that create ethical controversies, how to establish consensus is an important and difficult task. In addition, how to provide the condition for collaboration between the various participants in the research and development of nanotechnology and those individuals concerned with regulations, public policy and ethical standards is an important task that will require further analysis. Ultimately, however, more fundamental questions to raise are why do we want to, and by which means do we achieve particular goals through science and technology (i.e., bionanotechnology). Theses questions are not addressed specifically in this work due to its limited scope but will require careful attention for a responsible development of bionanotechnology.

Knowledge Production
in Nanotechnoscience

The World View of Nanotechnology – Philosophical Reflections

Andreas Woyke*

Abstract Philosophical implications of the current nanotechnology boom are not limited to possible problems in ethics. They result essentially from a new understanding of science, which is orientated toward applicability, and from an intensification of a technocratic view of nature in the context of visionary programs. For the integration of nanotechnological discovery in the great context of the history of science and ideas are relations prepared to synthetic chemistry and to the paradigm of progress in modern times. The instrumental comprehension of nature, which show us the leading-ideas of nanotechnology, leads to the question as to which influence they could have on existing views of science and world. Possible answers to this question are outlined by a look at the ontological character of nano-objects, and by references to the one-sidedness of the view of nature, which is founded in the nanotechnological scenarios of the future.

Keywords Nanotechnological visions, world views of nanotechnology, technological progress, philosophy of nature, chemistry, novelty, responsibility

Philosophical Aspects of Nanotechnology

Nanotechnology offers new technical possibilities and contexts of application through the manipulation of structures on the nanometer-scale ($1\,nm = 10^{-9}$ m). Crucial for the investigation of this field, which ranges between molecular physics and chemistry (>0.1 to <10nm) and the dimensions in physics and chemistry of solids (>100nm to <100μm), is the development of new microscopes, which can resolve measurements far beneath the wavelength of visible light.[1] Also, important

---

*Technical University of Darmstadt, Institute of Philosophy, Residenzschloss, D-64283 Darmstadt, woyke@phil.tu-darmstadt.de

[1] Cf. Chen (1993); Rubahn (2004, pp. 50–68); Hartmann (2006, pp. 58–82).

sections of basic scientific research, such as particle physics or neurophysiology, could not move forward without fundamental technological innovations, but the goals of the fields where these techniques are used are connected with principal questions underlying the scientific understanding of nature. In contrast, nanotechnology is booming because of radically new applications, which must be realized under innovation pressures in a globalized economy. Nanoscience and nanotechnology are internationally promoted with great political, institutional, and financial support, although it is currently unknown whether and how the intended possibilities could be realized.

The new understanding of sciences as *technosciences*, proclaimed by Donna Haraway,[2] Bruno Latour[3] and others subsequent to Martin Heidegger's "philosophy of technology", gets, in view of the results and visionary colorings of nanotechnological research, a new quality: The seemingly sharp boundaries between pure and applied research are blurred more than ever before. This point raises the question whether we can legitimately distinguish these general between science and technology. A philosophical interest in nanotechnological research therefore arises not only in reference to concrete ethical rules for the use of its results, but also from the possibility of discovering actual processes of the transformation of understanding and practice of science. We should reflect on the innovative potential of nanotechnology without taking it in the context of positively or negatively accentuated utopian schemes.[4] The way the "start into the nano-cosmos" is frequently represented and emphatically described conjures up pictures and metaphors that are used to illustrate the modern voyages of discovery and the progress in space travel. The image of the Cold War as the motivation behind the United States' and USSR's contest to "fly to the stars" no longer plays a role today. If America, Japan, and the European Union invest great sums into nanotechnological research it is primarily due to their economic competition. Closely connected is the relationship between nanotechnological research and more theoretically motivated questions. The popular brochure, *Nanotechnology - Shaping the World Atom by Atom*, published by the USA *National Science and Technology Council* (NSTC), makes clear that we are dealing with nanotechnology especially because of new technical possibilities for visualization, manipulation and application of atoms and molecules for functional contexts.

Existing knowledge about the nano-cosmos is drawn upon in many respects, and new theoretical insights result from the nanotechnological research, but a fundamental interest in the character of matter and similar intentions are not important. This also becomes clear from the fact that the scanning-tunneling-microscopic pictures from the nano-cosmos are interpreted in the sense of a naive realism, and by technically oriented questions like *"How are electrons going through molecules?"*,[5]

---

[2] Haraway (1995).
[3] Latour (1995).
[4] One-sided considerations in the line of technophobia or technophilia should be abandoned (cf. Fohler, 2003; Woyke, 2004, pp. 695–711); NSTC (1999).
[5] Nordmann (2004b, p. 52).

through which palpable pictures of classical physics are applied on the molecular level. While perceptions of quantum mechanics in manifold ways lead to new ways of thinking about man's place in nature,[6] in the visions of nanotechnology any integration of man into the natural order is lost. In traditional cosmologies, man's role is to be a mediator between the macro- and micro-cosmos,[7] while in the "cosmology" the NSTC-brochure's cover suggests he is completely absent. Nanoscience and nanotechnology are now the key-disciplines for a new conception of nature, which reaches from the molecules on the nanometer-scale (10^{-9} m) to the galaxies one billion light years away (10^{25} m). In this cosmology man is "without place" and he is able to look at, to understand and potentially to manipulate everything. The nanoscientific world order from the very small to the very big is represented, *atom by atom*, as a hierarchically structured complex system, which finds its foundation at the nanoscale:

> *People will be able to acquire a radically different instinctive understanding of the world as a hierarchy of complex systems rooted in the nanoscale.*[8]

In reference to a philosophical reflection of the phenomena of chemistry,[9] such euphoric use of the paradigm of self-organization could be emphasized, but there is a great difference between the orientation of nanotechnological research toward applicability and such theoretical considerations. The NSTC-brochure's title, *Nanotechnology – Shaping the World Atom by Atom*, shows that the world in the focus of nanotechnological visions isn't a given object of experience and reflection; instead, it is taken as material for new technological possibilities, through which it should be transformed. This program is based on the idea that there are no essential differences between natural and technologically planned and initiated processes:

> *Whatever nanotechnology brings about, it cannot violate nature since it works with nature's own principles.*[10]

This is an extremely superficial and instrumentally orientated understanding of nature, which does justice neither to an organismic view,[11] nor to the fullness of historical and present relations to nature beyond reductionism.[12] A philosophical reflection on paradigms and objects of nanotechnological research must improve the insufficient and problematic character of these one-sided perceptions, and confront them with a more balanced and holistic understanding of nature, in which the

---

[6] Kather (1998).
[7] Gloy (1996, 19ff.); Nordmann (2004a).
[8] Roco and Bainbridge (2002, p. 19).
[9] Müller (1998); Earley (2003). For a processualistic philosophy of nature in reference to Whitehead, Rescher and others cf. Woyke (2004, pp. 772–792); Stein (2004).
[10] Nordmann (2004a, p. 41).
[11] Gutmann and Edlinger (1994).
[12] Böhme (2002); Seel (2001); Spaemann and Löw (1991); Köchy (2003); Woyke (2004, pp. 86–105, 663–682).

technical plans and actions of man in view of new technological paradigms could also find a place.

Novelty in Chemistry, Modern Progress and Nanotechnology

"Technology", in a broad sense, is closely associated with the introduction of novelty through human practice. The examples reach from the discovery and use of fire, the development of the wheel, and the production of new materials, to the printing press, the use of heat engines, and the success of the computer technology. In spite of this general connection between technology and developments of novelty, there also are decisive differences in the quality of novelty. Chemistry as practical art and as science generates "the new" (in German: *"das Neue"*)[13] not only by the fact that what is already known is brought in a new order, where new functionalities become usable. Chemistry also produces, by processes of material change, authentically new materials with new qualities. In this respect, the potential of chemical synthesis, which has extended rapidly since the second half of the 19th century, stands for a qualitatively and quantitatively new dimension in the formation of "the new". Nanotechnology exhibits many relations to modern technologies of chemical synthesis and opens a "new degree of novelty": On the fundamental nanometer-scale, new functionalities and possible applications are to be developed. Therefore, a historical embedding of nanotechnology arises from its relation to the history of chemistry in the broader context of defining and evaluating "novelty" in the western history of ideas.[14]

This consideration of "novelty" leads at first to the essential difference between antiquity and Middle Ages on the one hand, and on the other hand that epoch which we generally call "modern times" (in German: *"Neuzeit"*). Second, it will become clear that especially the Christian-medieval ontology has shaped the constitution of modern thought. In so-called "antiquity", attention was directed at the permanence of the given through change – this is the antique cosmos or the medieval "creation".

---

[13] The German word *"das Neue"* is here meant in the sense of a completely abstract category to stress the different creative potentials of nature and the human mind against the classical view that all innovations can be attributed in the long run to well-known elements already given. Chemical phenomena and the possibilities of the chemical synthesis concretize this abstract term in a particularly eminent way and contradict, in this respect, all philosophical and scientific determinisms, which suppose a conceptually or a, by the given material elements and natural laws, closed universe. Hans Poser makes clear that innovations are not limited to the field of the cultural and technological development, but are, in accordance with the concept of evolution, a general characteristic of nature (H. Poser, 2004, 184ff.). The large pallet of new materials, which modern chemistry is able to synthesize, is in this sense not only a "work of man", but participate in the long run on the creativity of the natural sphere.

[14] Moltmann and Rath (1984).

In contrast, modern times esteem novelty as opposed to "the old", which has to be overcome. If the production of new things in antiquity is connected with the natural space of possibility to which, in the end, also humans are beholden in all their actions, in the Christian Middle Ages there is a shift toward the open space of possibility available to an omnipotent God who is able to make the world absolutely new also in the eschatological sense of apocalypse.[15] In historical perspective,[16] modern times go back exactly to this image of *Deus absolutus*, while man takes over the creative abilities of God in world's renewal. Modern consciousness tries to overcome the past, in all important fields, with new and better conceptions. In considering nanotechnological visions, it must be stressed that "the new" of modern times is associated quite substantially with the fact that new technical instruments like telescopes and microscopes open new possibilities for the perception of distant and, until now, unknown worlds. Philosophers like Francis Bacon[17] and René Descartes[18] deliver with their programmatic methodologies the new thinking for the scientifically and technologically founded progress.

In Georg Wilhelm Friedrich Hegel's dialectic philosophy of history,[19] the category of novelty attains an ontological significance going further out of such programs.[20] No matter how one judges in detail such ontology, it is certain that modern consciousness, along with modern scientifically shaped world views, lack a more basic philosophical foundation. Prevalent is the orientation toward purely instrumental programs, which we can find paradigmatically in examples of nanotechnological research. Friedrich Nietzsche's criticism[21] of an optimistically tinted progress of "the new" over and over again emphasizes its essential ambivalence and provides an important reference point for the general reflection about the one-sidedness of such paradigms.

In the history of philosophy and the general history of ideas, the category of novelty isn't only a negative category,[22] it is central for the understanding of our self-image and the world since modern times. However, as Joachim Schummer states, it reaches a special significance in the context of chemical synthesis of new substances which, until now, has barely been object of philosophical considerations. The basic dispositionality and context dependence of chemical qualities form the background of this

---

[15] e.g. Revelation 21,5.
[16] Hösle (1996).
[17] Bacon (1990, I 129, II 52).
[18] Descartes (1997, *Discours de la méthode* VI 2).
[19] e.g. Hegel (1988, Vorrede, 10).
[20] But the difficult relation between the spheres of "mind" and "nature" in the philosophy of Hegel constitutes a dichotomy between the creative innovativity of mind and the persistent conservativity of nature (cf. e.g. Hegel, 2005, p. 74)
[21] Cf. e.g. Nietzsche, *Die fröhliche Wissenschaft* 1, 4; KSA, Bd. 3, 376–377; *Jenseits von Gut und Böse* 192; KSA, Bd. 5, 113–114.
[22] Schummer (1996, pp. 290–296).

specific feature and limit our material knowledge; *"imperfection"* appears with it as an *"inherent character of the knowledge of material"*.[23] Ontological, epistemological and ethical conclusions,[24] which arise out of this, give an important context for a relevant investigation of nanotechnological products and processes and can lead to a better understanding of nanotechnological research, as well as deliver normative guidelines for the use of its results.

Does Nanotechnology Constitute a New World View?

In order to investigate the interaction between views of science and world views in the context of nanotechnology it is helpful to begin with the assumption[25] that a more or less obliging frame of reference for the experience and interpretation of the world can be assigned to all epochs of the western history of ideas that might be called "ontology".[26] This assumption is legitimized particularly by the fact that, for each of these epochs in western history, certain "forces" appear which have a constitutive influence on these ontologies.[27] To be sure, a frivolous and superficial use of this assumption becomes quickly implausible in view of historical details and contingencies.[28] Also in reference to the contrasts and inner conflicts of modern consciousness it loses basically its base.[29] To the extent that a "modern world view" (in German: *"ein modernes Weltbild"*) is determined by scientific theories and technological practices, we find ourselves in an extremely paradoxical situation: The functional connection between science, technology and economy plays an important role in structuring every day experience, but it is not powerful enough to shape the self-understanding of individuals and social actors which is characterized by a lack of orientation, randomness and aporeticity.

The position of the natural sciences in this context should be characterized in two significant ways: On the one hand, scientific theories contribute in an emphatically affirmative way to a rational-progressive relation to the world's interpretation and a steady enrichment of our knowledge about the world – beyond one-sided reductions. On the other hand, it comes to a programmatic coloring of scientific knowledge with a visionary view of the future that turns out to be very virulent in

---

[23] Schummer (1996, pp. 293–294); translation of the author.

[24] Schummer (1996, pp. 290–296, 2001a).

[25] Woyke (2004).

[26] Hübner (1985); Woyke (2004, pp. 119–124).

[27] Important are the archaic poetry, the antique philosophy, the Christian religion and theology, subjectivity philosophy and modern natural sciences and the modern "superstructure" between science, technology, industrial manufacturing and mass culture.

[28] I have tried to restrict this evident reproach by a much more detailed interpretation of the historical development having regard also to critical positions.

[29] In this view modernity has no ontology, which could be compared with the ontologies of times before.

nanotechnological research.[30] The basic scientific theories of the 19th and 20th century influence in a somewhat strong way that which, in an abstract sense, can be called the present world view, even if the theoretical crises of physics are still not present in general consciousness. Indeed, while the theory of relativity, quantum mechanics and theories of self-organization and chaos have found extremely important technical applications, they stand above all for the aforementioned world-view-donating contribution of the natural sciences. The knowledge of genetics and molecular biology belongs in this context too, but in the public consciousness it is present in particular by its possible and actual applications, where euphoric approval and fear of possible dangers meet. Here the "world-view-contribution" (in German: *"der Weltbild-Beitrag"*) already moves to the side of visionary hopes and fears. Nanotechnology and nanoscience, which also use the methods and interpretation patterns of genetic engineering, intensify this reduction of science to the socio-economic influence of technological products and procedures.

Indeed, nanoscientists can also use theoretical motivations as legitimizations for their field of research, however, these often flow *stante pede* into an orientation toward applicability, or are peripherally compared with the mainstream research being done. Neither the accentuation of the intersection between classical and quantum physics in the nano-cosmos, nor the talk of a new paradigm of the natural sciences regarding the theories of self-organization are satisfying theoretical legitimizations. It can be argued that nanotechnology isn't primarily orientated toward a better understanding for the intersection between quantum mechanics and classical physics, but on confirming the long-known theories again by pushing forward to the nano-cosmos and using it for technological applications.[31] Indeed, the second aspect has the potential to change our essential understanding of the natural sciences and to confront the traditional mechanical and genetic order with a new botanical or zoological order.[32] Nevertheless, this accentuation moves away from the practical direction of nanoscience and has no chance to play a decisive role for the orientation of the research process.

Which influence does this prevailing understanding of nature in nanoscience and nanotechnology have on the scientific, and on a general, world view, at least available in diffuse form? Does an increase of the general dominance of instrumental rationality and an orientation toward efficiency in the modern societies express itself in these world views? The overdrawn visionary colorings of nanotechnological research – which reach from the perspectives of the NSTC reports to the

---

[30] Bensaude-Vincent (2004). She refers to the fact that the "machine-metaphor" of the influential "nano-visionary" Eric Drexler has its historically roots in the mechanicism of René Descartes and others. The dynamic model of Richard E. Smalley opposing to this is orientated on complex chemical systems. But against the idea of Bensaude-Vincent we can't find two separated "cultures of nanotechnology": Most researchers in physics, chemistry and biology use the term "nano-machine" and make no important differences between natural and artificial "nano-machines" (cf. e. g. Bath & Turberfield, 2007; Hla, 2007; Jones, 2004; Moriarty, 2005; Seeman, 2003).

[31] A very popular example of this application is the possibility of a "quantum-computer" (cf. Rubahn, 2004, pp. 143–147; Edwards, 2006, pp. 191–196).

[32] Nordmann (2002, p. 3). For the vision of a generally systemic science cf. G. Khushf (2004).

techno-fantasies of Eric Drexler, Ray Kurzweil, William S. Bainbridge and others and the promises of the transhumanists – contribute substantially to the intensification of a technocratic understanding of nature. All images of a basic transformation of nature and humanity run inevitably against basic ontological and anthropological borders, which is why every attempt of realize them shows totalitarian characteristics. On the other hand, it is a matter of a more balanced position that illustrates the possibilities of nanotechnology more realistically and concentrates the attention upon the dynamism of nature and upon the character of humanity.

Our knowledge about substances and the material change is limited *per se* from what immediate ethical implications arise, in particular for the use of new synthesized substances.[33] If the central purpose of nanotechnological methods consists in generating new materials at the level of the nanoscale and in utilizing these materials for technological applications, it is evident that this comprehension in this field is very important.[34] Also, the point is not a rigorous relinquishment, but a little bit more care and responsibility by the choice of the purpose perspectives.

Looking for the Ontological Status of Nanotechnological Products

A philosophical consideration of nanoscience and nanotechnology is confronted with the question of how the relationship between science and technology is to be determined within this heterogeneous research field. In the Aristotelian philosophy, we can find the traditional ontological difference between "nature" (φύσις) and "technology" (τέχνη),[35] which leads to the question of the relevant ontological character of nanotechnological products and processes.

If we take seriously the nanotechnological image that nanotechnology can be understood as an imitation of nature that operates like a nano-engineer, the question

---

[33] Schummer (1996, pp. 290–296).

[34] For the discussion of environmental ethical aspects of nanotechnology cf. Jömann and Ach (2006, pp. 48–52).

[35] Schummer shows that this difference was interpreted in the historical reception in a way which isn't coherent to the original Aristotelian statements: Aristotle assumes neither that technology must be understood exclusively as an imitation of nature, nor negates the difference between the teleology of nature and the purposes of man. Also a reasonable differentiation between natural and artificial things which meets as a dichotomy between natural and synthetic substances in the context of chemistry is a product of later interpretations (cf. Schummer, 2001b). These statements are certainly right, but not new for everybody who is well acquainted with the *Corpus Aristotelicum*; the similar counts – at least conditioned – for Schummer's statement, that the Aristotelian differentiation between *physis* and *téchne* is only a little suitably for a modern philosophy of technology. Besides, he overlooks in my opinion the importance of the antique notion of *physis* for the foundation of a holistic philosophy of nature in the interaction between scientific knowledge and philosophical speculation.

above is abolished in view of an unreflected naturalism which also wipes all conceivable ethical and societal problems dealing with the use of nanotechnological products and processes. Nevertheless, one must critically stress the one-sidedness of this understanding of nature, and emphasize the relevance of the question about the place of nanotechnology in a generally taken philosophy of science and technology and a holistic philosophy of nature.

It is clear that the Aristotelian notions cannot be transferred immediately to the modern world, which is determined by scientific knowledge and technological practice.[36] Technical artifacts do not play a dominating role in the frame of ancient ontology: They can be characterized by their essential difference to natural things, their at least material dependence on the *physis* and by their causal, formal and final dependence on the plans and actions of man. In the modern world, technological systems have reached a "self-dynamism" and an autonomy that is completely new and leads to the fact that people alienate from these objects and perceive them as a "second nature".[37] The formulation of a "poietical philosophy", as Wolfgang Wieland demands, is maybe a suitable way to reflect the particularity of these technological dynamics, which distinguish themselves substantially from natural dynamics. However, these technological dynamics cannot be attributed completely to man's intentions of planning, constructing and applying.

For this purpose, it does not seem helpful to compare in an ahistoric way the Aristotelian differentiations between "science/knowledge" (ε'πιστμη') and "technology/art" (τε'χνη) with our concepts of "science" and "technology".[38] Even if the phenomena of chemistry are in many respects products of the art of chemical synthesis, synthetical chemistry uses the dispositionality of material qualities and is not able to form a completely open and free realm of determinable possibilities. The Aristotelian difference between *physis* and *téchne*, indeed, on the basis of different sources of movement and change, no longer fits into a universally naturalistic understanding of nature. However, this difference still represents a fertile principle of natural philosophy in the basic regard of a self-dynamic and resistant matter, as well as in the more special regard of the self-productivity of living beings. Like the scientific theories of self-organization[39] suggest, a modern creative concept of matter

---

[36] Wieland (2003).

[37] Guzzoni (1995).

[38] Schummer (1997). The differentiation between *epistéme* and *téchne*, which delivers Aristotle in his *Nicomacheun Ethics* (cf. EN VI 3–4), is not congruent with the stronger differentiating considerations to the relation of the "metaphysics" (πφω'τη φιλοσοφία) to the particular sciences in his *Metaphysics* and *Physics*. The accentuation of change and becoming in the sensuous-material universe is for Aristotle just the starting point for the development of a science of *physis* in the narrower sense (cf. e.g. Met. VI 1, 1026a; Met. XI 7, 1064a). That is why it seems to me too simple to say Aristotle has separated "science" and "technology" by the difference between unchangeable and changeable. In addition, an immediate comparison between the Aristotelian *epistéme* and the modern natural sciences is very ahistorically.

[39] Cf. e.g. Mainzer (Hrsg.) (1999); Kanitscheider (1993); Müller (1998); Prigogine and Stengers (1993); Lehn (1995, pp. 139–143); Jones (2004, pp. 88–125).

offers the possibility to connect animated and inanimate nature with each other and adapt the ancient idea of a general *physis* to the modern scientific knowledge.

It is not possible in modern times to plausibly determine the relationship between natural and technical things or beings in the widest sense by this easy dependence-relation, as it used to be for Aristotle and other ancient philosophers. But the degree of the mechanization and technologization of our social environment, intensified by nanotechnological products and processes, and the difficulty to distinguish at the conceptual level between natural and technical things should not lead to an uncritical acceptance of suitable technological and social developments. Even if the human body today is increasingly coupled with technological artifacts, and the world of the modern industrial societies, as Bruno Latour states, may appear as a universe of hybrid objects, I see in this connection no reason to totally blur the basic differences between the person and his technical productions, on the one hand, and natural and technical dynamics on the other hand.[40] Many subjects and objects of modern science like "the ozone hole", "the greenhouse effect", "*in vitro* fertilization", "the cancer mouse", "BSE" or "AIDS" cannot be more simply or clearly assigned to one or another area of science, for example, cultural, natural or technological sciences. Partly because they count as at least anthropogenic influenced, and partly because they are produced by specific technological manipulations and are connected with many very different implications for nature, technology and culture.[41] Much out of nanoscientific knowledge, nanotechnological practice and popular-scientific representations of nanoscience and nanotechnology can be characterized as "hybrid objects". The mixture between natural and social spheres, as Latour tries to make clear with his concept of the hybrid or quasi object, is certainly a relevant and new phenomenon, which cannot be concerned with simplistic area allocations. However, the priority of a self-dynamic nature, which cannot be reduced completely to the determinism of fundamental physical laws, is not negated by the presence of hybrid objects. The complexity of technological systems and social relations leads to the fact that they can emancipate themselves from a drastic control and rule of the subject and, in this respect, bear a "self-dynamism" as a "second nature". However, they remain, in the end, in one or another, still connected to a general natural sphere in the sense of the Aristotelian *physis*.[42]

---

[40] Fohler (2003, pp. 193–196). For a critical position cf. Woyke (2004, pp. 707–711).

[41] Latour (1995).

[42] Latour argues from the sociologically orientated science studies which represent a very constructivistic view of the objects of scientific knowledge; a general view of "nature" which encloses scientific knowledge, but tries to integrate elements of traditional philosophy of nature plays so far no role for him (cf. Latour, 1995, pp. 71–77). Latour considers the specifically modern separation between a natural and a social sphere as a starting point for the constitution and increase of quasi objects for which he names four essential springs of generation: *"naturalization, soziologisation, discursivation and, in the end, oblivion of the being."* (ibid., 91; translation of the author). The strategies to handle with all of these separations are one-sided and pursue an absolution of one aspect. It is certainly an important insight that everything what meets us in the world is in different ways "natural", "social", "discursive" and "being", but a general understanding of "nature" represents in my view, however, the "ontological *prius*" by which and in relation to these different dimensions should be unfolded.

Final Consideration

A central leading idea of nanoscientific and nanotechnological research and of the popular representation and marketing, is the picture of nature as a big nano-engineer which man only tries to copy in all of his technological actions. Every content-rich and non-instrumentally reduced understanding of nature, as well as the possibly problematic ecological, social, and all other results of nanotechnology are ignored here. It is certain that this leading idea is based on massive instrumental reductions of nature. It is unclear if this new quality of "technological reductionism"[43] will lead us to a new dimension of using scientific knowledge for nature's domination and control. How strong could its influence be on the public perception and evaluation of nanotechnological research? In the unbalanced relation between exaggerated visions, real applications and concrete dangers of nanotechnology *"a broad anti-scientific movement"*[44] could be generated, which argues above all on an emotional base. In contrast, we must try to defend a serious kind of scientific knowledge, an enlightened world view and a responsible form of acting in the world. The actual application of scientific knowledge in the context of nanotechnology and the so-called "converging technologies" needs a theoretical foundation and a connection to traditional motives of natural philosophy. Hans Jonas' philosophy of responsibility is still an important reference point for modern evaluations of technological innovations, which leads to a fertile relationship between natural philosophy and natural ethics. His principle, *"giving the evil prophecy more hearing than the good prophecy"*,[45] is undoubtedly a good guideline for the evaluation of nanotechnological visions of the future. Behind this principle stands the essentially philosophical view that every ethics needs a metaphysical base[46] if it doesn't want to lose itself in "adhoceries". Indeed, in post-metaphysical times no more metaphysics can be founded that is applicable to all ethical discourses, but a holistic philosophy of nature and an integrative anthropology are in my opinion the best arguments against a nanotechnological transformation of the world and a re-designing of humanity.[47,48]

---

[43] Schmidt (2004).

[44] Schummer (2004, p. 123).

[45] Jonas (1995, p. 70); translation of the author.

[46] Ibid., especially pp. 91–95. Jean-Pierre Dupuy agrees to this view of Jonas in principle and applies it to his philosophical analysis of nanotechnology and other "converging technologies" (cf. Dupuy, 2005).

[47] Dupuy names two essential elements for a holistic natural philosophy and an integrative anthropology, namely the finiteness of human life and the dynamic complexity of nature (cf. ibid.).

[48] For helpful comments and corrections I thank Fabrice Jotterand, Alfred Nordmann and Cade Warren.

References

Aristoteles (1970). *Ethica Nicomachea*. Griech. Text, recognovit brevique adnotatione critica instruxit I. Bywater. London: Oxford University Press.
Aristoteles (1989/1991). *Metaphysik, Bücher I – VI und VII – XIV*. Griech./dt., übers. und hrsg. von H. Seidl. Hamburg: Meiner.
Bacon, F. (1990). *Neues Organon/Novum Organum. Teilband I und Teilband II*. übers. und hrsg. von W. Krohn. Hamburg: Meiner.
Bath, J. & Turberfield, A. J. (2007). DNA Nanomachines. *Nature Nanotechnology*, 2, 275–284.
Bensaude-Vincent, B. (2004). Two Cultures of Nanotechnology?. *HYLE*, 10, 65–82.
Böhme, G. (2002). *Die Natur vor uns. Naturphilosophie in pragmatischer Hinsicht*. Kusterdingen: Die Graue Edition.
Chen, C. J. (1993). *Scanning Tunneling Microscopy*. New York: Oxford University Press.
Descartes, R. (1997). *Discours de la méthode pour bien conduire sa raison, et chercher la verité dans les sciences*. Franz./dt., übers. und hrsg. von L. Gäbe. Hamburg: Meiner.
Dupuy, J.-P. (2005). Aufgeklärte Unheilsprophezeiungen. Von der Ungewissheit zur Unbestimmbarkeit technischer Folgen (übers. aus dem Engl. von A. Hetzel). In G. Gamm & A. Hetzel (Hrsg.), *Unbestimmtheitssignaturen der Technik. Eine neue Deutung der technisierten Welt* (pp. 81–102). Bielefeld: transcript.
Earley, J. E. (2003). *Constraints on the Origin of Coherence in Far-from-Equilibrium Chemical Systems*. In T.E. Eastman & H. Keeton (Eds.), Physics and Whitehead: Quantum, Process and Experience. (pp. 63–73), Albany: State University of New York Press: New York.
Edwards, S. A. (2006). *The Nanotech Pioneers. Where Are They Taking Us?*. Weinheim: Wiley-VCH.
Fohler, S. (2003). *Techniktheorien. Der Platz der Dinge in der Welt der Menschen*. München: Wilhelm Fink Verlag.
Gloy, K. (1996). *Das Verständnis der Natur, 2. Bd: Die Geschichte des ganzheitlichen Denkens*. München: Verlag C. H. Beck.
Gutmann, W. F. & Edlinger, K. (1994). Organismus und Evolution: Naturphilosophische Grundlagen des Prozessverständnisses. In G. Bien, Th. Gil, & J. Wilke (Hrsg.), *Natur im Umbruch". Zur Diskussion des Naturbegriffs in Philosophie, Naturwissenschaft und Kunsttheorie* (pp. 109–140). Stuttgart/Bad Canstatt: frommann-holzboog.
Guzzoni, U. (1995). *Über Natur. Aufzeichnungen unterwegs: Zu einem anderen Naturverhältnis*. Freiburg/München: Alber.
Haraway, D. (1995). *Die Neuerfindung der Natur. Primaten, Cyborgs und Frauen*. Hrsg. von C. Hammer und I. Stieß. Frankfurt a. M.: Campus Verlag.
Hartmann, U. (2006). *Faszination Nanotechnologie*. München: Spektrum Akademischer Verlag.
Hegel, G. W. F. (1988). *Phänomenologie des Geistes*. Hrsg. von H.-F. Wessels und H. Clairmont. Hamburg: Meiner.
Hegel, G. W. F. (2005). *Vorlesungen über die Philosophie der Geschichte. Werke in 20 Bd., Bd. 12*. Frankfurt a. M.: Suhrkamp.
Hla, S.-W. (2007). Molecular Machines: Reinventing the Wheel. *Nature Nanotechnology*, 2, 82–84.
Hösle, V. (1996). Was sind die wesentlichen Unterschiede zwischen der antiken und der neuzeitlichen Philosophie? In V. Hösle (ed.), *Philosophiegeschichte und objektiver Idealismus* (pp. 13–36). München: C. H. Beck.
Hübner, K. (1985). *Die Wahrheit des Mythos*. München: C. H. Beck.
Jömann, N. & Ach, J. S. (2006). Ethical Implications of Nanobiotechnology – State-of-the-Art Survey of Ethical Issues Related to Nanobiotechnology. In J. S. Ach & L. Siep (eds.), *Nano-Bio-Ethics. Ethical Dimensions of Nanobiotechnology* (pp. 13–62). Berlin: LIT.
Jonas, H. (1995). *Das Prinzip Verantwortung. Versuch einer Ethik für die technologische Zivilisation*. Frankfurt a. M.: Suhrkamp.
Jones, R. A. L. (2004). *Soft Machines. Nanotechnology and Life*. Oxford: Oxford University Press.

Kanitscheider, B. (1993). *Von der mechanistischen Welt zum kreativen Universum. Zu einem neuen philosophischen Verständnis der Natur*. Darmstadt: WBG.
Kather, R. (1998). *Ordnungen der Wirklichkeit. Die Kritik der philosophischen Kosmologie am mechanistischen Paradigma*. Würzburg: Ergon Verlag.
Khushf, G. (2004). A Hierarchical Architecture for Nano-scale Science and Technology: Taking Stock of the Claims About Science Made by Advocates of NBIC Convergence. In D. Baird, A. Nordmann, & J. Schummer (eds.), *Discovering the Nanoscale* (pp. 21–33). Amsterdam: Kluwer.
Köchy, K. (2003). *Perspektiven des Organischen. Biophilosophie zwischen Natur- und Wissensch aftsphilosophie*. Paderborn: Schöningh.
Latour, B. (1995). *Wir sind nie modern gewesen. Versuch einer symmetrischen Anthropologie*. Aus dem Franz. übers. von G. Rossler. Berlin: Akademie Verlag.
Lehn, J.-M. (1995). *Supramolecular Chemistry. Concepts and Perspectives*. Weinheim: VCH.
Mainzer K. (Hrsg.) (1999). *Komplexe Systeme und nichtlineare Dynamik in Natur und Gesellschaft*. Berlin: Springer.
Moltmann, J. & Rath, N. (1984). Lemma "Neu, das Neue". In J. Ritter & K. Gründer (Hrsg.), *Historisches Wörterbuch der Philosophie, Bd. 6: Mo-O* (pp. 725–731). Darmstadt/Basel: WBG.
Moriarty, Ph. (2005). Nanotechnology: Radical New Science or Plus ça Change? *Nanotechnology Perceptions*, 1, 115–118.
Müller, A. (1998). Die inhärente Potentialität materieller (chemischer) Systeme. *Philosophia naturalis*, 35, 333–358.
National Science and Technology Council (NSTC) (1999). *Nanotechnology. Shaping the World Atom by Atom*. Washington, DC: NSTC. http://itri.loyola.edu/nano/IWGN.Public.Brochure.
Nietzsche, F. (1988). *Kritische Studienausgabe in 15 Bd*. Hrsg. von G. Colli und M. Montinari. München/Berlin: dtv/de Gruyter.
Nordmann, A. (2002). *Nanoscale Research: Application Dominated, Finalized, or "Techno"-Science?*. http://www.cla.sc.edu/cpecs/nirt/AN1.html.
Nordmann, A. (2004a). Nanotechnology's World View: New Space for Old Cosmologies. *IEEE Technology and Society Magazine*, 23/4, 48–54.
Nordmann, A. (2004b). Molecular Disjunctions: Staking Claims at the Nanoscale. In D. Baird, A. Nordmann, & J. Schummer (eds.), *Discovering the Nanoscale* (pp. 51–62). Amsterdam: Kluwer.
Novum Testamentum Tetraglottum (1981). *Archetypum Graecum cum versionibus Vulgata Latina, Germanica Lutheri et Anglica Authentica*. In usum manualem edundum curaverunt C. G. G. Theile (Nachdruck einer Ausgabe von 1858). Zürich: Diogenes.
Prigogine, I. & Stengers, I. (1993). *Dialog mit der Natur. Neue Wege wissenschaftlichen Denkens*. Übers. von F. Griese. München: Piper.
Roco, M. C. & Bainbridge, W. S. (eds.) (2002). *Converging Technologies for Improving Human Performance. Nanotechnology, Biotechnology, Information Technology and Cognitive Technology*. Arlington, VA: National Science Foundation. http://www.wtec.org/ConvergingTechnologies/Report/NBIC_report.pdf.
Rubahn, H.-G. (2004). *Nanophysik und Nanotechnologie*. Stuttgart: Teubner.
Schmidt, J. C. (2004). Unbounded Technologies: Working Through Technological Reductionism of Nanotechnology. In D. Baird, A. Nordmann, & J. Schummer (eds.), *Discovering the Nanoscale* (pp. 35–50) Amsterdam: Kluwer.
Schummer, J. (1996). *Realismus und Chemie. Philosophische Untersuchungen der Wissenschaft von den Stoffen*. Würzburg: Königshausen & Neumann.
Schummer, J. (1997). Challenging Standard Distinctions Between Science and Technology. *HYLE*, 3, 81–94.
Schummer, J. (2001a). Ethics of Chemical Synthesis. *HYLE*, 7, 103–124.
Schummer, J. (2001b). Aristotle on Technology and Nature. *Philosophia Naturalis*, 38, 105–120.
Schummer, J. (2004). Societal and Ethical Implications of Nanotechnology: Meanings, Interest Groups, and Social Dynamics. *Techne - Research in Philosophy and Technology*, 8/2, 56–87.

Seel, M. (2001). *Eine Ästhetik der Natur*. Frankfurt a. M.: Suhrkamp.
Seeman, N. C. (2003). Biochemistry and Structural DNA Nanotechnology: An Evolving Symbiotic Relationship. *Biochemistry*, 42, 7259–7269.
Spaemann, R. & Löw, R. (1991). *Die Frage Wozu? Geschichte und Wiederentdeckung des teleologischen Denkens*. München/Zürich: Piper.
Stein, R. L. (2004). Towards a Process Philosophy of Chemistry. *HYLE*, 10, 5–22.
Wieland, W. (2003). Poiesis. Das Aristotelische Konzept einer Philosophie des Herstellens. In Th. Buchheim, H. Flashar, & R. A. H. King (Hrsg.), *Kann man heute noch etwas anfangen mit Aristoteles?* (pp. 223–247). Hamburg: Meiner.
Woyke, A. (2004). *Die Entwicklung einer aprozessualen Welt- und Naturdeutung in der abendländischen Geistesgeschichte und ihre Bedeutung für die "Ausblendung des Prozessualen" in Chemie und Chemieunterricht*. Siegen. http://www.ub.uni-siegen.de/epub/diss/woyke.htm.

Nanomachine: Technological Concept or Metaphor?

Xavier Guchet[1] and Bernadette Bensaude-Vincent[2]

Abstract The molecular machines currently designed in nanotechonology laboratories are confronted with the molecular machines described in Eric K. Drexler's famous *Engines of Creation*. We argue that the new class of molecular machines coming both from chemistry and molecular electronics, does not fit within the two classical definitions of machines that are used in Drexler's molecular manufacture. Molecular machines are neither standard mechanical machines (Cartesian model), nor complex systems (Von Neumann's model). They consequently call for a new concept of machine more adequate to nano-objects, which is delineated in Richard Jones' *Soft Machines*. We conclude that nanotechnology still wants a technological approach to molecular machines.

Keywords Engineering, Molecular Machine, Nature and Artifact, Technology

Introduction

The term machine has been extensively used over the past decades both by biologists and by materials scientists (Bensaude-Vincent, 2004). DNA, RNA, ribosomes are described as machines performing specific tasks just as molecules or supramolecules are designed as machineries. Machines are everywhere, among living cells and among the ultimate units of matter. Their ubiquitous presence is undoubtedly a major feature of today's science and testifies for a close interaction between cognitive and engineering perspectives. However, the "technological turn" of today's research programs is grounded on rather vague definitions of machine. This lack of precise definitions of machine is particularly noticeable in nanotechnology. Indeed, "molecular

---

[1] Normes, Sociétés, Philosophies, Université Paris I Panthéon-Sorbonne, 12 place du Panthéon 75005 Paris

[2] Centre d'histoire et de philosophie des sciences, Université Paris X, 200 avenue de la république, F. 92001 Nanterre

machine" labels a special field of nanotechnology researches. Terms such as "artificial molecular motor", "molecular wheelbarrow",[1] "molecular rack-and pinion" and even "nano-car", have flourished in the scientific literature. What kind of machine do scientists have in mind when they use such labels? In this paper we try first to determine what sort of machines molecular machines are: are they classical mechanical systems such as Cartesian automata or something like complex machines? We will argue that both concepts are irrelevant to characterize molecular machines and that we need an alternative concept of machine. We will emphasize in particular that molecular machines designed and studied by scientists today do not fit in with Eric Drexler's concept of "nano-robot" (or nanobot).

Indeed, in his best-seller *Engines of Creation* (Drexler, 1986), Drexler based the so-called "coming era of nanotechnology" on the design of "Universal Assemblers" and "Replicators". Both are supposed to be molecular machines performing useful tasks much better than biological protein machines do. Emphasizing the "dark side" of molecular manufacturing, Drexler focused also on a terrific scenario related to the uncontrolled proliferation of "Universal Assemblers", eating the whole of biomass. This famous scenario labelled "Grey Goo", has been criticized by scientists, but it has undoubtedly played an important role in shaping the public ethical debate on nanotechnology. So the comparison between Drexler's nanobots and molecular machines that are actually designed by scientists may lead to major ethical issues.

It is often noticed that Drexler's vision of molecular manufacturing, focusing on the control of matter atom by atom, achieves the philosophical program defining modernity since the 17th century: to instrumentalize and to artificialize Nature in order to make it serve human goals. In this "bottom-up" approach to nanotechnology, Nature is exclusively viewed as a reservoir of available matter and energy that human beings are nowadays able to exploit at the nanoscale. It is considered a given sum of building blocks at the disposal of human technology. However, we argue that this representation of Nature is related to the kind of artefacts envisioned in this "bottom-up", "atom-by-atom" approach to molecular manufacturing. As emphasized by the French philosopher Maurice Merleau-Ponty in his lectures on the concept of Nature, our idea of Nature is always shaped and valued by the artefacts we build. According to him, we could not talk about Nature in the 1950s without talking about Cybernetics. Let us take this advice into account: could we say today that it is impossible to talk about Nature without talking about nanotechnology? To test this hypothesis, we have to analyse what kinds of artefacts are designed and built in laboratories.

This paper is an attempt to characterize the idea of Nature involved in the design and the building of molecular machines. We will argue that the conception of Nature related to this research field is not reducible to Drexler's vision of Nature as something to be purely exploited. The ethical debate surrounding the issue of molecular machines should be then reformulated.

---

[1] Rapenne et al. (2006a).

A Brief Genealogy of Molecular Machines

Looking back to the recent past, one can trace two possible origins of the field of molecular machines. The first one is chemistry, more precisely the synthesis of complex molecules such as catenanes (Fig. 1) (two or more interlocked macrocycles) and rotaxanes (a "dumbbell shaped molecule" threaded through one or several macrocycles) at the beginning of the 1980s. Indeed chemists did not use the word "machine" to qualify such molecules until they were able to control various kinds of motions inside these molecules, since the beginning of the 1990s. In the catenane, chemists can induce by means of electrochemistry the motion of one macrocycle with respect to the other. As soon as chemists put the molecule in its former electrochemical state, the macrocycle pivots again and returns to its first stable position. So the motion of the macrocycle is reversible. The concept of machine used by chemists in this context fits in with two major features: (1) chemists talk about machine as soon as they are able to control a reversible motion in the molecule. This definition of machine by motion seems to be very classical; (2) chemists talk about machine even though they do not address single molecules, but billions of molecules in solution.

The second origin of the research field of molecular machines is molecular electronics. The project to use single molecules as components in electronic circuits emerged in the mid 1970s and was reinvigorated in the 1980s by the invention of the Scanning Tunnelling Microscope (STM) in an IBM laboratory at Zurich. The STM opened up the fascinating perspective of moving single molecules and atoms with the tip of the microscope and prompted a number of experiments on the behaviour

Fig. 1 A catenane

of a single nano-object (electron transfer, molecular dynamics...). Physicists can today address single molecules laid on surfaces. They can design molecules able to perform useful tasks at the nano-scale, such as a molecular motor powered by means of electrochemistry or a molecular switch. Indeed, quantum physics predicted that a molecule laid on a surface would behave as a quasi-classical object. This idea opened the way to a new field of research labelled molecular mechanics (see for instance the design of a single molecular wheel rolling at the nanoscale by the researchers of the CEMES,[1] France, or the rack-and-pinion molecular mechanism (Fig. 2) designed by the same and their colleagues of the Frei Universität of Berlin, Germany[2]).

An example of molecular machine in the field of molecular electronics is the bistable biphenyl laid on a silicium surface (Fig. 3).[3]

The molecule is made of two benzene rings linked by a covalent bonding. By means of the tip of the STM, a current is applied inside the molecule. The molecule pivots. When a current is applied again at the same location, the molecule returns to its first position. It is actually bistable. Physicists of the Nanoscience Group of the LLPM (Laboratoire de Photophysique Moléculaire, Orsay, France), the laboratory leading this experiment, have noticed that the surface plays a very important role in the molecular dynamics of the bistable biphenyl. As a researcher noticed, "the surface is part of the molecule". The molecule is no longer described as benzene rings; its definition includes the interactions between the molecule and the surface itself. Such a description involves obviously an unfamiliar concept of machine, including the environment of the machine in the definition of the machine itself.

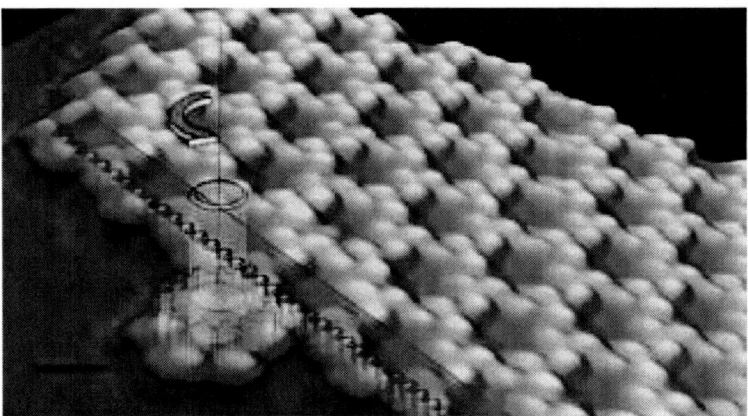

Fig. 2 Rack-and-pinion molecular mechanism

---

[1] Grill et al. (2006).

[2] Chiaravalloti et al. (2007).

[3] Mayne et al. (2004).

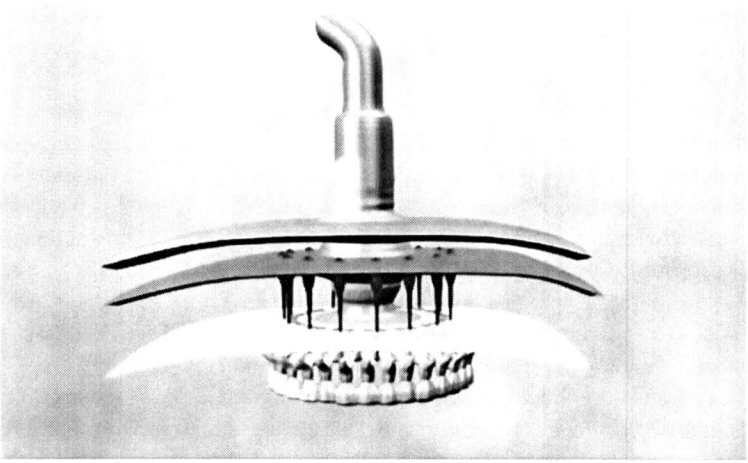

Fig. 3 Bistable Biphenylon Sio2

Two Standard Definitions

Such a molecular machine (bistable biphenyl) does not fit into the standard definitions of machine. Globally two major models of machines have been distinguished: the Cartesian model and the complex one. On the one hand, a Cartesian automaton is a multicomponent machine in which some parts are still, and some parts are in motion. Each part has been individually designed, to perform a specific task. Then all the parts have been assembled according to a specific design pre-existing in the mind of the craftsman or the engineer. Such a machine is perfectly understandable and predictable for its designer: its behaviour is completely and unambiguously determined by its components and their combination. The properties of the whole are deducible from the properties of the components, in a linear way.

By contrast, a complex machine cannot be deduced and does not pre-exist in the mind of its designer. The notion of complexity outlined by John Von Neumann at the Hixon symposium (Caltech) in September 1948, was a response to the model of neural network proposed by Pitts and McCulloch in their 1943 article (Dupuy, 2000, p. 68; Lafontaine, 2004). Von Neumann conceded that an automaton (natural or artificial) whose behaviour could be completely and unambiguously described in a finite number of words could be embodied in a neural network. In that case, it would be more difficult to describe the structure of the automaton than to describe what it is capable of. However, Von Neumann imagined an automaton whose behaviour is so complicated that it is impossible to describe it completely and unambiguously in a finite number of words. It would be simpler (infinitely simpler) to describe the structure of the automaton than its behaviour. As Dupuy and Grinbaum put it, "the threshold of complexity is the point at which the structure of an object becomes simpler than the description of its properties" (Dupuy & Grinbaum, 2004, p. 5). The behaviour of the whole is not a linear function of its components; it is unpredictable because the non-linear interactions between the parts generate effects that are not deducible from the parts. Instead of designing a structure to perform a specific task (the function determining the structure), you have to build the structure and then observe what it is capable of.

Thus, the contrast between classical machines and complex machines concerns the part/whole relationships. Classical machines are artificial totalities, for within these machines the parts exist prior to the whole and the whole is nothing but the sum of its components. By contrast, complex machines in Von Neumann's sense are more like natural totalities. They are made up of various elements interacting in loosely determined ways, and resulting in non-linear effects. From the interactions between the parts, a spontaneous order emerges. In return, the emergent order imposes constraints on the level of elementary interactions. "The whole and its elements therefore mutually determine each other" (Dupuy, 2000, p. 136).

Cartesian and complex machines are also differentiated by their ends. In classical machines, the aim of the machine is completely determined by the designer. Each component and the machine as a whole, are designed to perform a specific task. Such a machine can be said heteronymous, as its end lies in the designer's mind rather than being intrinsic in the machine. By contrast, a complex machine is autonomous because its behaviour is not strictly determined by the designer. The behaviour of such a machine is unpredictable and it escapes the control of its constructor. Therefore, the catastrophic scenario of mankind superseded by its own creation is no longer accidental in complex machines (if they are feasible). Far from being an accident, autonomy is their very essence, their *raison d'être*. Von Neumann himself prophesized that "the builders of automata would find themselves as helpless before their creations as we ourselves feel in the presence of complex natural phenomena" (Dupuy, 2000, p. 142).

According to Dupuy, Von Neumann' complexity is at the core of the metaphysical program underlying the convergence between nanotechnology, biotechnology, information technology and cognitive science, the NBIC program (Dupuy, 2004). Lack of control seems, thus, to be an essential feature of nanotechnology. "In keeping with that philosophy, the engineers of the future will not be any more the ones who devise and design a structure capable of fulfilling a function that has been assigned to them. The engineers of the future will be the ones who know they are successful when they are surprised by their own creations" (Dupuy & Gribaum, 2004, p. 6). Although the Grey Goo scenario has been discredited by chemists and even abandoned by its author, it can still be viewed as the metaphor of a major risk. "It will be an inevitable temptation, not to say a task or a duty, for the nanotechnologists of the future to set off processes upon which they have no control" (Dupuy & Grinbaum, 2004, p. 8).

Two Models of Machines in *Engines of Creation*

As a matter of fact, both standard models of machines can be found in Drexler's molecular manufacturing. Drexler defined molecular manufacturing as the fabrication of artefacts by manipulating isolated atoms or molecules with great precision. What kind of machine is able to that? Is it a Cartesian or a complex machine?

In *Engines of Creation*, Drexler quoted the definition of a machine from *The American Heritage Dictionary of the English Language*: "Any system, usually of rigid bodies, formed and connected to alter, transmit, and direct applied forces in a predetermined manner to accomplish a specific objective, such as the performance of useful work" (Drexler, 1986, p. 5). Three aspects are noticeable in this definition: (a) a machine is made on purpose out of rigid or stable components; (b) a machine is something which converts energy and transfers forces in a specific direction; (c) a machine is meant to produce work, to perform useful tasks. Nanomachines have to satisfy this definition. They are made of several parts, in motion or not, they "alter, transmit, and direct applied forces in a predetermined manner", and they have to be useful. Molecular machines are just small versions of the most familiar machines. However, despite this clear definition of what a (nano)machine should be, there is a persisting ambiguity in Drexler's use of the concept of nanomachine.

He relies on the Cartesian concept of automaton, a mechanistic and deterministic machine, in his description of the universal assembler. However, occasionally Drexler refers to Von Neumann, more precisely to his artificial self-replicating automaton (Bueno, 2004). How does he combine the two models?

First of all, Drexler claims that our modern technology relies on the same archaic principles that have prevailed since the stone era: from stone axes to electronic chips, there is no gap. Nanotechnology will break with the era of top down methods and "shape the world atom by atom". Drexler's confidence in the feasibility of his program lies in the existence of molecular machines operating in nature: biologists describe the variety of protein machines operating in the cell. Ribosomes, like factory machines, are solid; they assemble amino-acids according to the instructions given by RNA. Enzymes are capable of a wide range of tasks that they perform with great precision. Drexler acknowledges that chemistry has always been a nanoscience since chemical syntheses are molecular assemblies. However, chemists cannot control isolated molecules; they work in solution with crowds of molecules, in a messy manner. By contrast, molecular manufacturers will assemble complicated structures atom by atom according to a specific plan. This requires a convergence with biotechnology since proteins will have to be programmed to build second-generation nanomachines. "Genetic engineers are already showing the way", says Drexler. They have a precise control on assembling processes at the molecular level (for example, they can grow chains by bonding on nucleotides one at a time, in a programmed sequence). However, bio-engineering processes still rely on natural molecular tools. "When biochemists need complex molecular machines, they still have to borrow them from cells" (Drexler, 1986, p. 6). By contrast, tomorrow's nanoengineers will design artificial proteins which fold to make motors, bearings and so on, to build nanorobots capable of handling individual molecules. These tiny machines will be able to assemble parts and shape complex machines according to a specific plan, just as an automated machine tool programmed by punched tapes does. Protein machines will bond molecules with great precision (they will "combine the splitting and joining abilities of enzymes with the programmability of ribosomes"). Moreover, they will be made of a tougher stuff than the soft and weak molecular machines in the cell. Drexler calls these programmable artificial molecular machines "universal assemblers". "Enzyme-like-second-generation machines will be able to use as "tools" almost any of the reactive molecules used by chemists – but they will wield them with the precision of programmed machines" (Drexler, 1986, p. 14).

From Drexler's comparison between molecular machines in the cell and industrial machines in factories, it is clear that molecules are viewed as rigid building blocks similar to the elements of Lego construction sets. The functions performed by the various parts of molecular machinery are essentially mechanical. They position, move, transmit forces, carry, hold, and store. The assembly process itself is described as a "mechanosynthesis", positioning the components with a mechanical control. The assembler sounds like a robot arm in a factory: "the assembler itself looks like a box supporting a stubby robot arm a hundred atoms long. The box and arm contain devices that move the arm from position to position, and others that change the molecular tolls at its tip. Behind the box sits a device that reads a tape and provides mechanical signals that trigger arm motions and tool changes. In front of the arm sits an unfinished structure. Conveyors bring molecules to the assembler system. Some supply energy to motors that drive the tape reader and arm, and others supply groups of atoms for assembly. Atom by atom (or group by group), the arm moves pieces into place as directed by the tape; chemical reactions bond them to the structure on contact" (Drexler, 1986, p. 56).

Drexler's universal assembler is thus a very conventional machine, made up *partes extra partes* out of artificial parts performing precise functions. Drexler extends to the nanoworld our familiar notion of machinery: "Like all machines, they have parts of different shapes and sizes that do useful work. All machines use clumps of atoms as parts. Protein machines simply use very small clumps" (Drexler, 1986, p. 6). To be sure nanomachines do not work in Descartes's geometrical universe. As Drexler rightly emphasizes, "proposed molecular technologies likewise rest on broad foundation of knowledge, not only of geometry and leverage, but of chemical bonding, statistical mechanics and physics in general" (Drexler, 1986, p. 48). Nevertheless, Drexler shares a deterministic and mechanistic view of machines, far from the idea of a complex and self-replicating automaton. The universal assembler is not self-replicating ("assemblers will not replicate by themselves; they will need materials and energy, and instructions on how to use them"), even though Drexler himself sometimes uses ambiguous terminology.

However, the goal is to build a macroscale structure. How does Drexler plan to do it atom by atom? An isolated assembler operating at the atomic or molecular level would employ an infinite time to build such a structure, unless a great number of universal assemblers work in parallel. How would nanotechnologists obtain a great number of assemblers? Self-replication is the answer. Drexler thus introduces Richard Dawkins's notion of "replicator", a thing that can make a copy of itself (Dawkins, 1976; Drexler, 1986, p. 23). It is made of a reader, a tape, assemblers and other nanomachines. For instance, cells can be viewed as replicators, even though "some of these replicators will not resemble cells at all, but will instead resemble factories shrunk to cellular size" (Drexler, 1986, p. 56). The project is to build artificial cell-like replicators made of a more robust stuff than protein molecules.

Since universal assemblers are components of replicators, and will be replicated by them, the first problem to be solved is: how to design the first replicator? Drexler is rather vague on this point: "the chief requirement will be programming the first replicator, but AI systems will help with that" (Drexler, 1986, p. 96). He is convinced that "advances in automation will lead naturally toward mechanical replicators" (Drexler, 1986, p. 54). However, Drexler says nothing about how to make his replicators. In his vision of "radical nanotechnology", neither universal assemblers, nor replicators are proved feasible. Nowhere the reader can find a precise description of the operations performed within the machines or how replicators would work (if they were feasible). In noticing this silence, George Whitesides argues rightly that "the assembler, with its pick-and-place pincers, eliminates the many difficulties of fabricating nanomachines and of self-replication by ignoring them" (Whitesides, 2001, p. 81).

To sum up, Drexler's program makes use of two different notions of machines: universal assemblers are nanorobots operating like industrial robots. Replicators are more like Von Neumann's complex machines since they share two remarkable features with complex automata: autonomy and self-replication. But Drexler gives no clear view about the methods to make such complex machines.

His critics, George Whitesides, Richard Smalley, Philip Ball and Richard Jones pointed out a number of contradictions in his notion of molecular manufacture, and convincingly argued that they were neither plausible nor feasible. The core of these critics is that Drexler simply transferred the features of macro-machines to the nanoworld and neglected the special features of the nanoworld. Our conventional machines would not work at the molecular level, because of the unfamiliar features of the nanoworld, such as Brownian motion or surface forces (forcing nanoobjects to stick together). "I feel that the literal down-sizing of mechanical engineering popularized by nanotechnologists such as Eric Drexler – whereby every nanoscale device is fabricated from hard moving parts, cogs, bearings, pistons and camshafts – fails to acknowledge that there may be better, more inventive ways of engineering at this scale, ways that take advantage of the opportunities that chemistry and intermolecular interactions offer" (Ball, 2002, p. 16). Drexler's nanomachines, i.e. small-scaled versions of industrial macromachines, are not adapted to the special physics of the nanoworld. "Physics is different in the nanoworld, and the design principles that serve us so well in the macroscopic world will lead us badly astray when we try to apply them at these smaller scales" (Jones, 2004, p. 85).

Indeed, Drexler did not ignore the special physics of the nanoworld. He admitted that phenomena such as Brownian motion, stickiness or viscosity, had to be appreciated. But as Jones rightly pointed out, Drexler considered these phenomena as obstacles to be overcome by a number of tricks, instead of using them as positive "opportunities". It is quite clear, however that the various examples of molecular machines given above emphasize an alternative design strategy for molecular manufacturing. Indeed, in these examples, the environment of the machine (for instance the surface) is not an obstacle to be overcome, but a functional part of the machine itself. This lead to a new concept of machine.

"Soft Machines", An Alternative Concept for Nanotechnology

The controversy between Drexler and scientists such as Smalley, Whitesides, Ball and Jones focused on the feasibility of universal assemblers. The core of the debate is that Drexler failed to acknowledge the special features of the nanoworld: phenomena such as Brownian motion, stickiness, and viscosity prevent familiar mechanical engineering principles from being applied on a molecular scale to build nano-objects. The example of the nanoscale submarine is frequently quoted by critics. At the molecular level, a conventionally-designed submarine would be bumped by the numerous molecules in solution (Brownian motion), it would then be quite impossible to guide it. In addition the submarine would stick to other nano-objects.

However, Jones has to admit that Drexler's program of a "radical nanotechnology" is not contrary to the laws of nature. The real difficulties lie in the details. On his weblog Jones argues that "as soon as one starts thinking through how one might experimentally implement the Drexlerian program a host of practical problems emerge". According to Jones, Drexler is right when he claims that "nanotechnology, with sophisticated nanoscale machines operating on the molecular scale, can exist, because cell biology is full of such machines" (Jones's weblog). To a certain extent, Jones advocates the Drexlerian program of a "radical nanotechnology", applying a bottom up strategy at the nanoscale. The target of his criticisms is Drexler's underestimation of the difference between the nanoworld and the macroworld and his subsequent failure to acknowledge the special physics of the nanoworld. "Molecular machines of biology work on very different principles to those used by our macroscopic products of mechanical engineering" (Jones's weblog). The principles working at the nanoscale could not work in human engineering at the macroscale. "Big organisms like us consist of mechanisms and materials that have been developed and optimised for the nanoworld, that evolution has had to do the best it can with to make work the macroworld" (Jones, 2004, pp. 6–7). Biomaterials have been designed at the nanoscale, they work well at that level, and they are completely adapted to the special physics of the nanoworld, even though they are not always efficient at the macroscale. Drexler's program of Molecular Manufacturing is not *a priori* unfeasible, but it will succeed only by emulating the design principles of nature. Biology would be then the unique model for engineering at the nanoscale. Jones calls "soft machines" these tiny nanomachines designed on the biological principles. "The advantage of soft engineering is that it does not treat the special features of the nanoworld as problems to be overcome, instead it exploits them and indeed relies on them to work at all" (Jones, 2004, p. 127). Soft engineering will take advantage of Brownian motion combined with surface forces in the principle of self-assembly; the floppiness of small scale structures allows molecular shape changes. Together with Brownian motion they provide error correcting mechanisms that have no equivalent in conventional macroscale machines. Moreover, "there is another design principle that is extensively used in Nature, that nanotechnologists have not yet exploited at all. That is the idea of chemical computing – processing information by using individual molecules as logic gates, and transmitting messages through space using the random motion of messenger molecules, driven to diffuse by Brownian motion" (Jones's weblog). Self-assembly, molecular shape changes and chemical computing are the major design principles used by Nature that nanotechnology can exploit to build machines at the nanoscale.

As Jones argues, soft engineering principles will lead to complex nanoscale systems. The goal is actually to create complex systems that would be robust in the noisy and unpredictable nanoscale environment. "The ubiquity of noise in the nanoscale world offers a strong argument for using complex, evolved control systems" (Jones's weblog). Drexler and Jones advocate thus two different views of complexity. For Drexler, complexity means self-replication (in agreement with Von Neumann's understanding on self-reproducing automata); for Jones, complexity means robustness in a noisy environment. Drexler's mechanical nanoengineering is unfeasible because it does not emulate the complexity of biological soft machines: engineering at the nanoscale is equivalent to creating soft i.e., complex machines. Can we do that? Jones acknowledges that we still lack some essential tools for emulating biological complex systems in our soft engineering.

It is noticeable that we are not absolutely devoid of any concept of machine fitting to Jones' soft engineering. Indeed, changing obstacles into positive principles of work is exactly what French philosopher Gilbert Simondon (1924–1989) called "concretization" (versus "abstraction"). A machine is "abstract" when each part has been designed independently, each one for a definite and unique function. In the design process, the special features of the environment where the machine will operate are not included in the design itself: they remain exterior to it, they are possibly viewed as obstacles or problems to be overcome. Drexler's molecular manufacturing leads to abstract nanomachines. On the contrary, a concrete machine is more than the materialization or the application of scientific principles. It is not designed *partes extra partes* on the model of Cartesian machines. The environment where the machine will operate is not an external feature or a simple parameter that engineers have to take into account in the design process. The *milieu* is not something to which the machine will have to be adapted; it is an intrinsic dimension in the design of the machine. A concrete machine works precisely because of (and not despite) its association with a specific environment. It is invented straight off by envisioning, "imagining" the feedback loops between the machine and its *milieu associé*. Design and operation are not two independent tasks. A concrete machine is designed according to its operating conditions. The *milieu associé* becomes an integral part of the machine. Moreover, all the physical and chemical phenomena occurring in a concrete machine are functionalized.

Simondon exemplified this style of technological invention in describing a hydraulic power station, designed by the French engineer Guimbal.

Guimbal had to design an electric generator, small enough to be immersed into a water pipe. The major obstacle was the heat produced by the generator, which would cause its explosion at a critical point. Conventional engineers trained in "abstract" engineering would try to reduce the size of the couple turbine/generator to avoid explosion. The environment where the machine should operate (water) would not be taken into account in the design process: their machine would be an open air engine, modified to be immersed in the water pipe. By contrast, the "concrete engineer" will include the characteristics of the *milieu* in the design principles. The generator is in a container filled up with oil and is coupled to the turbine by means of an axis. The system turbine-generator is immersed into the pipe. Water will have a dual function : it supplies power to the turbine, and at the same time it exhausts the heat generated by the rotation of the turbine. Oil is also multifunctional: it lubricates the generator, it conveys the heat released by the generator to the surface of the container, which is cooled by the water, and it prevents water to come into the container, due to the difference of pressure between oil and water. The faster the turbine and the generator will rotate, the greater the agitation of oil and water (and subsequently the cooling of the system) will be. As G. Simondon emphasized, the water/oil *milieu* determined the design of the generator. The Guimbal turbine would explode in open air. The concrete machine is highly adapted to its environment (in this case, the couple oil and water). Simondon calls *individu technique* (a technological individual) such a self-conditioned machine.

Jones's soft machines share some of the major characteristics of Simondon's concrete and self-conditioned machines: the environment and its special features (Brownian motion, stickiness, viscosity, and floppiness) are included and exploited in the design process. In other words, such machines could not operate at the macroscale: they are designed to work in association with a specific *milieu*.

Soft Engineering in Nanobiotechnology

Current researches in the field of nanobiotechnology fit into Jones' approach to nanodesign in terms of "soft engineering". For instance, the Nanoscience Group of the Laboratoire d'Architecture et d'Analyse des Systèmes (LAAS, France) is leading an ongoing research focusing on the design and the building of an artificial analogon of the motor of a bacterial flagellum (E. Coli) (Fig. 4).[4]

The ultimate goal of researchers is to control the motion of a tiny vector that may be functionalized for a specific task, such as drug delivery in the human body. To reach this objective, scientists of the LAAS have considered three possible strategies.

The first one consists in shaping the motor "atom by atom", or rather "protein by protein", according to Drexler's vision of molecular manufacturing. However, this design strategy sounds impossible since scientists do not understand how nature arranges proteins to build the motor of the flagellum. They are convinced that if they want to succeed and build such a complex nanomachine, they will necessarily have to emulate the design principles of nature. A major objective of the research is thus to understand how nature arranges proteins step by step in its design process. The first conclusion is that scientists seem to fit in with Jones' analysis: natural mechanisms and materials are optimized at the nanoscale.

---

[4] See http://www.laas.fr/laas/1-5595-Nano-Moteur.php

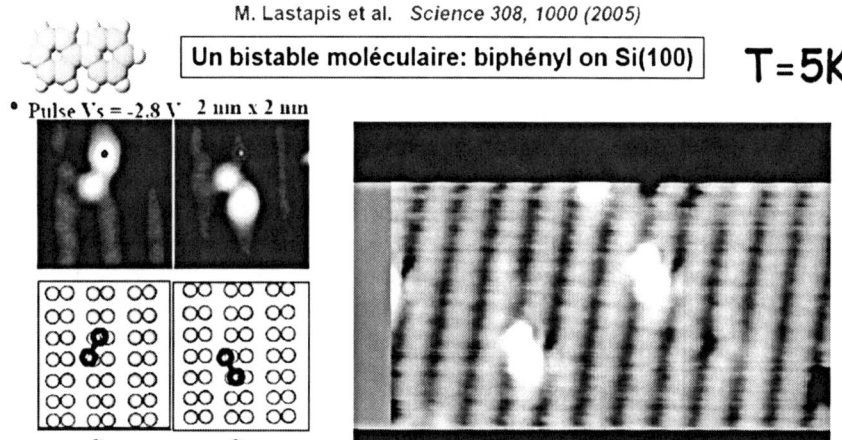

Fig. 4 The model of the nano-motor of the flagellum of E. Coli

The second strategy consists in avoiding the difficulty. Scientists do not aim at designing an artificial analogon of the biological motor. They take the whole bacteria and give it some interesting functions by means of genetic engineering. So they "instrumentalize" nature rather than mimicking it to design artificial nanomachines. Their goal is no longer to understand how nature proceeds, it is to exploit living beings and to make them perform useful tasks and serve human interests.

Thus, both the first and the second strategies have in common to consider nature as a reservoir of devices (elementary blocks of matter or living beings) at the disposal of engineers. On the contrary the third strategy, experimented by the Nanoscience Group of the LAAS, consists neither in designing the motor "protein by protein" according to a previous artificial plan, nor in harnessing living beings (bacteria). To be more precise, they adopted a hybrid strategy, which consists (i) in a "translation" of natural processes into technological, functional schemes, leading to a mechanical model of the biological motor; (ii) in the preparation of a surface as close to the natural membrane of the bacteria as possible, finally (iii) they lay the various proteins of the biological motor on such an artificial surface with the expectation that the proteins will self-assemble on the surface as they do in bacteria. The building process, if and when it occurs in these conditions, is followed step by step by means of an Atomic Force Microscope in order to validate (or falsify) the mechanical model of the motor. So this design strategy rests on a major assumption: to build an artificial analogon of a biological motor, scientists have to recreate its environment. The artificial motor can be built if and only if it includes the surface in its functional scheme. The surface is not external to the motor, it is part of it as in the example of the bistable biphenyl.

This approach to nanodesign resembles the "protein by protein" strategy because it implies an analysis of the whole process step by step; however the mechanical model of the motor is not an a priori program guiding the design process (as it is in Drexler's vision of molecular manufacturing with programmable universal assemblers): it is nothing but a mechanical hypothesis about the running of the motor, to be validated. Although the approach of the LAAS resembles the "harnessing living beings" strategy because it refers to the insuperable complexity of nature, it does not instrumentalize nature; the experiment consists in bringing some homogeneity between natural processes and human technology. The goal is to build up a model that is at the same time compatible with our mechanical understanding of phenomena and adequate to the effective running of the biological motor. Undoubtedly the motor of the LAAS fits in with Jones' approach to soft engineering and with Simondon's definition of concrete machines: the design strategy tested here actually aims at converting natural processes into functional, technological schemes. To be sure, the vision of nature involved in this experiment necessarily differs from both the "instrumentalized" vision of nature and the mechanistic vision of Drexler.

Conclusion

This critical survey of the concepts of machines used in the domain of nanotechnology and biotechnology emphasizes a major weakness in the understanding of nanotechnology. It has often been pointed out that the definition of nanotechnology suffered from the vagueness of the prefix «nano» which can refer to many different things at a length scale of the nanometer. However the term «technology» in the phrase «nanotechnology» is as vague and indefinite as the prefix nano, as long as we rely on a metaphorical notion of nanomachine.

The concept of "technology" (after the Greek *Technê* and *Logos*) means a rational approach to machines. It is a special research field devoted to machines, a field which has a rationality of its own, irreducible to scientific rationality as well to economical principles. In a technological perspective, machines are neither applications of established scientific principles, nor social objects defined by their utility.

In order to be a technology *stricto sensu*, nanotechnology should develop its own methods of analysis and classification of machines. It should be focused on operations rather than on structural aspects. Whereas Drexler's vision of a molecular manufacture seems to us a technological non-sense, Jones's description of soft machines with its emphasis on the operational conditions in the nanoworld points toward a technological understanding of molecular machines. Among the challenges that will have to be faced to develop a technological understanding of nanotechnology, let us mention the collapse of the traditional categories of nature and artefact and the construction of a "common world" for humans and machines.

References

Ball, P. (2002). Natural Strategies for the Molecular Engineer. *Nanotechnology*, 13, 15–28

Baum, R. (2003). Nanotechnology. Drexler and Smalley Make the Case For and Against Molecular Assemblers. *Chemical & Engineering News*, 81(48), 37–42

Bensaude-Vincent, B. (2004). Two Cultures of Nanotechnology?. *International Journal for Philosophy of Chemistry*, 10(2), 65–82

Bueno, O. (2004). Von Neumann, Self-Reproduction and the Constitution of Nanophenomena. In D. Baird et al. (eds.), *Discovering the Nanoscale* (pp. 101–118). Amsterdam: IOS Press

Chiaravalloti, F., Gross, L., Rieder, K.-H., Stojkovic, S. M., Gourdon, A., Joachim, C., & Moresco, F. (2007). A Rack-and-Pinion Device at the Molecular Scale. *Nature Materials*, 6, 30–33

Dawkins, R. (1976). *The Selfish Gene*. Oxford: Oxford University Press, reprinted in 1989

Drexler, K. E. (1981). Molecular Engineering. An Approach to the Development of General Capabilities for Molecular Manipulation. *Proceedings of the National Academy of Sciences*, 78(9), chemistry section, 5275–5278

Drexler, K. E. (1986). *Engines of Creation. The Coming Era of Nanotechnology*. New York: Anchor Books

Drexler, K. E. (1992). *Nanosystems. Molecular Machinery, Manufacturing and Computation*. Palo Alto, CA: Wiley

Dupuy, J.-P. (2000). *The Mechanization of the Mind. On the Origins of Cognitive Science*. Princeton, NJ: Princeton University Press

Dupuy, J.-P. (2004). Complexity and Uncertainty. A Contribution to the Work in Progress of the "Foresighting the New Technology Wave". Bruxels: High-Level Expert Group, European Commission. http://www.ulb.ac.be/penser-la-science/images/conf2/dupuy_complexity.pdf

Dupuy, J.-P., & Grinbaum, A. (2004). Living with Uncertainty: Toward the Ongoing Normative Assessment of Nanotechnology. *Techné*, 8(2), 4–25

Grill, L., Rieder, K.-H., Moresco, F., Rapenne, G., Stojkovic, S., Bouju, X. & Joachim, C. (2006). Rolling a Single Molecular Wheel at the Atomic Scale. *Nature Nanotechnology*, 2, 95–98

Guchet, X. (2005). *Les Sens de l'évolution technique*. Paris: Editions Léo Scheer

Jones, R. L. (2004). *Soft Machines. Naotechnology and Life*. New York: Oxford University Press

Jones, R. L. Weblog, access July 20, 2006

Lafontaine, C. (2004). *L'empire cybernétique. Des machines à penser à la pensée machine*. Paris: Seuil

Mayne, A. J., Lastapis, M., Baffou, G., Soukiassian, L., Comtet, G., Hellner, L., & Dujardin, G. (2004). Chemisorbed Bistable Molecule: Biphenyl on Si(100)−2×1. *Physical Review B*, 69, 045409

Merkle, R. (1992). Self Replicating Systems and Molecular Manufacturing. http://www.zyvex.com/nanotech/selfRepJBIS.html

Rapenne, G., Grill, L., Zambelli, T., Stojkovic, S., Ample, F., Moresco, F., & Joachim, C. (2006a). Launching and Landing Single Molecular Wheelbarrows on a Cu(1 0 0) surface. *Chemical Physics Letter*, 431, 219–222

Whitesides, G. M. (2001). The Once and Future Nanomachine. *Scientific American*, Sept, 78–83

No Future for Nanotechnology? Historical Development *vs.* Global Expansion*

Alfred Nordmann

Abstract Since there are few disruptive nanotechnological products and processes now, it would seem that ethical and societal deliberations concern what nanotechnology may bring in the future. This orientation towards the future is shown to be unnecessary and wholly inadequate to technoscientific research programs. Since these programs posit that there is something deficient or problematic about the present that will benefit from a nanotechnological solution, they posit not a future but an alternative world. Since nanotechnology is primarily a conquest of space, critiques of colonization and globalization may offer the most appropriate resources for the assessment of this alternative world.

Keywords Nanotechnology, technoscience, colonization, globalization, technology assessment, nanoscience and technology studies

The implications of nanotechnologies are often assumed to be like those of genetically modified organisms. One might also compare them to the introduction of plastics. Either way, nanotechnologies are said to be profoundly transformative. Whether one envisions the cure of cancer by 2015,[1] another industrial revolution, or a new renaissance,[2] much and perhaps everything will change and nothing remain as it is now.

But what does it mean and what should one do when told that everything will change? First, one might want to know with greater detail just how likely it is that this or that will actually change within our lifetimes or beyond. In particular, one might wonder how the envisioned changes affect our sense of self, in which ways

---

*This paper is a translation, somewhat updated and expanded revision of "Wohin die Reise geht: Zeit und Raum der Nanotechnologie," in Gamm and Hetzel (2005, pp. 103–123). It is profitably read alongside other contributions to this volume, especially those by Gerhard Gamm, Jean-Pierre Dupuy, Christoph Hubig, Jutta Weber, and Ingeborg Reichle.

[1] As does, for example, the brochure *Cancer Nanotechnology: Going Small for Big Advances*, US Department of Health and Human Services, National Institutes of Health, National Cancer Institute, January 2004. Only in March 2006, Sidney Wolfe of the Public Citizen's Health Research Group spoke out against the insensitivity of such promises to the hopes and fears of cancer patients.

[2] See especially the introduction to Roco and Bainbridge (2002).

Department of Philosophy, Darmstadt Technical University

and to what end they expand human powers, or how they might reconfigure society, nature, and technology. Second, one might want to resist the threat that is implied by the assertion that everything *will* change and that we better brace ourselves for what lies ahead. This implied threat of an ineluctable technological future has motivated publicly commissioned philosophical reflection and social science research of the ethics of nanotechnology and "human enhancement," in particular.[3] Such research thus operates in a paradoxically defined space as it explores degrees of freedom in the face of apparent inevitability. Against the unrelenting "*will* change" one might want to reclaim political space for the deliberation of a genuine choice in the matter.[4] The "*will* change" finally prompts a third response and it is the focus of the following reflections. Are we to imagine a more or less distant future for which we must assume responsibility now, or else, are we already implicated in this change, is it taking place as we speak?

On first sight, this third issue appears to be a matter merely of roadmaps and timelines. If expanded memory storage and technically enhanced computational capacity of the human brain will be achieved no sooner than 2100, this transformative change would seem to affect future generations and not the present. Accordingly, we would need to conceive our current research programs with a sense of responsibility towards the future. Beyond roadmaps, however, one might argue that this change is already happening. Experiments are being conducted now to explore the possibility of brain-machine interactions, the once fundamental distinction between organic and inorganic, living and dead matter has been undermined for some time, and some are already cursing the present and the shortcomings of their own bodies for being born too early – in light of what they envision for a merely hypothetical future.[5] Thus, whether we envision nanotechnological change in terms of the future or of the present is not a question of roadmaps and timelines alone.

After rehearsing different ways of conceptualizing the progress and future of nanotechnologies, the following reflections recommend a change of perspective, suggesting that the advance of nanotechnologies should be considered in terms of global expansion or the conquest of space, that is, as a process decidedly in and of the present.

The Futures of Nanotechnology

The English acronym "TA" for "technology assessment" has been rendered in German in a somewhat restricted manner that introduces a further dimension. Instead of translating the concept literally as "*Technikabschätzung*" or

---

[3] For a critique of such speculative ethics, see my "If and Then: A Critique of Speculative NanoEthics" (2007).

[4] This point was emphasized in discussions at the NanoEthics conference (Columbia, SC, March 2005) by Mickey Gjerris.

[5] Compare Dupuy (2007). Ray Kurzweil's desperate attempt to build a bridge towards the time at which nanotechnology will give him immortality is an example of this "promethean shame"- a term coined by Günther Anders, that is, a present feeling of deficiency in comparison to what we might make of ourselves in the future, see Anders (1980) and Kurzweil and Grossman (2004).

"*Technologieabschätzung*," German TA is conceived as the assessment of the implications, consequences or effects of technology ("*Technikfolgenabschätzung*"). When a German TA researcher like Armin Grunwald reflects the extent to which the development of nanotechnology is subject to social shaping, he must do so in the horizon of history and begin by probing conceptions of technology's relation to the future. Indeed, he juxtaposes three rivalling conceptions.[6] The *prognostic* approach to technology assessment assumes a knowable future, a future that is already given and thus impervious to our interventions. It accords with a view of technological determinism and the claim that what is technically possible will sooner or later become realized. The *constructive* or *social shaping* approach posits an open future that is up for grabs and yet to be determined by us. Only our decisions and actions make the future.[7] Finally, the *evolutionary* approach places the future in a genuinely historical perspective. In one respect at least, the future of technical development will be like its past: Historical analysis shows that it was never possible to predict, let alone derive the future from the past. This will surely hold for current attempts to predict the future, too. From the point of view of the past, the future is always open. History also shows, however, that the present is always indebted to the past, that the explanations of the present lie in the past, and that the present is hardly open to arbitrary shaping. From the point of view of the present or of the future, our lives are to a large extent determined by the past. An evolutionary understanding would therefore consider the future undetermined but also deny that we can shape it at will. The evolutionary approach to technology assessment will identify its underlying social dynamics and discover sites for intervention and debate. As for determining the future, it can influence the discursive landscape or economic environment in which further technical development unfolds. For example, the concept of 'sustainability' serves to frame but not to plan technological development, partly because it is itself subject to public contestation.

There is no clear-cut criterion for the correctness of one or the other conception of the future. Grunwald does not call for a choice between them but demands for the sake of transparency that pertinent presuppositions about historical development are rendered explicit in societal deliberations of technology – especially in the case of nanotechnology that thrives on representations of its imagined future.[8] He introduces a normative consideration where he emphasizes the distinction between

---

[6] Grunwald (2003), also Grunwald (2006).

[7] Grunwald does not present this as a characterization of "constructive technology assessment" which pays close attention, for example, to "emerging irreversibilities" and thus starts from an analysis of the space of actions and actors. In regard to nanotechnology, Rip's contribution to the European workshop on Social and Economic Research on Nanotechnologies and Nanosciences, Brussels, 14–15 April 2004.

[8] Here, Grunwald's position comes close to that of Rip who might consider the three conceptions of technological development as folk theories that may or may not be shared by enactors (typically actors who promote a technological development) and comparative selectors (typically their many publics who think of themselves as having a choice in the matter). A first step in constructive technology assessment (CTA) is for CTA actors to clarify the status and strategic role of such folk theories of technological development. See Rip (2005, pp. 15–24). Grunwald offers the term "vision assessment" for this process of clarification.

the perspective of actors and observers.⁹ While the three approaches to the future may appear equally credible or valuable to observers, actors have to follow what one might call an imperative of the political and must suppose that technical development is subject to shaping.¹⁰ A heuristics of precaution, of a prudential cost-benefit calculus, or of sustainability can set critical limits or establish positive goals, even if it were the case that the participants' faith in their ability to shape technological development will ultimately be exposed as illusionary.¹¹ Though only one or the other of Grunwald's three "futures" may prove adequate in any given instance, all three have this much in common: The future is in any case just that which will happen at a future time.

For Jean-Pierre Dupuy, in contrast, the future of nanotechnology is not what in the course of time turns out to be this way or that. The future of nanotechno*logy* is what corresponds to its *logos* and what is already contained in our conception of nanotechnology. In a sense, this future is already prefigured, for example, in the program of a "bottom up" approach that avails itself of principles of self-organization and that views our presently given world as an aggregate of attributes that can be manipulated at will.¹²

Dupuy introduces this conception of the future as "enlightened dooms saying" or "catastrophisme éclairé."¹³ Of the three approaches presented by Grunwald it has greatest affinity on first sight to the prognostic view in that it posits the future as already given and therefore not subject to prudential shaping. Grunwald comments

---

⁹Cynthia Selin complicates this picture in two essays. In discussions of the future of nanotechnology, there is not a definition of nanotechnology on the one hand and different interpretations of its future-orientation on the other. Different conceptions of time (evolutionary vs. revolutionary, the medium- or the long-term, a future that is inevitable or open to shaping, etc.) inform the dispute over what nanotechnology is in the first place. On the one hand, the competing conceptions of time co-exist indefinitely, on the other hand they enter into conflict over the proper conception of nanotechnology. Selin refers to Brown et al. (2000), compare Selin (2006, 2007).

¹⁰Here, perhaps, Rip and Grunwald part company. By distinguishing enactors and comparative selectors not in terms of social location but as engaging in two types of activity, constructive technology assessment questions the opposition of inside and outside perspectives. All parties are actors and for their actions they draw strategically on folk theories of technical development. As Rip points out, for example, enactors who promote a technical development often do not believe that the technology can be shaped but believe instead that the conditions of its reception (public acceptance) can be shaped.

¹¹Pace Selin (note 9 above), these critical limits and positive goals need not be oriented toward more or less general, more or less speculative or contested conceptions of nanotechnology. They engage the scope of technical action that at any time has already been constituted through instrumentally mediated interventions. Instead of defining nanotechnology as concerning everything molecular or everything in a certain size regime, the domain of nanotechnology is defined by Grunwald as that domain "where machines stand ready to analyze and manipulate at the nanoscale" (personal communication, 2005). Grunwald's conception owes to Peter Janich, e.g. his "Wissenschaftstheorie der Nanotechnologie," (2006).

¹²See Dupuy (2007, 2004).

¹³Dupuy (2002).

on this view: "The inner connection between prognosis and determinism leads to the absurd situation, that if an optimal prognosis were possible, it would have no use." It is a virtue of Dupuy's approach that he confronts this absurdity or paradox. In a sense, the paradoxical uselessness of saying what the future holds prompts an abhorrence that makes us shy away from this otherwise inevitable future.

According to Dupuy we cannot shop for a palatable future by choosing a preferred pathway according to the criterion of sustainability or the like. If one imagines the future as something that can be adapted to prudential considerations and if one imagines that there is always a choice between more or less sustainable but equally possible futures, one will never find credible the impending catastrophe. Moreover, we cannot even assign meaning to the word "future" if the future is as of yet undetermined, one of various scenarios that might become realized. Instead, there can be only *the* future, our one and only future, and it attains meaning precisely because it is what and who we will become. That one future is already determined, not however by being uniquely predicted through extrapolation from the inner logic of technical development. Instead, it is determined in the sense of a prophetically projected future that is envisioned or claimed by the *logos* of nanotechnology.[14] Such a projected future, Dupuy argues, is implicit not only in his negatively prophetic warning but also in attempts to create a positive feedback loop between projection and realization. In the latter case,

> it is a matter of obtaining through research, public deliberation, and all other means, an image of the future sufficiently optimistic to be desirable and sufficiently credible to trigger the actions that will bring about its own realization. It is easy to see that this definition can make sense only within the metaphysics of projected time, whose characteristic loop between past and future it describes precisely. Here coordination is achieved on the basis of an image of the future capable of insuring a closed loop between the causal production of the future and the self-fulfilling expectation of it.[15]

If the analysis of the *logos* of nanotechnology does not suggest a desirable image of the future, however, all we can hope for is a negative feedback loop such that the promise of a catastrophe can break the self-fulfilling cycle of expectation and causal production. Both, positive and negative feedback begins with an image of the future that does not need to be articulated in detail. On the one hand, the future of nanotechnology appears indeterminate; on the other hand we can know that it signifies a catastrophe. This only apparent tension is easily resolved by Dupuy in that he derives nanotechnology's catastrophic character immediately from its objective indistinctness or indeterminacy. This indeterminacy is objective because it does not depend on our presently and contingently limited state of information. It is not merely epistemic because it comes with systematic unpredictability at the nanoscale. We encounter this systematic unpredictability, for example, in the complex systems that are described by non-linear dynamics. These systems cease at so-called tipping points to change in a gradual manner that is strictly proportional to the

---

[14] This is how Dupuy's account differs from the prognostic view that was identified by Grunwald.
[15] Dupuy (2004, p. 91).

causal influences upon them. As they reach a tipping point, they suddenly shift into a new state of organization. In other words, at their tipping points these systems behave catastrophically. Dupuy describes this as follows:

> Beyond certain *tipping points*, they veer over abruptly into something different, in the fashion of phase changes of matter, collapsing completely or else forming other types of systems that can have properties highly undesirable for people. In mathematics, such discontinuities are called *catastrophes*. This sudden loss of resilience gives complex systems a particularity which no engineer could transpose into an artificial system without being immediately fired from his job: the alarm signals go off only when it is too late. And in most cases we do not even know where these tipping points are located. Our uncertainty regarding the behaviour of complex systems has thus nothing to do with a temporary insufficiency of our knowledge, it has everything to do with objective, structural properties of complex systems.[16]

This behaviour of complex systems occurs in nature most prominently where technical advance, excessive consumerism, exploitation, over-population and pollution overtax the resiliency of eco-systems. Such situations might also be produced through the creation of quasi-natural technical systems that enter into complex interactions which introduce further instabilities.[17] And precisely this appears to be implied by the program of nanotechnology. Its bottom-up approach aims to recruit principles of the self-organization of complex systems. Also, the pervasive integration of technical systems into the environment promises to increase complexity.[18] In both respects, the hybridization of technology and nature may produce an increase of objective unpredictability, ignorance, and catastrophic instability by way of 'complexification.'

According to Dupuy, if there is a way to avoid catastrophe, it does not consist in prudential measures like prevention and limitation, exercises of preparedness, improved sensors, strict legislative oversight, or the like. As perhaps with Heidegger and his students, a possible escape could only consist in a dramatic change of course that shies away from catastrophe. This would amount to a historical accident or singularity prompted by a negative feedback-loop:

> It is a matter of achieving coordination on the basis of a negative project taking the form of a fixed future that *one does not want*. [...] to obtain through scientific futurology and a meditation on human goals an image of the future sufficiently catastrophic to be repulsive and sufficiently credible to trigger the actions that will block its realization.[19]

Dupuy invests some work to establish the conceptual possibility of such a singularity. Historically, he finds it in the history of the arms-race where the prospect of an

---

[16] Dupuy 2004, pp. 80f.

[17] Compare my "*Noumenal* Technology: Reflections on the Incredible Tininess of Nano" (2005). It was inspired, in part, by Dupuy's critique.

[18] While the attempt to do so is clearly part of the program of nanotechnology, it is not at all clear yet whether science and technology will actually be able to harness these processes.

[19] Dupuy (2004, pp. 91f.) Dupuy goes on from here to resolve an apparent technical difficulty in his account of time: "*If one succeeds in avoiding the undesirable future, how can one say that coordination was achieved by fixing one's sights on that same future?*" (On Dupuy's account, of course, "that same future" is our one and only future.)

accidentally triggered "mutually assured destruction" may have prompted a turn towards arm-control.[20] With explicit reference to Hans Jonas, Günther Anders, and Hannah Arendt, Dupuy thus recommends a heuristics not of prudence or calculation, but of fear.

With explicit reference to Ernst Bloch, George Khushf finally pursues a heuristics of hope with regard to nanotechnology.[21] For him, the future is neither what will happen at some future time, nor is it what is already contained in the *logos* of nanotechnology. Instead, it is an anticipation, foreshadowing, or adumbration of a potential that needs to be realized responsibly. For Khushf, this potential is contained in the notion of a "technological convergence" that is rooted in nanotechnology, Bloch refers to this potential as "allied technology," that is, technology allied with nature [*Allianztechnik*]:

> Just like the final manifestation of history, so the final manifestation of nature lies in the horizon of the future, and toward this future are oriented the categories of mediation of the concrete technologies that we can safely expect. The formative powers of frozen nature will surely once again be unleashed to the extent that in place of a merely external technology an allied technology will become possible, that is, a technology that is mediated with the co-productivity of nature.[22]

While Dupuy warns of any technology that unleashes the self-organizing, formative, potentially catastrophic forces of nature – forces that have no interest in the human species as such –, Khushf bets on these forces and thus also on the ethical and developmental formation (*Bildung*) or co-evolution of humans and nature.[23] While Dupuy pursues a negative project that does not allow for precautionary prevention or anticipatory remediation but demands a radical break, Khushf's project is directed positively at the realization of a new world that is already announcing itself. And while Dupuy takes ethics to be rooted in an acknowledgement, even loving embrace also of human frailty that can radically challenge and doubt itself (Kant's warped wood from which man is made[24]), Khushf views ethics as an affirmation of freedom that consists in the reflection and bringing-forth of the good.

Khushf delineates the task of responsibly conceiving the future in a paper on "The Ethics of NBIC Convergence for Human Enhancement."[25] At any given time

---

[20] Jean-Pierre Dupuy, personal communication, 2006.

[21] The complementarity of Jonas's "heuristics of fear" and Bloch's "heuristics of hope" was identified by Wolfgang Bender (1996).

[22] Bloch (1973, p. 807) (my translation).

[23] Compare ibid., p. 810, where Bloch envisions a technology that accords with Kant's characterization of the arts and of a creative imagination that acts like nature and can be regarded as another nature. In Nordmann 2005 I identify the character of naturalness that is assumed by certain genetic or nanotechnologies and relate this seeming naturalness to the apparent uncanniness of these technologies.

[24] Kant (1983, p. 34).

[25] Khushf (2007). Khushf provides a more sustained analysis in (Khushf, 2003), also (Khushf, 2004). The following presentation of his approach is also based on a number of personal conversations, since 2002.

in history there obtains an equilibrium between technical capacities, conceptions of the good life, and ethical norms. Even continuous and gradual technological development can disturb this equilibrium, enable new conceptions of the good life and occasion new ethical norms. If technological development is radically discontinuous this does not signify a relatively smaller or larger departure from equilibrium. As indicated by theories of non-linear complex dynamics, it requires instead a spontaneous reorganization at a higher level. In the normal case where there is a mere departure from an equilibrium that needs to be restored, ethics lags behind as it identifies distributing issues and ties them back into traditional discourses. However, the visions of nanotechnology and of the convergence of nano-, bio-, and information technologies intend a radical transformation of the organization of research, of life and nature as subject to willful and creative shaping. They aim for an equilibrium at a higher level and thus challenge ethics to proactively take part in the creation of what shall be.

This ethical project unfolds in parallel at all the levels at which the formative powers and complex dynamics of nature are unleashed. On one level, nanotechnology aims to exploit the bottom-up self-organizing principles of nature. On quite another level the very same structural paradigm of self-organization brings about a new configuration of academic disciplines that no longer divide nature among themselves in a classically hierarchical manner (fundamental particles to social organizations). Finally, on the third level, the new technologies effect a profound reorganization of all aspects of human life, livelihood, economy, sociability and health. According to Khushf, ethics enters the game just as soon as one takes seriously the nanotechnological claim that it will radically transform the organization of knowledge and of society. One is then always already implicated in the possibly competing conceptions of how to "truly integrate humans with nature."[26] The same process unfolds on all three levels in parallel and this similarity or mutual mirroring serves to integrate the research process. This constitutes a kind of reflexive circle that incorporates and internalizes at a higher level what initially appears as an external disruption. Here the notion of physical and social formation explicitly links up with the educational notion of *Bildung*, that is, of an ethical development of the self. In the course of personal development, after all, traditional norms are first encountered as external rules and then incorporated into a mature self-understanding, a process that satisfies the general scheme of self-organization:

> There is an important difference between the way a child and adult approaches ethics. For children ethical norms are external impositions on wants and will. Rules prevent you from having candy, taking John's toy, or playing instead of going to school. For the adult those rules are internalized and become an expression of one's own life, sustaining vitality and orienting practice so that individual and communal flourishing coincide. The external rules about candy, stealing, and school are transformed into wisdom about how to eat, relate to others, and progress in knowledge and understanding. Adults transform the rules so they become an inner guidance for their life, the tools by which they craft in a responsible way their own future. In the face of the ethical challenges associated with

---

[26] Bloch (1973, p. 817).

> NBIC convergence, we need to enter maturity, developing that form of reflection that characterizes an adult. Only as adults should we enter the radically new world that opens up in front of us.[27]

In the self-education of humankind, the future thus appears in its infancy but already announces its mature personality. The shaping of this maturation process Khushf considers as a kind of formation or *Bildung* such that the outer formation of a radically new world is accompanied by implicit and explicit inner commitments also of the researchers and developers. The convergence of diverse technologies and disciplines is therefore quite distinct from the mere unleashing of science and industry's powers of productivity:

> Here the character of the task, and the opportunity to craft the future we now enter, all come into view. Appropriately understood, the NBIC initiative [of converging technologies] does not just drive into the future, with engines of science and industry running full throttle; even beyond this, NBIC convergence, with a newly developing form of ethical reflection, can responsibly lead into a future, where the engines of growth are also the engines of self-regulation, reflection, and mature governance.[28]

At the end of this survey, Armin Grunwald enters the conversation one more time with a strategy for dealing with "nanotechnology as a cipher of the future." If the future is a medium for the communication of and about nanotechnology,[29] one needs to consider closely how the future appears in these communications. Different representations of the future require different assessments. Constative statements that confidently pronounce what the future holds may prove more or less credible depending on how far-fetched or attainable this future is. In contrast, statements that posit a future in order to create positive or negative feedback-loops need to be judged perhaps in ethical or political terms, perhaps under the aegis of the precautionary principle. And yet again, a hypothetical mode of posing uncertain possibilities calls for public engagement in a process of shaping nanotechnological development. Grunwald concludes:

> This may sound sobering, but the function of nanotechnology as cipher of the future is *not* to show us the future. Its function is also not to display alternative futures that we can choose amongst as in the shelves of a supermarket. Instead, the function of ciphers of the future is to draw our expectations of the future into our current thinking, to reflect them there, to communicate and reach understanding about these reflections, and to finally render all of this fruitful for current actions and decisions – for, these require conceptions

---

[27] Khushf (2007). This passage indicates that, as opposed to Dupuy, Khushf considers the formative powers of nature to be subject to shaping. In his own work, Khushf seeks innovative ways to shape such processes, for example, by reframing research in collaboration with molecular biologists, geneticists, medical researchers, bioethicists such that the reflexive circle can become productive in the creation of new and shared concepts. – This perspective on the formation of a new generation of natural and engineering scientists is less pronounced but still present in Khushf's paper on "The Ethics of Nanotechnology" (2003). Here, too, this formation is said to contribute to the maturation of humanity.

[28] Khushf (2007).

[29] See Lösch (2006).

of the future. Nanotechnology is a cipher of the future – but by being precisely this, it casts us back upon ourselves and our present.[30]

Indeed, once we take this detour through the future, we need to adopt Grunwald's proposal and analyze various ways of talking of the future as a means to articulate our current expectations and demands. However, the following reflections challenge the notion that this detour is really necessary and that we need conceptions of the future to deliberate decisions at present. These decisions, one might say, concern what our world is like and how our world should be – quite irrespective of time or history. Accordingly, the critical analysis of nanotechnology as a cipher of the future should be supplemented, perhaps preceded by an effort to deflect talk of our nanotechnological future and to redirect it toward a consideration of the ways in which the various nanotechnologies lay claim on our world and our lives.

From Time into Space

If nanotechnologies take us into the future or confront us with an image of the future, that future eludes our grasp in a variety of ways. It remains unclear, first of all, whether our relation or orientation to this future should be conceived in Grunwald's terms as prognostic, constructive, or evolutionary, as projective in Dupuy's sense, or formative, even self-expressive as suggested by Khushf. Furthermore, according to all those ways of relating to the future it remains quite unclear what this future actually holds, which of our current programs will have been realized, how future developments ought to be taken seriously in today's policy decisions, or how they might engage our cultural conversations.[31] Even Dupuy's projective fixing of a catastrophic future construes our relation to it as ultimately metaphysical.[32] On all these conceptions, the future of nanotechnologies is a horizon of expectation in which something unheard-of or unspeakable will appear.[33] Indeed, all this confusedness regarding the future serves as a first argument for

---

[30] Grunwald (2006, p. 78).

[31] Valerie Hanson therefore warned in her presentation "The Role of Anticipatory Rhetorics in Discussions of Nanotechnological Ethics" firstly, that attention to the realization of future-oriented programs may obscure presently problematic aspects of nanotechnology, and secondly, that the training of ethical sensibility to emerging situations may preclude us from engaging seriously with present reactions to nanotechnology.

[32] On the one hand, there is a trust that nanotechnology can indeed realize its *logos*, for example, that it will actually become capable of technically exploiting processes of self-organization. On the other hand, the human condition is taken not merely as our frame of reference for making sense and assigning meaning, it is absolutized as the only conceivable frame of reference. Though it is indeed perverse to posit and desire an ill-defined post-, trans-, at any rate non-humanity, and though ethical and political debate is ill-served by speculations about future ways of being someone other than ourselves, there ought to be some other way out of this predicament rather than universalizing how we now find ourselves.

[33] Compare Kaminski (2004).

abandoning an orientation toward the future. This could be done by way of a critical analysis that refers us back to the present or, better yet, by way of ceasing to speculate altogether about the future of nanotechnologies.[34]

In respect to the unspeakable or unheard-of that lurks in the future, there is no Archimedean vantage-point that would allow us to pinpoint it or to subject it to a normative assessment. Hans Blumenberg characterizes this predicament by speaking of the coincidence of the "not yet" and the "no longer." Confronted with technology as promise and fulfilment, societies typically transition immediately from the sphere of mere promise where one cannot ask critical questions as of yet, to the sphere of accomplished fact in which one is already implicated and which one cannot question anymore.[35] On this side of technical innovation we do not know as of yet how it will challenge us culturally, politically, or ethically.[36] And on the other side of technical innovation we have already become different people and can no longer invoke standards that now belong to our technical and cultural past. The notion of (accelerating) progress does not allow for any middle ground between such a before and hereafter. As a matter of principle, however, political subjects ought not to place themselves into a temporal frame of reference that systematically deprives them of a decisive, real or merely assumed moment of possible intervention.

---

[34] This abandonment of an orientation toward the future concerns only a *historical* conception of time. In the course of history, the historical subjects (persons, nations etc.) undergo a change. This is not the case in a *technical-empirical* temporal succession. Of course, all "space travel" takes place in time, but its subject is thought to persist unchanged on its trajectory. This (admittedly, idealized) juxtaposition ought to be complemented by a concept of the future that no longer conceives of the future historically but as a space which we come to occupy in the course of our travel through time. It is just this dehistoricized "future" that makes for the ambiguity of the popular conception of a "nanocosm": What we see at first is only a conquest of space, but this space is supposed to represent also the future that awaits to be occupied and settled in the course of time. It requires an additional assumption that this future will historically transform the subjects who are embarked on this journey. Andreas Lösch (2006) has shown how nanotechnological visions conceive the future ahistorically as a space that opens up to "us" (as subjects whose identity remains untouched). Another analysis was recommended by Kate Marshall (2004). According to her, the spaces claimed by nanotechnology are endangered from the future. They are "risky spaces" since, in the risk society, everything present is a product of an imagined, possibly scary future. Again, this notion of a present as product of the future posits an entirely ahistorical future, one that is relevant only to the extent that it affirms and confirms our present existence. It may well be the case that citizens of today's societies can only countenance the ahistoric "futures" described by Lösch and Marshall. Indeed, with reference to Niklas Luhmann's "Beschreibung der Zukunft (1993)," Sabine Maasen 2005 made a compelling case how this holds for visual anticipations and verbal descriptions of the nanotechnical future. I take this diagnosis not as a final verdict, however, but as an argument for purposes of policy and deliberation – that is in immediate relation to my considerations on the "If and Then" (2007) and to the notion of "entanglement" (see my "Knots and Strands: An Argument for Productive Disillusionment" (2007)).

[35] Blumenberg (1963). I owe this reference to Christoph Hubig.

[36] Reinhart Koselleck construes this as a characteristic difference between sphere of experience (*Erfahrungsraum*) and horizon of expectation (*Erwartungshorizont*), thus of a spatial discontinuity within the modern conception of historical progress, see his *Vergangene Zukunft* (1989, pp. 349–375).

If they did so, they would abandon the possibility of politics and the necessary illusion that they can express their values or shape their societies. This matter of principle motivates the proposal to imagine the advance of nanotechnologies not as a progression into the future but as a conquest of space, and thus as a journey that, empirically speaking, takes time but throughout which our moral point of view remains intact.[37]

The proposal to stop relating ourselves to the future appears outrageous and unheard-of in its own right: How could this even be achieved? It is worth recalling, however, that Science Studies, the Philosophy of Science, and the technosciences themselves have achieved this already. Science and Technology Studies has shown that technoscience differs from classical science precisely in that it is oriented not to the future but to space. For hypothesis-testing science and traditional philosophy of science, the truth was thought to lie in a remote future. According to Max Weber, Charles Sanders Peirce, or Karl Popper science approaches but never reaches this truth as it keeps postulating and testing hypotheses. If it advances further and understands more, this is because it builds upon the work of its predecessors and thus stands on the shoulders of giants. And for that very reason scientists must hope that their findings do not last but will be superseded in the course of progress.

Aside from the idea of progress, that of objectivity is also conceived by traditional science in historical terms. The main threat to objectivity is seen in historical or cultural contingency. The truth will have to be eternal and must therefore be cleansed of idiosyncrasies of personality, context of discovery, or cultural background. In the words of Paul Feyerabend, objective knowledge depends on the "separability assumption" and thus on the separability of a scientific claim from the historical conditions under which it was produced.[38]

None of this holds for technoscience.[39] The difference is apparent already in its conception of objectivity. Instead of looking to dehistoricize claims, technoscience is said to delocalize phenomena. The object of technoscience is not to gradually approximate eternal truth. Instead, it concerns the acquisition and spread of

---

[37] Especially Dupuy's analysis could be reformulated accordingly. Instead of positing the future as it is presently given with the *logos* of nanotechnology, he could argue more simply and without invoking the future: Nanotechnology aims to create a paradise on earth, I show you this paradise, see for yourself whether you really find it so enticing. This would make explicit an implicit assumption of Dupuy's (one that is not shared by Grunwald and Khushf), namely, that he is judging nanotechnological visions from the point of view of today's human being who acknowledges his mortality, his lack of technical perfection. At this point one should also consider the approach of the chemist George Whitesides. Refraining from predictions of the future, he identifies some core assumptions that underwrite our contemporary culture and form of life (following Wittgenstein, Andreas Kaminski and Barbara Orland refer to these assumptions as the "hinge propositions" since all other propositions and our form of life hinges upon them; they include propositions like "humans are mortal"). Whitesides (2004) goes on to show how these assumptions have become questionable, for example by way of nanotechnological visions.

[38] Compare, for example, Weber (1946), Merton (1965) and Feyerabend (1999)

[39] See, for example, Nordmann "Was ist Technowissenschaft?" (2004).

capabilities.[40] Its goal is, first of all, to produce a phenomenon in the laboratory. One then needs to establish that the phenomenon does not exist under the special local conditions of the laboratory alone, but that it is stable enough to be transported to other laboratories and, finally, into society at large. This delocalization requires on the one hand that the production of phenomena becomes routinized, scaled to production, etc. It requires on the other hand that the external world is assimilated to laboratory conditions, that it becomes homogenized, standardized, sanitized. Technical or scientific advance therefore does not pursue an ideal of perfectibility towards the future, it marks no transcendence of past limitations. Instead, it is an advance quite concretely outward into the world. It expands territorially. First it may conquer inner space at the nanoscale, then it structures our daily actions in a pervasively technologized environment, and finally it pervades technically less developed cultures.

Space Travels

The term delocalization appears in a paper by Peter Galison that shows his proximity to as well as his distance from Bruno Latour.[41] This final section will compare three conceptions of technology. All three view technology as permeating or conquering space. Despite the differences among them, Galison and Latour represent the first of these, Gerhard Gamm the second, and my own third proposal will be sketched very briefly only.

Galison's and Latour's Science Studies notion of delocalization makes the beginning. According to Latour, the laboratory is no longer a locally bounded space for experimentation. The presumed difference between its inner life and a societal environment has evaporated.[42] It has evaporated because society at large is implicated in various ways in the biopolitical experiments of genetic and agricultural engineering, of nano- and biotechnology. First of all, these experiments are undertaken by an alliance of stakeholders and not a specialized scientific community. Inversely, the experiments are performed on all of society and not on a more or less self-selected sample population. The social benefits as well as environmental or health risks of new technologies are determined not in advance but only in the course of such large-scale experiments, namely by observing their diffusion and appropriation. The divide between the inner workings of science and the outer social order has evaporated also because technologically significant facts need to be sustained through the co-production of innumerable human and non-human actors,

---

[40] This is reflected also in recent philosophy of science with its emphasis on modelling practices (local models, fitting models to phenomena and *vice versa*) and the specification of mechanisms.

[41] Galison (1997).

[42] Compare, for example, Bruno Latour's notion of collective experimentation (2004, pp. 196–200).

and through a distributed effort that involves a continuous societal effort.[43] Finally, the boundary has vanished because technical interventions are negotiated and appropriated in social settings that include advocates of the economy and the environment, professional and lay cultures.[44]

When Peter Galison speaks of delocalization, he places the emphasis somewhat differently. Against Latour's image of a global network that is everywhere local and that, as a whole, sustains the facts, Galison asks how things move between local cultures and thus how phenomena travel from one laboratory to another, from there to industrial production and from there into our households. Delocalized objectivity owes to the objects. Highly idiosyncratic local cultures build conceptual bridges and stable practices as they develop the instruments for the representation and manipulation of objects.[45] Just like the tools and instruments, the results of this engagement with the object are not transported merely in the form of writing. They do not simply travel as it were on the rails provided by a shared theory or conception of reality. Instead, they must be carried from place to place by persons. In a sense, objectivity is spread by adventurers, explorers, missionaries, and developers – the kind of people who are celebrated in books about the great seafaring "discoverers." While Latour tends to equate product and process (the global network coordinates local practices and owes its existence to their coordination), Galison emphasizes the effort involved in overcoming divisions and local contingencies: Initially, the worlds of the laboratory and of the environment at large are still divided and it requires work to universalize the laboratory phenomenon. Only if that work is successful, a new process or artefact will serve to coordinate practice in a global network of knowledge and industry.[46]

This difference between Galison and Latour can be related to current nanotechnological developments. Carbon nanotubes are presently manufactured in more or less cumbersome ways, in greater and smaller quantities with considerable variance among their properties. Standards of production and characterization are emerging only slowly, and everyone is still awaiting whether they will live up to their promise as universal building blocks for global solutions (chip architectures, video display technologies, etc.). The global network of artefacts and practices based on nano carbontubes remains programmatic. All the while, researchers are working to

---

[43] One of Latour's prime example is the *Pasteurizaton of France* (1988).

[44] Here, one of the most compelling case studies was provided by Steve Epstein (1996).

[45] Galison quotes Latour's conception that instruments as simple as a clock can "travel very far without leaving home." He contrasts this with his own view that "meanings, values and symbols often stay home or switch identities when scientific theories and instruments travel," compare Galison (1997, pp. 677, 679). Latour speaks of global networks that are everywhere local (1993).

[46] As opposed to Latour, Galison exercises rather more restraint when it comes to telling global stories about the social fabric. The previous reconstruction extrapolated from his analyses of the scientific interactions within and among large physics laboratories, (1998).

bridge different laboratory cultures so that these artefacts might travel more smoothly among and between them. The program that motivates this work – Dupuy would call it the metaphysical program of nanoscale research – cannot be represented by Latour's networks that are already all-encompassing. These networks are a spatial equivalent to Grunwald's evolutionary view of the future: They cannot be governed, they are neither determined nor subject to shaping. In contrast, Galison paints a picture of piece-meal constructions that might be open to shaping but that offer no Archimedean point for a social intervention that could globally orient technological development.

Aside from the image of nanoscale research as (inner) space travel and aside from a research process that moves from local tinkering to global solutions, there are numerous further indications that nanotechnologies are engaged in a conquest of space.[47] The first of these is the very label "nanotechnology" that refers to a region of space and the intermediary realm between classical and quantum physics. For the surprises that it holds, this highly complex world has been called an "exotic territory."[48] The first goal of nanoscale research was and is to find one's bearing or orientation and to act in this space. After learning to see and to move single atoms, one writes the name of one's lab in molecular script, acquires the capability to build a wire, to produce some effect, etc. The second goal of nanotechnology is to create great effects from small things. Here, miniaturisation gives way to the project of advancing from the nanoscale to larger scales. Thus, nanotechnologies advance in space by taking nanoscale processes and nanostructured materials to construct larger technical artefacts and systems. As opposed to traditional (outer) space travel or colonial conquests of discovery, however, nanotechnologies do not aim to inhabit this or that particular corner of the world. The nanocosm is presumed to extend everywhere where things consist of molecules. When nanoscale research seeks to control the molecular domain, it literally claims the space of everything. Silicon chips and nerve cells, proteins and pharmaceuticals used to belong to ontologically separate realms of organic and inorganic nature and technology. Now, nanotechnology considers all of these as aggregates of molecules that might be recombinable at will. With the unification of previously separate realms, nanoscale research extends its reach to biotechnology, information and communication technology, and other disciplines.[49] In the language of atoms and molecules, everything becomes nanotechnologically malleable.

Nanotechnology may thus appear to be a paradigm case of what Gerhard Gamm calls technology as medium: "Technology is like language or money a circulatory

---

[47] For a more comprehensive account see my "Design Choices in the Nanoworld: A Space Odyssey" (2007); for rather more detailed accounts see my "Molecular Disjunctions" (2004), and "Nanotechnology's Worldview: New Space for Old Cosmologies" (2004).

[48] See Roukes (2001).

[49] This aspect of nanotechnology's expansiveness underwrites the so-called NBIC-convergence, that is, the convergence of "nano, bio, info, cogno." In a soberingly deflationary manner, the etc-Group speaks of the "little BANG" that combines bits, atoms, neurons, and genes.

system in modern society."⁵⁰ Considered not as means to an end but as an indeterminate site for mediation, Gamm's medium permeates space and lies "at the limits of time." When Gamm distinguishes between transcendental and immanent indeterminacy of technology, this distinction can be applied to nanotechnology.

> Transcendental indeterminacy aims for a fundamental transformation of the modern age. This is to occur as technical agency inscribes itself [...] into the emptiness of a non-stereotyped productivity. For this, there is in principle neither an inner nor an outer limit.⁵¹

This "emptiness of non-stereotyped productivity" has been characteristic for the visions of nanotechnology ever since the "crazy" engineer Eric Drexler imaginatively claimed Richard Feynman's room at the bottom.⁵² Immanent indeterminacy, in contrast, is based on the gap between technical function and use and thus on the seemingly unlimited adaptability of technical functions to different contexts of use. An example of this was offered above – the case of carbon nanotubes that may produce the next generation of computer chips, new textiles, video displays, medical and environmental sensors, etc.

Obviously then, nanotechnology can be conceived as a medium that fills space and within which wholly original functions and uses can be imagined productively. However, one might reject this conception of technology as a medium for the same reasons that one might prefer Galison's over Latour's concept of delocalization. When Gamm's medium is compared to language, money, the circulation of blood, and the matrix of being, it proves to be too thin-bodied, immaterial, and rare. This subtle medium has already spread everywhere. Its spatial expansion can no longer be felt as a material claim, conquest, colonization or annexation. According to Gamm, this medium becomes apparent only when it becomes a form of reflection:

> By harbouring within it the *logos*, technology is essentially a medium for the *disclosure of self and world*. [...] [It] refers to the horizon from within which we invent the world and from within which with increasing insistence we technically reinscribe the image of our selves [...] It is the medium in which human beings become transparent in their artefactual character.⁵³

This notion of technology as a medium therefore serves well to describe how we find ourselves in regard to technology that is routinized and normalized through use. It characterizes a way of thinking that conceives of every problem first and foremost as a technical problem. The notion of technology as a medium finally captures the spirit of specific technological visions like "ambient intelligence" or "ubiquitous computing" – visions that aim for a deeply pervasive technological environment as a kind of second nature. However, like Latour's networks, the notion of technology as a medium does not capture how such visions need to be asserted and materially implemented.

---

[50] This and the following quotations are from Gamm, (2000, pp. 275–287)
[51] Gamm (2000, p. 279).
[52] See Feynman (1960).
[53] Gamm (2000, p. 285).

Moreover, in light of Marcuse's one-dimensional technological culture or Heidegger's "*Gestell*," Gamm's conception asserts all too optimistically an emancipatory dynamics. It advances all too quickly and with the semblance of necessity from thought in the medium of technology to a technological indeterminacy as a form of reflection that discloses self and world. Technology as a medium is supposed to lead us not only to reflect our artefactual character but at the same time to reveal indeterminacy as a norm that establishes a critical relation to technology. According to Gerhard Gamm, the indeterminacy of technology leads us to perceive

> another aspect of openness that is qualified and not arbitrary and almost always overlooked. It includes a normative significance that can be summarized in the form of a maxim: to probe actions, projects, decisions, plans for the future as to whether openness will be preserved also after the realization of the projects. In regard to the implementation of risky technologies this amounts to the question whether the decision for or against it includes the possibility of a reversal, whether technologies open spaces for action in which errors and mistakes do not lead to irreparable, that is, catastrophic consequences.[54]

In contrast to Latour and Galison, Gamm has thus achieved a normative point of view that permits an assessment of technological programs. This is an important achievement, but it comes at a high price: Delocalization comes at the expense of dematerialization with the added assumption that we necessarily advance on a conceptual path that will lead us to reflect and evaluate the pervasive, yet indeterminate presence of technology.

In contrast, I would like to finally suggest that delocalization is quite literally the program of technoscience. On this account it is nothing but territorial expansion pure and simple – and it is human beings and societies who are the ultimate object of appropriation, that is, human bodies and everything that structures human decisions and actions. By way of the technosciences, in general, and nanotechnologies, in particular, we engage in a project of self-colonization.[55] This process of self-colonization takes place at a variety of levels. It begins with the break-down of the boundary between laboratory and society, and continues with the ways in which promoters and critics alike are drawn into the affirmative project of "responsible development of nanotechnology." The project of self-colonization also involves the creation of systems of total information and control that would allow us to use our material resources far more efficiently. Such systems of control are pursued for environmental monitoring, medical imaging, or manufacturing, and they forge new constellations between material processes and human agency as one cultivates on the one hand an attitude of surprise toward the bottom-up emergence of novel

---

[54] Gamm (2000, p. 226).

[55] Here emerges another point of contact with Dupuy's work on the self-mechanization of the mind in AI-research. He shows Dupuy (2000) that self-mechanization or self-colonization are perfectly coherent technological projects that involve us in a paradoxical relation to ourselves as developers of these technologies.

properties, and on the other hand the expectation of precision control that reaches beyond human powers of imagination and understanding.[56]

In the age of technoscience we pursue projects of delocalization and self-colonization to develop solutions to currently identified problems. This "we" is not challenged by or obliged to a historical process but views the future merely as the place where technical possibilities will be realized. At the same time, this "we" is historical in that it is contingently given with its world, its values and traditions. Without arrogating to itself an entirely fictitious view from eternity, this "we" can only claim to be a citizen of its presently given world and it decidedly does not represent a persistent human nature against which the future can be measured. This historically contingent subject of its own world is aware of its contingency and therefore at odds with an ethics of responsibility for the future as postulated, for example, by Hans Jonas.[57] Our world is indeed, as George Khushf put it, only a particular equilibrium of nature, technology, society, and individual. We do not know whether we have any right to pass judgment or to act on behalf of future generations whose values or sense of self may be quite different from ours. At the same time, however, we are obligated to act in accordance with our values, to assert our cultural sense of body and self. On the one hand, therefore, we have no right to paternalistically judge in the name of future generations the transhuman cyborg, for example, as deficient, perverse, or alienated. For, if such cyborgs were to have a self, they would be no more or less alienated from themselves as we are (and if cyborgs have no selves, the problem takes care of itself). As hybrids of humans and machines, cyborgs will also find themselves in an equilibrium of values and physical facts – there will be no need for them to relate the conception of machine to that of a human being, since to them the machine represents no alien otherness. On the other hand and at the same time, we cannot do otherwise but to experience the technological transformations of the human body in the terms of invasion and heightening of self, as a violation or alienation of physical being. When artists like Stelarc place their bodies in experimental situations of extreme technological control, the significance of their work consists in the fact that they apply a discourse about the future of the human being to the body of the present human being. They draw technological visions into the horizon, values, and evaluations of the present. The troubling immediacy of transition, from the time when critical questions can "not yet" be asked to the time when they can "no longer" be raised, is addressed by artists like Stelarc: He asks "already now" what is effected by the technical penetration of his body.[58]

---

[56] These themes cannot be developed here in any detail. They constitute a major strand of ongoing inquiries (in close collaboration with Astrid Schwarz) on naturalized technology, the limits of knowledge and understanding in nanotechnologies, the enhancement of material nature, the nano- and ecotechnological discovery of unlimited possibility beyond the limits of growth, and the seductive power of technoscience. To be sure, the currently popular discussions of human enhancement should also be placed in the context of technoscientific projects of self-colonization (they constitute the tip of the iceberg, so to speak).

[57] Jonas (1984).

[58] Stelarc thus experimentally reunites what Koselleck diagnosed as the modern separation of the sphere of experience (*Erfahrungsraum*) and horizon of expectation (*Erwartungshorizont*), see notes 35 and 36 above. See also Hanson (2005).

Paradoxically perhaps, it is therefore the descriptive enterprise of Science and Technology Studies with its emphasis on the spatial orientation of the technosciences that enables normative critiques of technology, licensing even attempts to halt certain technological trajectories in light of current and contingent values. Its analyses link up with political and ethical discourses of the present precisely in that they surrender the concern for future generations in favour of critiques of colonialism and globalization. This linkage has only begun to be developed and explored within the context of Science and Technology Studies – and must therefore remain programmatic here.[59]

References

Anders, Günther (1980). *Obsolescence of the Human*, London: Radius.
Bender, Wolfgang (1996). "Zukunftsorientierte Wissenschaft – Prospektive Ethik," in Anna Wobus et al. (eds.), *Stellenwert von Wissenschaft und Forschung in der modernen Gesellschaft* in *Nova Acta Leopoldina*, Neue Folge 74: 297, Heidelberg, pp. 39–51.
Bloch, Ernst (1973). *Das Prinzip Hoffnung*, Frankfurt: Suhrkamp Verlag.
Blumenberg, Hans (1963). "Lebenswelt und Technisierung unter Aspekten der Phänomenologie," *Sguardi su la philosophia contemporanea* Vol. 51, pp. 3–31.
Brown, Nik, Rappert, Brian, Webster, Andrew (Hg.) (2000). *Contested Futures – A Sociology of Prospective Techno-Science*, Aldershot: Ashgate.
Dupuy, Jean-Pierre (2000). *The Mechanization of the Mind: On the Origins of Cognitive Science*, Princeton, NJ: Princeton University Press.
Dupuy, Jean-Pierre (2002). *Pour un catastrophisme éclairé: Quand l'impossible est certain*, Paris: Seuil.
Dupuy, Jean-Pierre (2004). "Complexity and Uncertainty," in European Commission, Community Health and Consumer Protection, *Nanotechnologies: A Preliminary Risk Analysis*, Brussels: European Commission, published electronically at www.europa.eu.int/comm/health/ph_risk/documents/ev_20040301_en.pdf, pp. 71–93.
Dupuy, Jean-Pierre (2007). "Some Pitfalls in the Philosophical Foundations of Nanoethics," *Journal of Medicine and Philosophy* 32:3, pp. 237–261.
Epstein, Steve (1996). *Impure Science: AIDS, Activism, and the Politics of Knowledge*, Berkeley, CA: University of California Press.
Feyerabend, Paul (ed.) (1999). "Realism and the Historicity of Knowledge," *The Conquest of Abundance*, Chicago, IL: University of Chicago Press, S. 131–146.
Feynman, Richard (1960). "There's Plenty of Room at the Bottom: An invitation to open up a New Field of Physics," in: *Engineering and Science* 23:5, S. 22–36.
Galison, Peter (1997). "Material Culture, Theoretical Culture and Delocalization," in John Krige and Dominique Pestre (eds.), *Science in the Twentieth Century*, Amsterdam: Harwood, pp. 669–682.
Galison, Peter (1998). *Image and Logic: A Material Culture of Microphysics*, Chicago, IL: University of Chicago Press.
Gamm, Gerhard (2000). *Nicht Nichts*, Frankfurt: Suhrkamp Verlag.
Gamm, Gerhard (2005). "Unbestimmtheitssignaturen der Technik," in Gerhard Gamm and Andreas Hetzel (eds.), *Unbestimmtheitssignaturen der Technik: Eine neue Deutung der technisierten Welt*, Bielefeld: transcript Verlag, pp. 17–35.

---

[59] See note 56 above. A different kind of beginning was offered by Meaney (2006); see also Marshall (2004).

Gamm, Gerhard & Andreas Hetzel (eds.) (2005). *Unbestimmtheitssignaturen der Technik: Eine neue Deutung der technisierten Welt*, Bielefeld: transcript Verlag.

Grunwald, Armin (2003). "Die Unterscheidbarkeit von Gestaltbarkeit und Nicht-Gestaltbarkeit der Technik," in Armin Grunwald (ed.), *Technikgestaltung zwischen Wunsch und Wirklichkeit*, Berlin: Springer, pp. 19–38.

Grunwald, Armin (2006). "Nanotechnologie als Chiffre der Zukunft," in Alfred Nordmann, Joachim Schummer, and Astrid Schwarz (eds.), *Nanotechnologien im Kontext: Philosophische, ethische, gesellschaftliche Perspektiven*, Berlin: Akademische Verlagsgesellschaft, pp. 49–80.

Hanson, Valerie (2005). "The Role of Anticipatory Rhetorics in Discussions of Nanotechnological Ethics," Presentation at the Nanoethics Conference in Columbia, South Carolina, March.

Interagency Working Group on Nanoscience, Engineering and Technology (1999). *Nanotechnology – Shaping the World Atom by Atom*, Washington, DC: IWGN.

Janich, Peter (2006). "Wissenschaftstheorie der Nanotechnologie," in Alfred Nordmann, Joachim Schummer, and Astrid Schwarz (eds.), *Nanotechnologien im Kontext: Philosophische, ethische und gesellschaftliche Perspektiven*, Berlin: Akademische Verlagsgesellschaft, pp. 1–32.

Jonas, Hans (1984). *The Imperative of Responsibility: In Search of an Ethics for the Technological Age*, Chicago, IL: University of Chicago Press.

Kaminski, Andreas (2004). "Technik als Erwartung," *Dialektik*, 2, pp. 137–150.

Khushf, George (2003). "The Ethics of Nanotechnology – Visions and Values for a New Generation of Science and Engineering," *Emerging Technologies and Ethical Issues in Engineering*, Washington, DC: National Academy of Engineering, pp. 29–55.

Khushf, George (2004). "Systems Theory and the Ethics of Human Enhancement: A Framework for NBIC Convergence," *Annals of the New York Academy of Sciences*, 1013, pp. 124–149.

Khushf, George (2007) "The ethics of NBIC Conuergence." *Journal of Medicine and Philosophy* 32, pp. 185–196.

Koselleck, Reinhart (1989). *Vergangene Zukunft*, Frankfurt: Suhrkamp Verlag.

Kurzweil, Ray & Terry Grossman (2004). *Fantastic Voyage: Live Long Enough to Live Forever*, Emmaus: Rodale.

Latour, Bruno (1988). *The Pasteurization of France*, Cambridge, MA: Harvard University Press.

Latour, Bruno (1993). *We Have Never Been Modern*, Cambridge, MA: Harvard University Press.

Latour, Bruno (2004). *Politics of Nature: How to Bring the Sciences into Democracy*, Cambridge, MA: Harvard University Press.

Lösch, Andreas (2006). "Antizipationen nanotechnischer Zukünfte: Visionäre Bilder als Kommunikationsmedien," in Alfred Nordmann, Joachim Schummer, and Astrid Schwarz (eds.), *Nanotechnologien im Kontext: Philosophische, ethische und gesellschaftliche Perspektiven*, Berlin: Akademische Verlagsgesellschaft, pp. 223–242.

Luhman, Niklas (1993). "Beschreibung der Zukunft," in Rudolf Maresch (ed.), *Zukunft oder Ende*, Munich: Boer, pp. 469–478.

Maasen, Sabine (2005). "Comments on a Presentation by Andreas Lösch: Antizipationen nanotechnischer Zukünfte, Visionäre Bilder als Kommunikationsmedien," workshop Medien der Wissen(schaft)skommunikation: Erprobungen analytischer Konzepte am Fall 'Nanotechnologie', Darmstadt Technical University, Darmstadt, Germany, November.

Marshall, Kate (2004). "Future Present: Nanotechnology and the Scene of Risk," in Katherine Hayles (ed.), *Nanoculture: Implications of the New Technoscience*, Bristol: Intellect Books, pp. 147–159.

Meaney, Mark (2006). "The Glorified Body: A Reterritorialization of Death," abstract for the XXth European Conference on Philosophy of Medicine and Health Care, August 23–26, Helsinki, Finland, http://www.espmh.cm-uj.krakow.pl/files/AbstractHelsinki2006_0.doc (accessed November 22, 2007).

Merton, Robert (1965). *On the Shoulders of Giants*, New York: Harcourt, Brace, & Jovanovich.

Nordmann, Alfred (as rapporteur for the expert group "Foresighting the New Technology Wave") (2004). *Converging Technologies: Shaping the Future of European Societies*, Luxemburg: Office for Official Publications of the European Communities, 63pp.

Nordmann, Alfred (2004). "Molecular Disjunctions: Staking Claims at the Nanoscale," in Davis Baird, Alfred Nordmann, and Joachim Schummer (eds.), *Discovering the Nanoscale*, Amsterdam: IOS Press, pp. 51–62.

Nordmann, Alfred (2004). "Was ist TechnoWissenschaft? – Zum Wandel der Wissenschaftskultur am Beispiel von Nanoforschung und Bionik," in Torsten Rossmann and Cameron Tropea (Hg.), *Bionik: Aktuelle Forschungsergebnisse in Natur-, Ingenieur- und Geisteswissenschaften*, Berlin: Springer, pp. 209–218.

Nordmann, Alfred (2004). "Nanotechnology's Worldview: New Space for Old Cosmologies," *IEEE Technology and Society Magazine*, 23:4, pp. 48–54.

Nordmann, Alfred (2005). "Noumenal Technology: Reflections on the Incredible Tininess of Nano," *Techné* 8:3, pp. 3–23, also in Joachim Schummer and Davis Baird (eds.), *Nanotechnology Challenges: Implications for Philosophy, Ethics and Society*, Singapore: World Scientific Publishing, 2006, pp. 49–72.

Nordmann, Alfred (2007a). "Knots and Strands: An Argument for Productive Disillusionment," *Journal of Medicine and Philosophy*, 32:3, pp. 217–236.

Nordmann, Alfred (2007b). "If and Then: A Critique of Speculative NanoEthics," *NanoEthics*, 1:1, pp. 31–46.

Nordmann, Alfred (2007). "Design Choices in the Nanoworld: A Space Odyssey," in Marian Deblonde, Lieve Goorden, et al. (eds.), *Nano Researchers Facing Choices (The Dialogue Series*, vol. 10), Universitair Centrum Sint-Ignatius Antwerpen, Antwerpen, pp. 13–30.

Rip, Arie (2005). "Constructive Technology Assessment of Nanotechnology" and "Folk Theories of Nanotechnogy," both in the pre-publication Arie Rip: Occasional Paper # 2005-01: Positioning and Assessing Nanotechnology, Twente: Technology Assessment/nano ned, pp. 38–40.

Roco & Bainbridge (eds.) (2002). *NBIC Converging Technologies for Improving Human Performance*, Arlington, VA: National Science Foundation, Department of Commerce.

Roukes, Michael (2001). "Plenty of Room, Indeed," *Scientific American*, September, pp. 48–57.

Selin, Cynthia (2006). "Time Matters: Harmony and Dissonance in Nanotechnology Networks," *Time & Society*, 15:1, pp. 121–139.

Selin, Cynthia (2007). "Expectations and the Emergence of Nanotechnology," *Science, Technology & Human Values*, 32:2, pp. 196–220.

Weber, Max (1946). "Science as a Vocation," in H.H. Gerth and C. Wright Mills (Translated and edited), *From Max Weber: Essays in Sociology*, New York: Oxford University Press, pp. 129–156.

Whitesides, George (2004). "Assumptions: Taking Chemistry in New Directions," *Angewandte Chemie International Edition*, 43, pp. 3632–3641.

Ethics and (Bio)Nanotechnology

Bionanotechnology: A New Challenge for Ethical Reflection?

Christoph Baumgartner

Abstract This paper addresses the rapidly developing field of biomedical nanotechnology from an ethical perspective. It first provides an overview over important ethical issues of biomedical nanotechnology all of which require a thorough and comprehensive analysis. The question whether these ethical concerns are novel or specific to biomedical nanotechnology is cautiously answered in the negative: most of the questions that are considered ethically relevant in the context of biomedical nanotechnology are already discussed in several fields of applied ethics. However, some of these ethical issues are accentuated specifically by biomedical nanotechnology. Finally, the article includes considerations concerning basic requirements of a comprehensive ethical investigation into biomedical nanotechnology and gives arguments for an inclusion of visionary projects in an analysis of the ethical aspects of nanotechnology.

Keywords Bionanotechnology; NBIC-Convergence; Ethical Issues; Ethical Analysis; Nanoethics

Introduction

Although still a relatively young field of research, nanotechnology has already secured its position as one of the key technologies of the twenty-first century. Given the current pivotal place that it occupies across a diverse range of fields, in all likelihood the influence of nanotechnology, both within science and beyond, will only increase further in the years to come. There is considerable optimism over the future applications of nanotechnology with proponents claiming that its impact on the environment, as well as the health and wealth of people, will be equivalent to the combined influences of microelectronics, medical imaging, computer-aided engineering and synthetic polymers in the twentieth century (see Smalley, 1999). Biomedicine is considered one of the areas

---

Utrecht University, Faculty of Humanities, Department of Theology, PO Box 80105, 3508 TC Utrecht, The Netherlands
Email: c.baumgartner@uu.nl

where nanotechnology could deliver the most impressive results, contributing to significant improvements in both diagnosis and treatment.

However, as with most emergent technologies, questions of the ethical implications of nanotechnology inevitably arise. The title of this paper, "Biomedical Nanotechnology: A New Challenge for Ethical Reflection?", demonstrates the ambiguous nature of attempting to disentangle the ethical issues at stake. Responding to such a question is not merely a case of identifying and responding to the ethical problems that nanotechnology generates, but also of launching an investigation into the challenges that the field could create for ethics as a whole. The purpose of this article is not to provide an exhaustive investigation into the ethical issues surrounding biomedical nanotechnology, nor will it attempt to formulate a definitive set of moral norms and principles to guide nanotechnology's application. Rather it has the more modest – and ultimately more fundamental – goal of responding to both possible readings of the title; looking both at ethical issues that arise from nanotechnology and whether these issues should be considered novel or unique within the larger ethical debate.

I will begin by naming some of the actual and (currently) potential applications of biomedical nanotechnology. I will subsequently address six clusters of ethical issues that nanotechnology engenders. Finally, I will investigate whether such ethical concerns are novel or specific to nanotechnology. This will include an assessment of the requirements necessary for a comprehensive ethical investigation into nanotechnology.

Although nanotechnology is multidisciplinary, involving such diverse areas of technology as agriculture, environment, information, communication and space, this article will focus on the field of biomedicine, which is especially ethically sensitive. It goes without saying that the exclusion of non-medical nanodevices does not indicate a lack of ethical issues outside the realm of *biomedical* nanotechnology.

Biomedical Nanotechnology and NBIC-Convergence

Processes essential for living organisms occur largely at the nanoscale: the elementary building blocks of most biological objects such as DNA, proteins or cell membranes are nano-sized. Nanotechnology (or nanobiotechnology) aims at both the improved understanding and the specific manipulation of nano-sized biological objects. Within the field of biomedicine, the miniaturization of technology down to the nano-scale promises to become an essential feature of processes and devices in the post-genomics era. While the possibilities of nanotechnological products and procedures in biomedicine may not always be entirely original or groundbreaking, much of its potential lies in improving technologies and devices that already exist at the macro- or micro-level. These improvements and new developments within the field of biomedical technology are expected to result from the convergence of different disciplines and technologies such as biotechnology, information technology, and cognitive science. This melting-pot approach, largely fostered by developments in nanotechnology, has been coined *NBIC-convergence* (see Roco & Bainbridge, 2003; Nordmann, 2004). The expectations and hopes connected with nanotechnological developments in biomedicine are immense. Therefore, just a few of the many envisaged or developed products will be dealt with in this article (see Baumgartner et al., 2003; Paschen et al., 2004; EGE, 2007).

Nanotechnology is expected to enable the development of improved or innovative methods of medical diagnosis. This applies both to the diagnosis of viruses or bacterial infections and to genetic diagnoses, all by optimizing the precision of devices such as diagnostic biochips. This should facilitate earlier identification of infections and diseases, thereby making it easier to treat them. Nanotechnology is expected to open up new vistas of genetic diagnosis that are currently only hypothetical. In twenty years time a complete genetic map might be a standard test, much like a blood test is today (see Williams & Kuekes, 2001). Developments like this could aid significant advancements in the field of individualized medicine, for example personalized pharmacogenomics.

It is expected that nanoparticles in the near future will serve as *drug delivery* and *drug-targeting systems*. Due to their size, nanoparticles can move through the human body very easily. They are not detected by the immune system and they are able to penetrate biological barriers, such as cell membranes or the blood-brain barrier. These properties make nanoparticles potentially valuable as a means for transporting drugs to specific target areas in the human body. They may also be used to deposit biologically active substances and release them over a controlled period of time or at the precise moment they are needed – either by response to the body's own signals or to external signals (such as those from a physician). It is expected that the successful employment of nanoparticles as drug delivery systems and other such developments will open up new pharmaceutical possibilities and significantly decrease undesirable side effects from current drugs.

Another anticipated result of nanotechnological research in biomedicine is biocompatible *transplants* such as artificial skin and new – more complex – artificial organs. Furthermore, extensive research is being conducted into the field of human-machine interfaces as the basis for direct and bidirectional exchanges of information between the human body and artificial devices. One notable example, neuroprosthesis, would make the restoration of lost or damaged mental and sensory faculties possible. Some research projects even aim at *improving* human performance by means of NBIC-convergence technologies (see Roco & Bainbridge, 2003).

Bordering on the utopian, one of the more visionary applications of nanotechnology was proposed by K. Eric Drexler, who described nanorobots or "immune machines" actively roaming the blood stream like tiny submarines and eliminating unwanted concretions and pathogens. According to Drexler, similar cleanup machines could effectively apply nanotechnology to oral hygiene, whereby "A mouthwash full of smart nanomachines could identify and destroy pathogenic bacteria while allowing the harmless flora of the mouth to flourish in a healthy ecosystem…" (Drexler, 1991, p. 207).

Summarizing the possible impact of nanotechnology in biomedicine, it can be claimed that the use of nanotechnology aims to improve almost all fields of biomedicine – diagnosis as well as treatment and even the "improvement" of human performance. Nanotechnology is said to be an *enabling* technology because it makes further developments possible and opens up new technological horizons, which would have remained out of reach without nanotechnology. The prospective impact that nanotechnology and, even more so, NBIC-technologies will have in biomedicine is evidently anything but nano-sized. They are expected to initiate a genuine transformation of both biomedical technology and medical health care systems.

The enormous potential influence of biomedical nanotechnology does not, however, only open up promising new possibilities for biomedical treatment. This relatively new technology also raises questions and poses a challenge for the ethical analysis and evaluation of biomedicine. Some of these ethical issues, both actual and potential, will be discussed in the following section.

Ethical Issues in Biomedical Nanotechnology

In 2003, a Canadian group of ethicists published one of the very first articles explicitly dealing with the ethical implications of nanotechnology (Mnyusiwalla et al., 2003). They pointed out that there is a significant gap between the rapid development of nanotechnology in different fields including biomedicine, on the one hand, and an apparent lack of related ethical investigation on the other hand. In their own words: "As the science leaps ahead, the ethics lags behind" (Mnyusiwalla et al., 2003, R9). In the years since their article was published, the situation has changed and biomedical nanotechnology is now widely considered an important research topic for ethicists. There are an increasing number of articles in academic journals and volumes addressing the ethical, legal, and social issues surrounding the emergence of biomedical nanotechnology, and also more fundamental philosophical issues in the context of biomedical nanotechnology.[1] Furthermore, non-governmental organizations, such as the etc group (action group on erosion, technology and concentration) and Greenpeace, have picked up on the topic of ethical and societal implications of nanotechnology (see e.g. ETC-Group, 2003; Arnall, 2003). In the meantime, there have also been a number of survey-based studies by bioethical committees (see e.g. Royal Society, 2004; CEST, 2006; Dutch Health Council, 2006; UNESCO, 2006; EGE, 2007) focusing predominantly on the ethical and social issues arising from *biomedical* nanotechnology. However, even the most recent studies, such as the report issued by the European Group on Ethics and New Technologies in January 2007, stress that there is still an urgent need for

---

[1] The Journal *NanoEthics: Ethics for Technologies that Converge at the Nanoscale* started in 2007. Recent volumes that include articles on the ethical, legal and social aspects of biomedical nanotechnology are Schummer and Baird (2006) and Weckert et al. (2007).

interdisciplinary research on the ethical, legal, and social implications of biomedical nanomedicine (EGE, 2007, p. 60).

In the following section I will identify some of the ethical issues considered relevant to the field of biomedical nanotechnology and critically address the ethical implications, perhaps even ethical *problems*, of nanotechnology in medicine. I do not wish to imply that there are no positive aspects of biomedical nanotechnology that are also ethically relevant. Reducing the negative side effects of drugs, for example, or improving the chances of fighting diseases such as cancer are certainly desirable ends not only from a medical but also from an ethical perspective. These aspects will not be discussed, however, because they are largely self-evident and non-contentious consequences of medical advancement.

As biomedical nanotechnology is still a nascent and rapidly developing technology, any attempt to provide a comprehensive list of ethical implications would inevitably fail. For this reason, the following list is intended to function as an overview of some of the most important ethical issues in biomedical nanotechnology. Future developments in specific fields are unpredictable, and it is more than likely that nanotechnology will bring surprises – some beneficial and some problematic or even harmful – to scientific or technological fields and corresponding ethical discussions (see Robison, 2004). In order to lay the foundations for a response to the question posed in the title of this article, the central ethical issues of biomedical nanotechnology and NBIC-convergence will be numerated in this section.

1. *Privacy and control* (see Dutch Health Council, 2006; Jömann & Ach, 2006; Moor & Weckert, 2004). Any nano-technological developments in the field of medical diagnosis will probably exacerbate current problems associated with the ethical evaluation of such technologies as biochips, genetic tests, and individualized medicine. It is not difficult to see that the envisioned *nanotechnological* devices and procedures will make personal (e.g. genetic) information easily, quickly, and cheaply accessible. Such information could be used not only to draw conclusions related to health, but also about lifestyle, biological origin (parentage), or the "genetic fate" of an individual. The pertinent ethical issues in this context are quite obvious. Firstly, the availability of privatized tests that can garner personal information outside of the formal – and thus somewhat protected – relationship between patient and physician increases the avenues to abuse (quasi-)medical tests. It would be possible and relatively easy to acquire biological data from people, using nano-related test-kits, without having first obtained consent. However, abuse does not necessarily imply acting without consent of the tested person. Biomedical (especially genetic) testing raises ethical problems even before it approaches the domain of abuse. Until recently, there was a broad consensus that such tests should focus on disorders that can be prevented or treated effectively. However, this starting point is coming more and more under pressure; the debate is influenced by the widening gap between possibilities for relatively easy and early detection of biochemical data by means of biomedical nanotechnology and the availability of treatments (Dutch Health Council, 2006, p. 73).

The participation in medical tests is usually bound to informed consent. "Informed consent", however, is under threat and could become a well-meant, but inapplicable principle if it is not possible to transform myriads of biomedical data and statistic probabilities into information that is meaningful for those participating in the test. Furthermore, biomedical information could be gained unintentionally; for example, information about disorders for which people did not want to be tested could be inadvertently obtained. How should this – potentially significant – information be dealt with? Do patients enjoy any "right not to know", if there is data suggesting a genetic disposition for a specific disease? Questions like these are certainly not specific for biomedical *nano*technology, and they are not even intrinsically connected with it: nano-enabled diagnostic tests are not ethically problematic *per se*. Nevertheless, the problem of privacy described here must be taken into account in every discussion of the ethical issues of medical nanotechnology. The developments nanotechnology fosters will only further aggravate the problems already present in the fields of "pre-nanotechnological" (genetic) diagnosis and individualized medicine.[2] It must also be kept in mind that access to information often involves power relationships and control.

Regarding biomedical developments in the field of NBIC-convergence, there are also controversies surrounding devices and implants that not only make personal information available, but can also function as data transmitters that can be connected together (see the pertinent contributions in Roco & Bainbridge, 2003). In this context, surveillance and manipulation are the keywords for the ethical issues that may arise. James Moor and John Weckert point out that nanotechnological implants (which serve the purpose of getting personal biological information) might turn over control of an individual's body and mental phenomena to others. For them, it is "not speculation [...] that with the advent of nanotechnology invasions of privacy and unjustified control of others will increase" (Moor & Weckert, 2004, p. 306).[3]

2. *Machinization of humans – anthropological issues* (see Dupuy, 2007; Grunwald & Julliard, 2007; Jömann & Ach, 2006). Medical nanotechnology mirrors and promotes a trend favouring technological advancement in biomedicine that raises several philosophical questions. Progress in the field of medical diagnosis is not only important, but also necessary, for what we consider to be effective medicine. However, this does not necessarily mean that a more or less permanent surveillance of the human body, including "remote monitoring" to guard against all deviations from biochemical norms, is desirable. Practices like these could become readily available through the use of nanotechnological devices. As ethicists, philosophers and theologians, we are interested, for instance, in the dominant concepts of health, disease, medicine, patient-physician relationship that guide biomedical developments, and in the anthropological self-conceptions that underpin our efforts in the

---

[2] See Wade L. Robinson (2004) who discusses the issue of invasion of privacy by means of nanotechnological devices more extensively, but with a wider focus than *biomedical* nanotechnology.

[3] The problem of delegation of agency to nano-enabled technology in biomedicine is addressed by Tsjalling Swierstra and Arie Rip (see Swierstra & Rip, 2007).

field of research and development of biomedical technologies. An analysis of texts such as the reports from the National Nanotechnology Initiative in the United States reveals a presumption of "technological feasibility" that sometimes slides, especially in the field of biomedical nanotechnology, into fantasies of omnipotence in the "rhetoric of nanotechnology" (David Berube). This applies even if we refrain from referring to clearly visionary projects, such as K. Eric Drexler's notion of nanorobots or self-replicating molecular assemblers capable of reproducing any object by taking atoms from its environment and arranging them atom by atom to assemble the object (see Drexler, 1986). How we conceptualise issues and challenges on the theoretical and even rhetorical level strongly influences how research and development approaches the content, and thus the future, of medical technologies. Particularly in the context of NBIC-convergence, and regarding nanotechnology-enabled man-machine interfaces, questions arise that are not explicitly ethical, but must be addressed on a more fundamental philosophical and anthropological level. Researchers in the field of NBIC-technologies assume (and invest substantial effort in) the idea that in time more and more functions of the human body could be executed by NBIC-technological implants. It is already possible to identify a research-and-development tendency, which could lead to a (currently hypothetical) point where a development that is well known in the debate about artificial intelligence is inverted. Whereas philosophical debates on artificial intelligence usually focus on the topic of a possible "humanization" of the machine, this not the issue here, but rather vice versa, namely the mechanization or "machinization" of humans. This could eventually occur through successive biomedical NBIC-implants, artificial organs and other technologies. Whereas philosophical and theological traditions used to ponder the concept of a human being by defining the boundaries to God on one side and animals on the other, the current challenge is defining the boundaries between human and machine (see Böhme, 2002), and the borders traditionally separating the two are coming under increasing pressure from both sides. Smith has even gone so far as to ask the question of "How much nano-prosthesis will make one non-human?" (Smith, 2001, p. 269). What becomes evident with such questions is that a possible introduction of NBIC-technologies into biomedicine urges us to adopt new anthropological perspectives. We have to ask whether being distinct from the realm of technology is still an important part of what defines us as human. Anthropological questions like these are closely connected with explicitly ethical questions. Does the merging of man and machine through biomedical nanotechnology endanger people's autonomy and humanity or, on the contrary, does it strengthen them? Can such mechanized interventions be deemed morally acceptable? For example, it is difficult to know how NBIC-implants in the human brain might influence mental phenomena, such as thoughts and emotions. If the implants become essential for the capacity to act, or for emotions and moods, as envisaged in some NBIC-projects, there would be important ramifications for the philosophical concept of responsibility. In the context of law and jurisdiction, these questions are important for accountability and liability.

Against this backdrop of anthropological and ethical issues it is imperative to confront the questions of who we want to be and what we want to be able to control.

Enhancement is one of the most pressing developments in biomedicine that relates to both of these questions.

3. *From medicine to anthropotechnologies: Enhancement* (see Dutch Health Council, 2006; Preston, 2005; Royal Society, 2004). Especially in relation to long-term developments of nanotechnology and within the context of NBIC-convergence, projects of biomedical nanotechnology often aim to improve people's physical, mental and sensory performances. A new synthesis of the biotic and the a-biotic is considered especially promising; ambitious projects in the field of NBIC-technologies pursue the goal of human-machines hybrids that are capable of far exceeding current human capabilities (Preston, 2005).

Such developments, which aim at improvements in the capacities of "normally functioning" and generally healthy human beings, and therefore not at the treatment of diseases or the restoration of partially or entirely lost capacities, are denoted by the term "enhancement". However, it does not necessarily follow that every form of non-therapeutic enhancement of human performance is ethically problematic. On the contrary, even the notion of non-therapeutic enhancement can be questioned because, ostensibly, what is the difference between improving human capabilities by means of nano-enabled "non-therapeutic enhancement" on the one side, and increasing capabilities through education or exercise on the other? Nevertheless, to make a distinction between therapeutic medical practices and non-therapeutic enhancement is undoubtedly reasonable; it even proves necessary when confronting the need to organize and implement effective and fair healthcare systems and health insurance policies. In the context of enhancement, at least two categories of ethical issues can be identified.[4] Firstly, there are concerns regarding particular forms of non-therapeutic enhancement, which are considered ethically problematic as such. Arguments against enhancement, in this context, are hinged on notions such as an assumed (or claimed) unnaturalness of artificial improvements to human capabilities and the attitude of mastery, control of nature and "men playing God" (see Kass, 2003). Such concerns regarding the "intrinsic moral wrongness" of enhancement are considered especially important in the field of merging man and machine. The second group of ethical issues is not directly linked with enhancement as such, but with research on enhancement technologies in view of questions of justice, equality, freedom, coercion, and priorities in biomedical research and development. While these two groups of ethical issues should be distinguished, they must not be completely separated from one another. Analyzing particular conceptions of improvements in human performance one recognizes that the two are indivisibly intertwined with each other. Should such conceptions ever come to fruition, there would be significant consequences for our understanding of medicine, disease, disability and "normality", each of which is vital for maintaining society's cohesive fabric? In the context of reproductive medicine and prenatal diagnosis, we can already see that many people are faced with increasing

---

[4] Often mentioned as one of the goals of biomedical nanotechnology, concerns resulting from extreme longevity could be seen as a third group of ethical issues. For a discussion of the ethical issues concerning longevity and nanotechnology, see Moor and Weckert (2004).

societal pressure to standardize human beings, both their bodies and capabilities. This "overt and subtle coercion," in the words of Leon Kass, results from both particular images of what constitutes a "good" or even "ideal" human being (often highly influenced by economic interests), and the availability of specific biomedical techniques (which again are developed by a market driven biomedical industry). It is reasonable to assume that the implicit social pressure to make use of such techniques in order to conform to particular images or ideals will only increase with the further availability of NBIC-enabled enhancements. This leads to the very serious question of whether people can "opt out" of a society in which the medical sector has been transformed into an anthropotechnological industry (Smith, 2001, p. 271). If the (presently) utopian visions of an *Improvement of Human Performance* are realised by means of NBIC-technologies, scenarios such as Alan Buchanan's and others' so-called "Genetic Communitarianism" become conceivable (Buchanan et al., 2000, pp. 2, 177–179). For it seems possible that different groups could choose to pursue eugenic "self-enhancement" according to their own particular ideas of the perfect human being and the good life, and therefore develop in potentially disparate directions from other individuals or groups. Buchanan and others write,

> [W]e can no longer assume that there will be a single successor to what has been regarded as human nature. We must consider the possibility that at some point in the future, different groups of human beings may follow divergent paths of development through the use of genetic technology. If this occurs, there will be different groups of beings, each with its own 'nature,' related to one another only through a common ancestor (the human race), just as there are now different species of animals who evolved through random mutations and natural selection. (Buchanan et al., 2000, p. 95)

Reading many of the visionary reports on NBIC-technologies one gets the impression that what Buchanan and his fellow authors considered a (presently hypothetical) societal outcome of *genetic* enhancement could become increasingly likely through the convergence of nanotechnology with bio- and information technology and cognitive science. While the ethical issues at stake here are not difficult to imagine, Jürgen Habermas has outlined one of the potentially more dangerous. He points out that such a scenario would jeopardize the unity of human nature as the basis for all people to mutually recognize one another as equal and autonomous individuals (Habermas, 2003, p. 121). Social and even conceptual inequalities concerning issues such as moral rights between the different descendants of human beings (as we use the term today) would almost automatically be the result. Scenarios like this are utopian, however, and are not intrinsically linked with nanotechnology.

4. *Patenting* (see CEST, 2006; EGE, 2007; ETC-Group, 2005; Smith, 2001). As with almost all new technologies, in nanotechnology there is a race to get key patents, which affect practically all fields of technology, including biomedicine. It is regarded as fact that the protection of intellectual property through patents on (particular) nanotechnological developments will provoke a lively and controversial ethical debate, especially in the field of biomedical nanotechnology (see Weil, 2001). Whether or not traditional arguments for the protection of intellectual property rights are valid for particular nanotechnological developments is already under discussion. Granting patents on developments important for broad fields of

research can make scientists dependent on patent holders. Such dependencies can impede the process of research and development instead of stimulating and supporting it, which is one of the main purposes of the patent system.[5] This problem is especially relevant to new and emerging disciplines like biomedical nanotechnology; lessons learned in fields such as biotechnology and pharmaceuticals show that a direct, unaltered application of the classical patent system contributes to strengthening inequalities, and often creates unfair initial opportunities between the research groups at big companies and small-scale enterprises and universities. Particular problems arise in the field of nano*bio*technology where biological components have essential functions for a particular development. Should products that merge man and machine be eligible for patenting just like purely mechanical inventions? Or does the fact that one has to make use of *natural* components not made from technology, coupled with the aforementioned ethical considerations relating to justice and freedom of research, necessitate a *sui generis* system for the protection of intellectual property in the field of nanotechnology? Within the framework of such questions, an ethical debate on the patentability of developments in biomedical nanotechnology becomes essential, one indeed comparable to the current debate on the patentability of biotechnological developments.

Even if such concerns are not *specific* to nanotechnology, there are certainly characteristics of biomedical nanotechnological advancements that raise doubts over whether the classical patent system is an adequate means to reward the work of researchers and to protect intellectual property. Biomedical nanotechnology is highly market driven (see Robison, 2004), it is a new and burgeoning technology, and last but not least it is an enabling technology, which not only makes progress in existing devices and technologies possible, but even opens up entirely new technological fields. Therefore, granting patents on fundamental developments of biomedical nanotechnology could be tantamount to blocking access to new fields of biomedical research, instead of opening up new possibilities.

5. *Justice in access and opportunity – the nano divide* (see Jömann & Ach, 2006; Arnall, 2003). Among the most important ethical challenges to arise from biomedical nanotechnology are issues concerning access to nanotechnological developments and the pursuant problems of justice, equality, and participation. It is likely that many developments that incorporate biomedical nanotechnology will be prohibitively expensive (see e.g. Litton, 2007; Paschen et al., 2004). Therefore, only a relatively small group will have access to them. This will almost certainly result in ethically problematic

---

[5] Michael Heller and Rebecca Eisenberg coined the term "Tragedy of the Anticommons" for this phenomenon in the context of medical research: While a resource is used too extensively as in the case of the well known "Tragedy of the Commons" (Garrett Hardin) because of *free* use, exactly the opposite is the case in the scenario described by Heller and Eisenberg. Because of the privatization of research, and especially because of the extensive patenting to be expected, there is scientific evidence that valuable biomedical research resources are not used much and not effectively, because the holders of patents can exclude others from using important or even decisive parts of the whole resource (see Heller & Eisenberg, 1998).

social inequalities. Those capable of making use of nanotechnological developments will have better diagnostic and therapeutic devices and procedures at their disposal, whilst others will remain excluded from these advances. Nanotechnology is expected to have the potential to make qualitative improvements to existing medical techniques possible, as well as to *transform* whole technological areas such as biomedical technology. A strongly market-driven development of biomedical nanotechnology could result in a scenario where the experiences and the chances of different groups are divergent and to a large extent separated from each other.[6] It is not difficult to see why the "digital divide" served as a conceptual model for this separation, and this gap between "haves" and "have-nots" has already been appropriately coined the "nano divide" (Jömann & Ach, 2006; Smith, 2001) This presents fundamental problems for justice on the regional and global levels, fair chances, and participation both within a society and between different societies. Since nanotechnological research and development in medicine necessitates a highly developed and differentiated technological infrastructure, to be found primarily in industrialized countries, it is likely that the "nano divide" will increase the existing gap between technologically developed societies and those less developed in this respect (see Paschen et al., 2004, p. 367). In contrast to the digital divide, the scope and the depth of a nano divide might not necessarily be limited to the external environment; rather, the presently fictitious fracturing of human nature, as in the previously mentioned "NBIC-shaped bio communitarianism," could also constitute such a divide.

However, there is no consensus on whether nanotechnology in general, and nanomedicine in particular, would really contribute to the intensification of current inequalities and dependencies between differently developed societies. Whilst several authors fear that the rapid progress in the field of nanotechnology will fuel the nano divide, a group of Canadian bioethicists assert exactly the opposite thesis as a feasible outcome (see Court et al., 2004). They argue that developing countries could be the principal beneficiaries of nanotechnology and that nanotechnology could, if adequately supported, reduce the gap between industrialized and developing countries. Accordingly, these authors call for a debate on societal and ethical issues of nanotechnology that does not focus primarily on the risks and morally problematic aspects of nanotechnology, but rather on its potential benefits.

6. *Self-replication: The "grey goo" problem* (Joy, 2000; Moor & Weckert, 2005; Preston, 2005). The final ethical issue of nanotechnology I wish to address is one of the most popular problems in nanotechnology – the so-called "grey goo" problem. K. Eric Drexler (1986) first identified this problem, raising the possibility of molecular assemblers running out of control. If the self-replication of nanorobots runs out of control, Drexler warns, everything in their environment – the scope of which

---

[6] In a sense, a development like this represents exactly the opposite of what John Rawls intended with his difference principle: balancing out unfair inequalities resulting from the "lottery of nature and social origin" by creating institutions according to the principle that social and economic inequalities must be arranged so that they are both (a) to the greatest benefit of the least advantaged persons, and (b) attached to offices and positions open to all under conditions of equality of opportunity (see Rawls, 1971).

would increase with the number of self-replicating nanorobots – would be a potential resource for the production of new nanorobots. Accordingly, the biosphere would (hypothetically) be transformed into a dust of "grey goo" in a matter of days.

The "grey goo" problem is in no way specific for *biomedical* nanotechnology, and the problems at stake are fairly self-evident, since "[a]n uncontrollable, environment-consuming goo is obviously undesirable for reasons of self-interest" (Preston, 2005).[7] Despite such claims, there are many practical doubts regarding even the theoretical possibility of molecular assemblers and the reality of a "grey goo" threat. Most scientists familiar with nanotechnology deny the physical possibility of nanorobots and even K. Eric Drexler agreed during a working group of the Royal Society and The Royal Academy of Engineering that "this [the grey goo problem] isn't what we should be worrying about" (see Royal Society, 2004, p. 109). Nevertheless, the issue deserves mention here due to its pre-eminence in the public debate on nanotechnology.

Biomedical Nanotechnology: A New Challenge for Ethical Reflection?

How should we approach an evaluation of the aforementioned ethical issues of biomedical nanotechnology? Do we need a separate ethical investigation of this emerging technology, including developments resulting from the convergence of different technologies such as nanotechnology, biotechnology, information and communication technology, and cognitive science?

It is evident that nanotechnology certainly raises important ethical questions, the answers to which require a thorough interdisciplinary ethical reflection. There are issues that challenge deeply held moral beliefs, and the values and concepts at stake far surpass just medical decision-making. Hence biomedical nanotechnology and NBIC-technologies are certainly valid objects for ethical investigation and evaluation. But are they unique; is there something specific that entitles us to call for a new and specific branch of applied ethics and to coin phrases like "nanoethics"? This would at least imply the claim that there is a need for a new kind of ethics, distinct from what has been used to investigate the ethical aspects of such areas as biotechnology, reproductive medicine, and pharmacogenomics.

Almost all of the ethical issues tackled in the section above are topics that were relevant to biomedical ethics long before the development of biomedical nanotechnology. This suggests that when it comes to nanotechnology there is "nothing new under the sun of ethical reflection." Such a conclusion implies that biomedical nanotechnology and NBIC-convergence are just components of biomedicine, and that there is nothing specific about them which would require a separate manner of ethical investigation replete with new ethical principles and criteria (see e.g. Litton,

---

[7] Preston (2005) provides an interesting and more extensive analysis of the "grey goo" problem from the perspective of environmental ethics.

2007). There is some evidence to support this thesis, but some qualifying remarks are necessary. These qualifications relate to some distinctive features that characterize biomedical nanotechnology at both the scientific and the technological level. Nanotechnology is considered an enabling technology, making significant progress and even new uses for known products and procedures possible, including merging different technologies whose biomedical applications have previously been distinct from each other. It is also considered a transformative technology, capable of changing entire areas both of technology and society. Both of these features are salient for the requirements of a comprehensive ethical investigation of biomedical nanotechnology. Nanotechnology could force some issues that currently only exist in the realms of fantasy or hypothesis onto the agenda of biomedical ethics as genuine problems. Biomedical nanotechnology could also facilitate the convergence of different ethical issues into a single debate.

Against this background it is reasonable, perhaps even necessary, that the ethical investigation of biomedical nanotechnology goes beyond the methodological and conceptual framework of biomedical ethics. This is especially true if the concept of biomedical ethics is strongly focused on an individual-oriented perspective and on issues such as the moral status of particular living beings, integrity and self-determination of the patient, and the patient-physician relationship. All of these are valid and necessary, but a thorough and comprehensive investigation of the ethical aspects of nanotechnology obviously requires the integration of different perspectives such as social ethics (including questions of justice in opportunity and "overt and subtle coercion" with regard to biomedical choices), but also perspectives not explicitly ethical at first glance, such as philosophical and theological anthropology.[8] Merging biological and technological structures by means of nanobiotechnology may also make the inclusion of concepts, distinctions, and criteria developed in environmental ethics pertinent (see Preston, 2005). Furthermore, it is worth noting that one of the most important issues in discussing far-reaching and highly transformative technologies, like nanotechnology and NBIC-technologies, concerns the character of the questions we ask. Do we discuss moral questions in a "strongly normative" sense, referring to moral rights and duties, or should we respond to questions on our ideas and visions of a good human being? Probably both of these categories of questions will require contemplation; indeed they are closely connected to each other but must be delineated in order to provide reliable and precise answers.

Biomedical nanotechnology and NBIC-convergence promise to open up new horizons for medicine and for practices that appear to be medicinal (as we conceive it), but may differ from it in their guiding visions. Even if one shares the view of Laurie Zoloth[9] that the *conditio humana* is not defined by suffering but rather by healing, the question of whether particular biomedical visions are to be considered

---

[8] This conclusion concurs with the considerations presented by Jean-Pierre Dupuy (see Dupuy, 2007).
[9] The following considerations are inspired by Laurie Zoloth's lecture at the conference "Biomedicine within the Limits of Human Existence", Utrecht University, April 2005.

utopian in a desirable sense, or rather dystopian, still has to be answered. These visions guide the development of technologies, which have until this point aimed mostly at healing or restituting lost bodily functions or sensory capacities. The emergence of nanotechnology, and even more of NBIC-convergence, seems to bring into reach developments that could make significant improvements and perhaps even changes in human performance possible, at least according to the claims of the representatives of NBIC-initiatives (see Roco & Bainbridge, 2003). It is important to see that these developments are not coincidentally linked with certain technologies. Biomedical science has a strongly normative component, especially in a market-driven society. Individual biomedical visions advance particular motives and tendencies to specific research projects and developmental efforts in the field of biomedical nanotechnology. As already mentioned above, it is possible that biomedical nanotechnology, and particularly NBIC-convergence may foster the transgression of boundaries, such as from treatment to non-therapeutic enhancement and the boundary between medicine and anthropotechnology. Authors like Laurie Zoloth claim that medicine is not about the world as we find it, but about the world as we imagine it could be. There is indisputably some wisdom in this view. But we have to take a closer look and examine the nature of the normativity that is linked with medicine in Zoloth's opinion. Above all we must investigate and assess its desirability and legitimacy.

Within this context, the ethical investigation of biomedical nanotechnology and NBIC-convergence must not limit its purpose to promoting public acceptance and preventing the derailment of nanomedicine, which is an identifiable expectation in some articles and reports stating the need for reflection on the ethical, legal, and social aspects of biomedical nanotechnology. In this regard the European debate on genetically modified food serves as a deterring example for several authors (see e.g. Mnyusiwalla et al., 2003; Rolinson, 2002). Ethics is a critical practice, always analyzing different possible alternatives for action and different ethical criteria and norms considered to be candidates for enabling responsible decisions without being biased in favour of any one of the numerous alternatives. A thorough ethical evaluation of any technology needs to take all perspectives into account that are concerned with the technology in question, and all important issues at stake have to be investigated thoroughly and critically. This applies equally to the strongly visionary aspects. It may be quite probable that it will never be possible to produce self-replicating nano-robots or devices that enable us to communicate directly with computers, but all the same it would be irresponsible to pay no attention at all to such seeming "impossibilities," as Nicolas Agar clearly points out in the context of genetic engineering:

> Science so often confounds the best predictions, and we should not risk finding ourselves unprepared for the genetic engineer's equivalent of Hiroshima. Better to have principles covering impossible situations than no principles for situations that are suddenly upon us. (Agar, 1999, p. 172)

There is a second argument for the inclusion of presently only visionary and utopian aspects of biomedical nanotechnology in ethical investigation: they are constitutive parts of the process of nanotechnological research and development itself. Even if there is convincing evidence that they are not, and never will be, "real possibilities," they still have an active effect as "potentialities." Visions of nano-robots in the blood stream, self-replicating molecular machines or NBIC-based Improvement of Human Performance and brain-to-machine-networks deeply influence the public perception of nanotechnology, as well as funding policies. An example of this has been the deliberate utilization of enthusiasm for optimistic, futuristic visions as a means of promoting technology development in parts of the United States (Paschen et al., 2004).

Conclusion

A cursory examination of the actual, prospective, and even merely envisioned developments in the field of biomedical nanotechnology makes it evident that significant ethical issues are at stake. Diverse ethical concerns must be considered, among them issues of privacy and control, problems concerning patents on nanotechnological devices, issues of distributive justice, ethical aspects of enhancement and, related to this, fundamental anthropological questions concerning the (potentially changing) self-understanding of humanity. Even this brief enumeration clearly indicates that biomedical nanotechnology is not an entirely new challenge for ethical investigation. Most of the ethical issues raised by nanotechnology are familiar from bioethical debates, including those focused around, for example, biotechnology, genetic diagnosis and pharmacogenomics. This even applies, to a certain extent, to the anthropological questions raised. This twofold result of the considerations above has several implications.

First, a thorough and comprehensive ethical investigation of biomedical nanotechnology is urgently needed. This research should include not only actual developments, but also prospective and long-term, future-envisioned enterprises and trends. This does not mean, however, that the potentially problematic ethical implications of imagined self-replicating assemblers or life-prolonging nanorobots should direct the ethical evaluation of medical nanotechnology and its future development and support. Yet at the same time, a wholesale exclusion of these seemingly farfetched "visions" within the field of biomedical nanotechnology and NBIC-convergence would ignore not only their potent influence on public debate but also their suggestive normative power within the images that frame such debates.

A second implication of the twofold result of this contribution is that an ethical investigation of medical nanotechnology does not have to start from a theoretical blank slate. Given that nearly almost all of the central ethical issues are to a certain extent familiar, it is possible to make use of insights and methods developed in these other, closely related fields of biomedical ethics. Conversely, it has to be acknowledged that several of the familiar ethical issues are sharply accentuated by nanotechnology, necessitating an investigation from fresh and innovative perspectives. The emergence of biomedical nanotechnology – coupled in part with its convergence with biotechnology, information technology, and cognitive science – forces together ethical issues that remain separate from one another in other fields of ethical investigation. In addition, the assumed transformative power of nanotechnology, not only in medicine but also with regard to society and even humanity, results in a significantly different framework for ethical issues. Therefore, a thorough and comprehensive investigation into the ethically relevant implications of biomedical nanotechnology does not primarily require a further "hyphen-ethics" – like bio-ethics, environmental-ethics, gene-ethics, and so on – but an integrative approach that cuts across diverse ethical disciplines. And in so doing, it should not only be the particular advancements in nanotechnological devices that are addressed, but also the hidden yet powerful models that shape cultural understanding. On the one hand, this multi-perspective ethical approach would incorporate and benefit from insights from different ethical discourses. On the other hand, it would be expected that a thorough investigation of the ethical aspects of nanotechnological research and development could in turn generate methodological insights which would be fruitful and important to ethical analysis in other fields of scientific research as well. This give-and-take exchange would be particularly useful for both the methodological integration of anthropological considerations and issues of international distributive justice in bioethical discourse. Thus biomedical nanotechnology and NBIC-convergence present not only challenging issues for ethical reflection but also significant possibilities to achieve an "Improvement of Ethics' Performance."

References

Agar, N. (1999). Liberal Eugenics. In H. Kuhse & P. Singer (eds.), *Bioethics* (pp. 171–181). London: Blackwell.
Arnall, A.H. (2003). *Future Technologies, Today's Choices. Nanotechnology, Artificial Intelligence and Robotics; A Technical, Political and Institutional Map of Emerging Technologies. A Report for the Greenpeace Environmental Trust*. London. Available online: http://www.greenpeace.org.uk/MultimediaFiles/Live/FullReport/5886.pdf
Baumgartner, W., Jäckli, B., Schmithüsen, B., & Weber, F. (2003). *Nanotechnologie in der Medizin. Studie des Zentrums für Technologiefolgen-Abschätzung beim schweizerischen Wissenschafts- und Technologierat*. Bern. Available online: http://www.ta-swiss.ch/www-remain/reports_archive/publications/2003/TA_47_Nano_Schlussbericht.pdf
Böhme, G. (2002). Über die Natur des Menschen. In A. Barkhaus & A. Fleig (eds.), *Grenzverläufe: Der Körper als Schnitt-Stelle* (pp. 233–247). München: Wilhelm Fink Verlag.
Buchanan, A., Brock, D.W., Daniels, N., & Wikler, D. (2000). *From Chance to Choice: Genetics and Justice*. Cambridge: Cambridge University Press.
CEST (2006). Commission de l'éthique de la science et de la technologie. *Ethics and Nanotechnology: A Basis for Action. Position Statment*. Québec. Available online: http://www.ethique.gouv.qc.ca/Ethics-and-nanotechnology-a-basis.html
Court, E., Daar, A., Martin, E., Acharya, T., & Singer, P.A. (2004). *Will Prince Charles et al. Diminish the Opportunities of Developing Countries in Nanotechnology?* Available online: http://nanotechweb.org/articles/society/3/1/1/1
Drexler, K.E. (1986). *Engines of Creation. The Coming Era of Nanotechnology*. Oxford: Oxford University Press.
Drexler, K.E., Peterson, C., & Pergamit, G. (1991). *Unbounding the Future. The Nanotechnology Revolution*. New York: William Morrow.
Dupuy, J.P. (2007). Some Pitfalls in the Philosophical Foundations of Nanoethics. *Journal of Medicine and Philosophy* 32, 237–261.

Dutch Health Council (2006). *Health Significance of Nanotechnologies.* The Hague. Available online: http://www.gr.nl/pdf.php?ID = 1417&p = 1

EGE (2007). The European Group on Ethics in Science and New Technologies. *Opinion No 21 (17 January 2007): Opinion on the Ethical Aspects of Nanomedicine.* Available online: http://ec.europa.eu/european_group_ethics/activities/docs/opinion_21_nano_en.pdf

ETC-Group (2003). *The Big Down. From Genomes to Atoms. Atomtech: Technologies Converging at the Nano-scale.* Available online: http://www.etcgroup.org/documents/TheBigDown.pdf

ETC-Group (2005). *Nanotech's "Second Nature" Patents: Implications for the Global South.* Available online: http://www.etcgroup.org/upload/publication/54/02/com8788specialpnanomarjun05eng.pdf

Grunwald, A. & Julliard, Y. (2007). Nanotechnology – Steps Towards Understanding Human Beings as Technology? *NanoEthics* 1, 77–87.

Habermas, J. (2003). *The Future of Human Nature.* Cambridge: Polity Press.

Heller, M.A. & Eisenberg, R.S. (1998). Can Patents Deter Innovation? The Anticommons in Biomedical Research. *Science* 280 (1 May 1998), 698–701.

Jömann, N. & Ach, J.S. (2006). Ethical Implications of Nanobiotechnology – State-of-the-Art Survey of Ethical Issues Related to Nanobiotechnology. In J.S. Ach & L. Siep (eds.), *Nano-Bio-Ethics. Ethical Dimensions of Nanobiotechnology* (pp. 13–62), Berlin: LIT-Verlag.

Joy, B. (2000). Why the Future Doesn't Need Us. *Wired* 8 (April), 238–262.

Kass, L. (2003). *Beyond Therapy: Biotechnology and the Pursuit of Human Improvement. Prefatory Note to Council Members of the President's Council on Bioethics.* Available online: http://www.bioethics.gov/background/kasspaper.html

Litton, P. (2007). 'Nanoethics'? What's New? *Hastings Center Report* 37, 22–25.

Mnyusiwalla, A., Daar, A.S., & Singer, P.A. (2003). Mind the Gap: Science and Ethics in Nanotechnology. *Nanotechnology* 14 (2003), R9–R13. Available online: http://www.utoronto.ca/jcb/home/documents/nanotechnology.pdf

Moor, J. & Weckert, J. (2004). Nanoethics: Assessing the Nanoscale from an Ethical Point of View. In D. Baird, A. Nordmann, & J. Schummer (eds.), *Discovering the Nanoscale* (pp. 301–310). Oxford: IOS Press.

Nordmann, A. (2004). *Converging Technologies. Shaping the Future of European Societies. Report from the High Level Expert Group "Foresighting the New Technology Wave".* Available online: http://europa.eu.int/comm/research/conferences/2004/ntw/pdf/final_report_en.pdf

Paschen, H., Coenen, C., Fleischer, T., Grünwald, R., Oertel, D., & Revermann, C. (2004). *Nanotechnologie. Forschung. Entwicklung. Anwendung.* Berlin: Springer.

Preston, C.J. (2005). The Promise and Threat of Nanotechnology. Can Environmental Ethics Guide Us? *Hyle – International Journal for Philosophy of Chemistry* 11, 19–44.

Rawls, J. (1971). *A Theory of Justice.* Oxford: Oxford University Press.

Robison, W.L. (2004). Nano-Ethics. In D. Baird, A. Nordmann, & J. Schummer (eds.), *Discovering the Nanoscale* (pp. 285–299). Oxford: IOS Press.

Roco, M.C. & Bainbridge, W.S. (eds.) (2003). *Converging Technologies for Improving Human Performance. Nanotechnology, Biotechnology, Information Technology and Cognitive Science.* Dordrecht, The Netherlands: Kluwer.

Rolison, D.R. (2001). Nanobiotechnology and Its Societal Implications. In M. Roco & R. Tomellini (eds.), *Nanotechnology. Revolutionary Opportunities and Societal Implications* (pp. 89–90). Available online: ftp://ftp.cordis.lu/pub/nanotechnology/docs/nano_lecce_proceedings_05062002.pdf

Schummer, J. & Baird, D. (eds.) (2006). *Nanotechnology Challenges: Implications for Philosophy, Ethics and Society.* Singapore: World Scientific Publishing.

Smalley, R.E. (1999). *Nanotechnology. Prepared Written Statement and Supplemental Material. US Congress Testimony, 22 June 1999.* Available online: http://www.nano-and-society.org/NELSI/document/congressional_testimony.html

Smith, R.H. (2001). Social, Ethical, and Legal Implications of Nanotechnology. In M. Roco & W. Bainbridge (eds.), *Societal Implications of Nanoscience and Nanotechnology* (pp. 257–271). Dordrecht, The Netherlands: Kluwer.

Swierstra, T. & Rip, A. (2007). Nano-Ethics as NEST-Ethics: Patterns of Moral Argumentation About New and Emerging Science and Technology. *Nanoethics* 1, 3–20.

The Royal Society & The Royal Academy of Engineering (2004). *Nanoscience and Nanotechnologies: Opportunities and Uncertainties. Report.* Available online: http://www.nanotec.org.uk/finalReport.htm

UNESCO (2006). *The Ethics and Politics of Nanotechnology.* Paris. Available online: http://unesdoc.unesco.org/images/0014/001459/145951e.pdf

Weckert, J., Moor, J., Linn, P., & Allhoff, F. (eds.) (2007). *Nanoethics. The Ethical and Social Implications of Nanotechnology.* Hoboken, NJ: Wiley.

Weil, V. (2001). Ethical Issues in Nanotechnology. In M. Roco & W. Bainbridge (eds.), *Societal Implications of Nanoscience and Nanotechnology* (pp. 244–251). Dordrecht, The Netherlands: Kluwer.

Williams, R.S. & Kuekes, P.J. (2001). We've only just begun. In M. Roco & W. Bainbridge (eds.), *Societal Implications of Nanoscience and Nanotechnology* (pp. 103–107). Dordrecht, The Netherlands: Kluwer.

Nanoparticles: Risk Management and the Precautionary Principle

Armin Grunwald

Abstract The production, use, and disposal of products containing nanomaterials may lead to their appearance in air, water, soil, organisms or the human body. There might be adverse effects on human health or on the environment. In analysing this situation with respect to conclusions for risk management strategies, careful normative reflection and careful analysis of the state of the knowledge available is required as well. Because of large knowledge gaps and scientific uncertainty concerning possible adverse effects of nanoparticles the precautionary principle comes into the game. The aim of this paper is to analyse the current situation with respect to the applicability of the precautionary principle and possible consequences of its application.

Keywords Risk, precautionary ethics, uncertainty, nanoparticles, nanotechnology, ethics of technology

Introduction and Overview

Nanotechnology as an emerging technology field (Schmid et al., 2006) attracts more and more public attention. On the one hand, there are far-reaching expectations with respect to potential contributions of nanotechnologies[1] to welfare, health, sustainable development and to new information and communication technologies. Often, nanotechnology is even regarded as key technology of the new century, sometimes even as the kernel of a third Industrial Revolution. On the other, however, people are aware that there might be a "dark side" of the accelerating technological progress.

Institute for Technology Assessment and Systems Analysis (ITAS), Research Center Karlsruhe, Germany

[1] 'Nanotechnology' is an umbrella term covering a large diversity of different developments with the only common characteristics of exploring or making use of specific new and size-dependent effects and properties at the nanoscale (Schmid et al., 2006). Because of this, I prefer using the plural 'nanotechnologies'.

Concerns about possible risks and side-effects of nanotechnologies have been expressed in public debate and in the media for some years. A lot of ELSA (ethical, legal and social aspects) activities have been undertaken (e.g. Nanoforum, 2004; Coffrin & MacDonald, 2004; Wood et al., 2003), and many expert groups worldwide are still debating about risk analysis, risk assessment, public perception and regulation of nanotechnologies (recently: Renn & Roco, 2006).

Currently, special attention in the public risk debate is being paid to synthetic nanoparticles.[2] A vast potential market for nano-based products is seen in this field. New products, based on new properties of nano-materials can be brought about in admixtures or specific applications of nanoparticles, for instance, in surface treatment, in cosmetics, or in sunscreens. Furthermore, nanoparticles and nanostructured materials offer new ways of designing and controlling catalytic functions, including the provision of enhanced activity and selectivity for target reactions. Since the activity and selectivity of catalyst nanoparticles are strongly dependent on their size, shape, and surface structure, as well as on their bulk and surface composition, being able to synthesize particles at the nanoscale with defined physical and chemical properties is an important step to achieve the goal of catalysis by design. This is just one of the many fields where nanotechnology will enable green chemistry and offer a route to a "green" nanotechnology.

But all of these potential solutions may come at a price. The production, use, and disposal of products containing nanomaterials may lead to their appearance in air, water, soil, or even organisms. There might be adverse effects (Colvin, 2003). Synthetic nanoparticles could be released into the environment or could enter the human body. This might occur via emissions during the production process or by the daily use of nanotech products. Their ways of spreading and interacting with other particles, their impacts on health and on the natural environment, in particular their possible long-term effects, are largely unknown at present. Possible hazardous impacts on human health or the environment are under scientific consideration, but there is still no comprehensive knowledge available. In this situation of large knowledge gaps and scientific uncertainty, the precautionary principle comes into the game. The aim of this paper is to analyse the current situation with respect to the applicability of the precautionary principle and possible consequences of its application.[3] It can be learned from the asbestos story (Gee & Greenberg, 2002) that the current diagnosis of "no evidence of harm" caused by nanoparticles must not be misunderstood in the sense of an "evidence of no harm".

---

[2] Synthetic nanoparticles are artificially designed particles at the nanoscale (like fullerenes, nanotubes, or the titaniumdioxide particles used in sunscreens). In this paper, I will concentrate on synthetic nanoparticles of this type and not consider nanoparticles possibly emerging from non-intended side-effects of production or incineration processes.

[3] This work has partially been performed within the expert group "Nanotechnology. Assessment and Perspectives" of the European Academy Bad Neuenahr-Ahrweiler (see Schmid et al., 2006). I would like to thank the group members for a lot of fruitful discussions. More specifically, I am deeply indebted to Harald Krug who introduced the knowledge of human and eco-toxicology of nanoparticles into these discussions.

In analysing this situation with respect to conclusions for risk management strategies, careful normative reflection is required (Schomberg, 2005). More precisely, thinking about the precautionary principle implies the absence of a "standard situation" in moral, in epistemic and in risk respect (Grunwald, 2005). Instead, ethical reflection is needed to shed light on the normative premises of the options at hand as well as on the criteria of decision-making. Such an ethical "enlightenment" is a necessary precondition for deliberative procedures within which society could identify adequate levels of protection, threshold values or action strategies. Questions of the acceptability and comparability of risks, the advisability of weighing up risks against opportunities, and the rationality of action under uncertainty are, without doubt, of great importance in the field of nanoparticles (cp. for the general challenge Rescher, 1983; Shrader-Frechette, 1991). A new field of *application* is developing here for the ethics of technology, where close cooperation with toxicology, social sciences, and jurisprudence is necessary (Renn & Roco, 2006).[4] Against this background, this paper can only provide first rough ideas.

At the beginning, a brief look at the short history of the risk debate on nanotechnology is given in the second section in order to provide a well-illustrated idea of the various conceptual proposals for dealing with this challenge. In a second step, a brief introduction into the precautionary principle and its differences compared to risk management strategies is given in the third section. Prepared in this way, it is possible to directly address the central question of this paper: should the precautionary principle be applied to the field of nanoparticles and which implications would follow from its application (fourth section)?

Risk Debate on Nanoparticles

Nanotechnologies were perceived as seemingly non-risk technologies for a long time. Public perception was, in the 1990s, low. The prefix "nano", however, and this is a strong indication for a positive perception, was used in the media – not in mass media but, for instance, in science magazines – as a synonym for "good" science and technology. In contrast to large power plants or big chemical factories – which are frequently related to a negative image in the public – nanotechnologies promised a better, clean and smart technological future based on the very positive appreciation of "the small". The "nano-hype" in science and in political communication about nanotechnologies (Paschen et al., 2004) was also a hype of public perception and media interest in nanotechnologies as an outstanding example of a "positive story" with respect to scientific and technological advance.

---

[4] This does not imply that principally new ethical questions have to be solved. There are considerable similarities to the challenge of dealing with new chemicals, for example by defining environmental standards or safety standards (Schmid et al., 2006, sect. 6.2.1). However, a new application field will mostly induce new conceptual and intellectual challenges.

This situation changed radically in 2000. The positive utopias of nanotechnologies, based on a technical access to "the small", were inverted to horror scenarios, based on the same "small" technologies (Joy, 2000). The ambivalence of technology-based visions became obvious in a dramatic way (Grunwald, 2007). The public risk debate on nanotechnology emerged around issues of visionary and more speculative developments. The topics of "grey goo", "nanobots" and "cyborgs" became well-known to many people within few months (Schmid et al., 2006, chap. 5). Concerned groups began to think about analogies and parallels between nanotechnologies and technology lines with a specific history in the public risk debate: nuclear technology and biotechnology (ETC Group, 2003). Newspapers put nanotechnology in the category of risky technologies. Re-assurance companies followed this line of thought (Munich Re, 2002) but became quickly aware of the risks of nano-materials and related governance questions (Swiss Re, 2004).[5] Thus we could witness, in the mirror of public perception and mass media communication, the fall of nanotechnology from a synonym of "good" scientific and technological progress to a technology line that is expected to bear a lot of still unknown risks as well.

At this point, the societal experience with technology risks *in general* enters the game. Technological risks in modern societies show certain characteristics that influence the approaches to their anticipative investigation and evaluation through risk research and technology assessment as well as their public perception. Among these characteristics are, in particular (Grunwald, 2004):

- Large spatial range of adverse effects up to the global dimension (for example, by distributing aerosols through the atmosphere or the oceans)
- Increased temporal scope of the consequences of technology (for example, because of the persistence of chemicals)
- Immense enlargement of those possibly affected by hazards, up to all of present and future humanity
- Delayed effects: often, perceptible damage appears only decades after its cause (examples are the CFC story and the asbestos case, Harremoes et al., 2002)
- Difficulties in ascertaining the chain of causes in view of highly complex and hardly reproducible causal relations (example: the mad cow disease BSE)
- Poor, insufficient or not existing perceptibility of the risks with normal human sensory organs (e.g. in the case of radioactive radiation)
- Diffuse distribution of responsibilities due to complex cause-effect chains and multi-actor constellations (cp. the climate change issue)
- Irreversibility of risks (for example: genetically manipulated organisms, once released, can no longer be completely retrieved from the natural environment)
- Absence of precise knowledge of the probability of adverse effects or the extent of possible damage

It is not difficult to imagine the relevance of many of these characteristics in the field of synthetic nanoparticles, or of nanotechnologies in general. For example,

---

[5] At the side of science this risk debate has mostly been regarded as a threat to the further development of nanotechnologies, remembering the histories of nuclear power technology and of biotechnology.

there have already been fears fed by the fact that nanoparticles or nanorobots could invade human bodies without humans being able to observe them. We are neither able to see or feel them nor to smell or taste them. Consumers and citizens can easily come in contact with nanoparticles already today, and the probability of directly having contact with synthetic nanoparticles will increase considerably in the next years because of the expanding market of respective products. Furthermore, the spatial distribution of nanoparticles probably could not be restricted by containment strategies, and, once released to the environment there wouldn't be any chance of retrieval. Rational risk management strategies must be grounded by knowledge about ways of spreading, behaviour in the atmosphere or in fluids, lifetime of nanoparticles *as* nanoparticles until agglomeration to other (larger) particles, their behaviour in the human body and in the natural environment, etc. Such knowledge, however, is currently not available to an extent which would allow for classical risk management strategies (Schmid et al., 2006, sect. 5.2). In this situation, the precautionary principle has to be considered (fourth section).

A specific item contributed considerably to the "fall" of nanotechnology in public perception. The asbestos story served as case study which had shown what could result from the intensive use of materials without a careful previous impact analysis. Some voices pointed to possible analogies of synthetic nanoparticles to asbestos: "Some people have asked whether the ultra-small particles and fibres that nanotechnology produces, such as carbon nanotubes, might become the new asbestos" (Ball, 2003). In fact, there are almost no analogies in physical or chemical respect, or with respect to particle size or shape between asbestos fibres and synthetic nanoparticles of today. The relevance of the asbestos story to the nanoparticle risk debate does originate more in the dramatic case of what could happen in case of no or low precaution. Because of miraculous attributes of asbestos in respect of engineering its exploitation and use had developed very fast. Adverse health effects (asbestosis) had been observed rather early and led to some workplace regulations already in the 1930s. Other relevant pieces of knowledge (concerning lung cancer and mesothelioma caused by asbestos fibres), however, had been ignored or even suppressed. No comprehensive data collection and assessment had been performed prior to the 1960s (Gee & Greenberg, 2002). The asbestos story demonstrating severe health and economic disasters served as strong motivation to ask for more precautionary approaches in the field of nanoparticles.

The emergence of the risk issue in combination with the fact of having nearly no knowledge available about side-effects of nanotechnology led to severe irritations and to a kind of helplessness in the early stage of that debate. Statements from that time waver between an optimistic "wait-and-see" strategy (Gannon, 2003) on the one hand, and strict precautionary and sometimes "alarmistic" approaches on the other: "The new element with this kind of loss scenario is that, up to now, losses involving dangerous products were on a relatively manageable scale whereas, taken to extremes, nanotechnology products can even cause ecological damage which is permanent and difficult to contain. What is therefore required for the transportation of nanotechnology products and processes is an organisational and technical loss prevention programme on a scale appropriate to the hazardous nature of the products" (Munich Re, 2002, p. 13).

The still most famous position on nanoparticle regulation probably is the postulate of the ETC group for a moratorium: "At this stage, we know practically nothing about the possible cumulative impact of human-made nanoscale particles on human health and the environment. Given the concerns raised over nanoparticle contamination in living organisms, the ETC group proposes that governments declare an immediate moratorium on commercial production of new nanomaterials and launch a transparent global process for evaluating the socio-economic, health and environmental implications of the technology" (ETC Group, 2003, p. 72). The ETC work gave a significant push to nanotechnology regulatory debates in many countries but also increased the fears at the side of nanotech researchers against a broad public front of rejection and protest.

A completely different but also far-reaching recommendation aims at "containing" nanotech research: "CRN has identified several sources of risk from MNT (molecular nanotechnology), including arms races, grey goo, societal upheaval, independent development, and programmes of nanotech prohibition that would require violation of human rights. It appears that the safest option is the creation of one – and only one – molecular nanotechnology programme and the widespread but restricted use of the resulting manufacturing capability" (Phoenix & Treder, 2003, p. 4). This containment strategy would imply a secret and strictly controlled nanotech development which seems to be unrealistic and unsafe as well as undemocratic. Furthermore, this recommendation is irritating regarding the ideal of an open scientific community.

All of these different proposals have enriched (and heated) public and scientific debate. Seen from today's perspective, these proposals are documents of a very specific situation. Nanotechnology, itself still in an embryo state, has experienced itself, more or less suddenly, as a subject of public risk debate. Nobody seemed to be prepared for this case. The situation was characterised by severe challenges: while high expectations of benefits still dominated, there was no reliable knowledge about possible side-effects of nanotechnology available. Even the principal feasibility of some of the specific developments under criticism like the Molecular Assembler (Drexler, 1986) and its possible running out of control (Joy, 2000) could not be assessed in a transparent manner.

Against this background, it is understandable that the first years of the nanotech risk debate may be characterised mainly by mere suspicions and uncertainties rather than by knowledge-based and rational deliberation. In the meantime, however, things have changed considerably. An intensive debate on nanotechnology issues took place in many countries rather early. The German Parliament debated in 2003 about nanoparticles, informed by a Technology Assessment study (Paschen et al., 2004). The study of the Royal Society and the Royal Academy of Engineering (2004) resulted in a lot of statements and recommendations aiming at closing the knowledge gaps and at minimising the risks possibly resulting from production and use of nanoparticles by a preventive approach. Following that report, the UK Government promised to promote more research on nanotechnology risks, to conduct reviews of existing regulations with respect to their applicability to nanotechnology, and to enforce public debate (UK Gov, 2005). Also in the United States and in Germany ambitious research programmes on possible side-effects of nanoparticles have been started in the meantime.

In spite of these activities, public uneasiness concerning nanoparticles did not disappear. New postulates for a moratorium have been brought forward (Friends of the Earth, 2006). Empirical research on the public perception of nanotechnology shows considerable concern. Though the recent "MagicNano" case in Germany – a product of this name had to be recalled because of adverse health effects (in spite of its name no nanotech was inside, according to the company's statement) – did not lead to a scandalisation of nanotech in the media, an enormous sensitivity at the side of nanotech researchers has been the consequence. There is still no equilibrium or stable state concerning public attitude to nanoparticles. In this situation, I will concentrate the further analysis on the question of the application of the precautionary principle and its possible impacts.

Risk Management and Precautionary Principle

Risk management strategies accompanying the implementation of new technologies and the introduction of new materials are standing in a long tradition. In earlier times, often a "wait-and-see" approach has been taken. New substances have been introduced assuming that probably either no negative developments and impacts would occur at all or that, in case of adverse effects, *ex post* repair and compensation strategies would be appropriate. Experiences with hazards to human health or the environment caused by new materials, by radiation, or by new technologies have led to risk regulations in different fields, in order to prevent further negative impacts on health and the environment. Important areas are (following Schmid et al., 2006, chap. 5.1):

- Regulations for workplaces with specific risk exposures (nuclear power plants, chemical industry, aircrafts, etc.) to protect staff and personnel
- Procedural and substantial regulations for nutrition and food (concerning conservation procedures, maximum allowed concentrations of undesired chemicals like hormones, etc.) to protect consumers
- Environmental standards in many areas to sustain environmental quality (concerning ground water quality, maximum allowed rate of specific emissions from fabrication plants, power plants, heating in households, etc.)
- Safety standards and liability issues to protect users and consumers (in the field of automobile transport, for power plants, engines, technical products used in households, etc.)

There are established mechanisms of risk analysis, risk assessment and risk management in many areas of science, medicine, and technology, for example in dealing with new chemicals or pharmaceuticals. Laws like the Toxic Substances Control Act in the United States (Wardak, 2003) constitute the framework for dealing with such situations. Such "classical" risk regulation is adequate if the level of protection is defined, and the risk can be quantified as product of the probability of occurrence of the adverse effects multiplied by the assumed extent of possible damage. In such situations, thresholds can be set by law, by self-commitments, or following participatory procedures, risks can be either minimised or kept below a certain level, and also precautionary measures can be taken to keep particular effects well below particular thresholds by employing the ALARA (as low as reasonably achievable) principle. Insofar such mechanisms are able to cover challenges at hand to a sufficient extent there is a "standard situation".[6]

If the conditions required for the "classical" risk management approach are no longer fulfilled, uncertainties, controversies and ambivalent situations are the consequence. This is the case if, at the one hand, scientific knowledge concerning possible adverse effects is not available at all or if it is controversial and hypothetical, or if empirical evidence is still missing. On the other, classical risk management might not be applicable if adverse effects could have catastrophic dimensions with respect to the extent of possible damage, also in case of (nearly) arbitrary small probabilities of their occurrence. This type of situations motivated, for example, Hans Jonas (1979/1984) to postulate a "heuristics of fear" and the obligation to use the worst scenario as orientation for action. There has been a lot of criticism against Jonas' approach. Especially the naturalistic premises and the supposed teleology of nature but also the arbitrariness of conclusions resulting from running into aporetic situations have been major arguments. It became clear that Jonas' approach might be very appropriate to raise awareness with regard to precautionary situations but would be completely inadequate to be operationalised by regulation. Jonas' approach completely missed a legitimate procedure for deciding about the applicability and adequacy of precautionary strategies.

However, Jonas' approach put – with high justification – emphasis on the relevance of normative reflection in cases where classical risk management would no longer be adequate. Such situations are often welcomed entry points for ideology and interest-driven statements. In fact, it is very difficult to identify what a "rational" approach to dealing with this situation could be and in which way it could be proven to be rational. Neither ideological positions nor an arbitrary choice of the risk management strategy could be accepted as "rational" (Decker & Grunwald, 2001). The implementation of the precautionary principle in the European Union might be interpreted as a step-by-step approach to a rational risk management strategy in cases of low knowledge about possible damages.

The observation that in many cases severe adverse effects in the course of the introduction of new materials had not been detected in an early stage but rather led to immense damage on human health, on the environment, and also on economy (cp. impressive case studies in Harremoes et al., 2002) motivated debates about precautionary regulation measures. A wide international agreement on the precautionary principle was reached during the Earth Summit (United Nations Conference on Environment and Development UNCED) in Rio de Janeiro 1992 and became

---

[6] The notion of a "standard situation" with regard to risk can be used quite analogously to the notion of "standard situations" in moral respect (Grunwald, 2000/2005): in both fields it is decisive that the respective normative framework is able to cover the situation at hand (for criteria of this "ability" see Grunwald, 2005).

part of Agenda 21: "In order to protect the environment, the precautionary approach should be widely applied by States according to their capabilities. Where there are threats of serious or irreversible damage, lack of full scientific certainty shall not be used as a reason for postponing cost-effective measures to prevent environmental degradation" (as stated by principle 15 of the Rio Declaration). The precautionary principle has been incorporated in 1992 in the Treaty on the European Union: Article 174 postulates: "Community policy on the environment shall aim at a high level of protection taking into account the diversity of situations in the various regions of the Community. It shall be based on the precautionary principle [...]."

The precautionary principle thus establishes a rationale for political action: it substantially lowers the (threshold) level for action of governments (see Schomberg, 2005 for the following). It considerably changes the situation compared to the previous context in which politicians could use (or abuse) a persistent dissent among scientists as a reason (or excuse), simply not to take action at all. In cases for which the accumulation of relevant scientific evidence can take decades, this implies that political action could always be postponed with the argument that scientific knowledge would still have to be completed. In this way, political action could simply come much too late. Table 1 shows the different situations of risk management: classical risk management in a "standard situation" is to be applied in situations with

Table 1 Overview of the state of affairs in science and the possible corresponding responses by risk management (Modified after von Schomberg, 2005)

Circumstances	State of affairs in science	Policy framework/regulatory action/examples
Risk (classical approach)	Known effects, quantifiable probabilities, remaining uncertainties may have statistical (e.g. stochastic) nature	Risk management by defining thresholds on the basis of the chosen level of protection, exercising prevention, minimalisation of risk: applying the ALARA principle, etc.
Unquantifiable risk, lack of knowledge	Known effects/unknown or uncertain cause-effect relations, therefore unknown probabilities	Examples: antibiotics in feeding stuff; protection of the North Sea. Invocation of the precautionary principle and preventive measures to eliminate the possible causes can be justified
Epistemic uncertainty: scientific controversies, lack of knowledge	Unknown scope of effects, degree and/or nature of their "seriousness" (in relation to the chosen level of protection) can only be estimated in qualitative terms	Examples: genetically modified organisms (GMOs); climate change; ozone depletion. Invocation of the precautionary principle is justified
Hypothetical effect/ imaginary risk	Arguments on the basis of a fully conjectural knowledge base, no scientific indication for their possible occurrence	Invocation of the precautionary principle is not justified

quantifiable and well-known risks. In the opposite case, if instead of any knowledge only imaginary fears or hypothetical assumptions about damage are available there is no rational ground for risk management. In the Jonas approach one could follow the worst thinkable hypothesis, but this strategy would not provide a legitimate and "rational" strategy of action. Between the extremes there are two lines where certain pieces of knowledge are available but not sufficient knowledge for classical risk management. These two cases – unquantifiable knowledge and epistemic uncertainty – are the fields where the precautionary principle has to be considered.

It is, however, a difficult task to make legitimate decisions about precautionary measures without either running into the possible high risks of a "wait-and-see" strategy or overstressing precautionary argumentation with the consequence of no longer being able to act. The following characterisation of the precautionary principle shows – in spite of the fact that it still does not cover all relevant aspects – the complex inherent structure of the precautionary principle:

> Where, following an assessment of available scientific information, there is reasonable concern for the possibility of adverse effects but scientific uncertainty persists, measures based on the precautionary principle may be adopted, pending further scientific information for a more comprehensive risk assessment, without having to wait until the reality and seriousness of those adverse effects become fully apparent. (von Schomberg, 2005, modified; Schmid et al., 2006)

Thinking about the application of the precautionary principle therefore generally starts with a *scientific examination*.

There is a need to have an assessment of the state of the knowledge available in science and of the types and extents of uncertainties involved. Drawing the borderline with classical risk management practice or the situation of a purely conjectural risk involves making normative choices that need to be made explicit (Rescher, 1983). In assessing the uncertainties involved, *normative* qualifiers come into play while applying the precautionary principle (Schomberg, 2005). It has to be clarified whether there is "reasonable concern" in this situation of uncertainty. The qualifier "reasonable concern" as employed by the EC guidelines relates to a judgment on the quality of the available information (von Schomberg, 2005). Therefore, the assessment of the knowledge available including its uncertainties enters the centre of precautionary reflections.

The Precautionary Principle in the Field of Nanoparticles

Argumentation Chain

Following the preceding analysis, the central questions in the current situation concerning the use of nanoparticles and the knowledge about possible impacts are:

1. Is there a precautionary situation characterised by *epistemic uncertainty* or *unquantifiable risk* (see Table 1)?

2. Is there "*reasonable concern* for the *possibility* of adverse effects" in the field of synthetic nanoparticles to legitimate the application of the precautionary principle?
3. If yes, what would follow out of this judgment with respect to adequate precautionary measures?

The first two questions concern the diagnosis of the current state of affairs, especially with respect to an assessment of the current state of knowledge, while the third question is directed towards the conclusions resulting from that diagnosis in combination with the application of the precautionary principle.

The first question has to be answered *positively* (e.g. Schmid et al., 2006, chap. 5.2; Haum et al., 2004). There are unknown or uncertain cause-effect relations and still unknown probabilities of risks resulting from production, use and proliferation of nanoparticles. The scope of possible effects, their degrees and the nature of their "seriousness" (in relation to the chosen level of protection) can currently, even in best cases, only be estimated in qualitative terms. Therefore, we are witnessing a typical situation of uncertainty where established risk management strategies are not to be applied (in the same sense also Royal Society, 2004, summary, p. 4; Haum et al., 2004).

The answer to the second question, whether there is "*reasonable concern* for the *possibility* of adverse effects" in the field of synthetic nanoparticles is also to be answered positively. There are first toxicological results from the exposure of rats to high concentrations of specific nanoparticles which showed severe and lethal consequences. Because the exposure concentrations have been extremely high and the transfer of knowledge achieved by the exposure of rats to the situation of humans is difficult, these results do not allow for the conclusion of an evidence of harm – but they support a conclusion that there is "reasonable concern for the *possibility* of adverse effects" caused by synthetic nanoparticles. Please note that in order to legitimate the application of the precautionary principle the evidence of the *possibility* of adverse effects is sufficient. This is, on one hand, an important difference to the imperative of responsibility (Jonas, 1979/1984), because in the Jonas approach no *evidence* of the possibility of adverse effects is required to reject a certain technology, but the hypothetic imaginability of adverse effects is regarded as sufficient. On the other, there is a strict difference to classical risk management situations where *evidence of adverse effects* is required, not evidence of their *possibility*.

The third question is the most difficult one. With regard to resulting measures the precautionary principle regularly has an impact at two levels (Haum et al., 2004, p. 26): at the level of improving the knowledge base in order to enable "classical" risk management strategies as well as at the level of elaborating regulatory mechanisms to be better able to deal with the uncertainties involved as long as the first level could not be reached. What this could mean in detail is, however, difficult to assess. In the following, I will first describe selected existing regulatory systems against the background of nanoparticles, and then summarise the normative analysis recently elaborated (Schmid et al., 2006).

Current Regulatory Situation

The Toxic Substances Control Act (TSCA) as the U.S. regulatory act for chemical substances (passed in 1976) aims at regulating production and use of chemicals with risk or risk potential to the environment (Wardak, 2003). Main element is the TSCA Chemical Substance Inventory which is a database of about 80,000 chemical substances from which about 50,000 have been reviewed and about 5,000 have been subject to rigorous testing (Wardak, 2003). Ideas of the precautionary principle are not implemented. Regulatory action can only be undertaken if scientific uncertainty about possible hazards has been resolved. The "burden of proof" applied requires that there is clear scientific evidence for the existence of adverse effects – a *"reasonable concern* for the *possibility* of adverse effects"* (see above) does not legitimate any regulatory action. With regard to nanotechnologies there are three types of possible relevant exemptions: the Low Volume Exemption, the Low Release and Exposure Exemption, and the Test Marketing Exemption. The Low Volume Exemption frees a manufacturer from full reporting in accordance with the TSCA regulations if less than 10t of a particular substance are produced per year. Probably, many nanomaterials will fall below this exemption condition. Also the Test Marketing Exemption might apply to some nano developments because in many cases small amounts will be produced in order to exploit its market potential (see Wardak, 2003).

In the future European REACH system a different perspective is taken, according to a stronger precautionary approach. It aims at testing all known chemical substances and including the data in a central database while new chemicals will only be reviewed optionally (Haum et al., 2004, p. 36). This *prima facie* strange approach is motivated by the diagnosis that new substances are only a small portion compared to the complete set of existing chemicals – whose features, side-effects and risks are often only partially known. In case of nanoparticles this situation is ambivalent: on the one side, there is a chance to carefully check the impacts of nanoparticles based on well-known chemicals (differing only with respect to particle size and/or morphology), but, on the other side, there might be a gap if such nanoparticles were declared as new chemicals – with only optional review within REACH. Currently, it is not clear whether nanoparticles will be covered by REACH.

The existing regulatory systems for chemicals have been optimised to large volumes, and they do not consider specific attributes of nanoparticles (maybe that REACH will change this point in the course of recent initiatives in the European Parliament but this is still an open process under political pressure). It is well-known that the toxicity of fine and ultra-fine particles depends strongly on size and form of the particles. Regulation of chemicals, however, does not take these attributes into account (indeed, they are not relevant in the field of usual chemicals). Against this background, a lot of reflection activities on nano-regulation are currently taking place. The first workshops and conferences on nanotechnology regulation have been organised, and an international community is establishing itself around these questions.

Normative Analysis

In the precautionary situation given by the positive answers to the first two questions mentioned in the section on Argumentation Chain the challenge is to identify a "rational" way of action. An important judgment has to be made (with good reasons):

- Is there reasonable concern about adverse effects of synthetic nanoparticles, or
- Is there reasonable concern about the *possibility* of adverse effects?

In the first case, there would be *prima facie* a sound basis for implementing strict measures. The diagnosis of a reasonable concern about adverse effects of nanoparticles could serve as a legitimating reason for very strict measures like a moratorium on nanoparticle use in specific products (e.g. cosmetics, Friends of the Earth, 2006), a strict regulation of the release of nanoparticles (ETC Group, 2003), or even a complete ban on nanoparticles in research and development. In the second case, weaker instruments of precaution would be adequate.

Recalling that the precautionary principle requires a scientific assessment of the state of the art, and that the "quality of the information available" (Schomberg, 2005) serves as qualifier in order to assess the "reasonability of concern", it becomes clear that the assessment of the state of knowledge has a key function in deciding among the two alternative options. This scientific assessment of the knowledge available has been provided in Schmid et al. (2006, chap. 5.2) in a comprehensive manner, reaching the frontier of present research.[7] The result is that there are indications for nanoparticle risks for health and environment in some cases. However, based on the state of the art, there is *no reason for serious concern* about adverse effects but serious concern about their possibilities.

This result is supported by other studies: "Taking into account our present-day knowledge, there is, with regard to nano-specific effects (excluding self-organisation effects and cumulative effects of mass production), no reason for particularly great concern about global and irreversible effects of the specific technology "per se", with it being on a par with the justifiable apprehension concerning nuclear technology and genetic engineering" (Haum et al., 2004, p. 16). The mere possibility of serious harm implied by a wider use of nanoparticles, however, does not legitimate to use the precautionary principle as argument for a moratorium or other prohibitive measures. It will always be possible to create speculative scenarios with highly adverse effects but following those speculations would not reach the "rationality" of the precautionary principle (as described in the third section). The conclusion in this respect is therefore that there is no reason for a moratorium or similar strict measures according to the present state of knowledge. Because the state of knowledge permanently changes, a continuous monitoring and assessment of the

---

[7] Usually, such a comprehensive analysis is not included in ELSA studies on nanotechnology, cp. Haum et al. (2004); Royal Society (2004); Nanoforum (2004) which endangers the strength and evidence of the recommendations given.

knowledge production concerning impacts of nanoparticles on human health and the environment is urgently required.[8]

The next question is what other (and weaker) types of measures are required to face the precautionary situation. These could, for example, be self-organisation measures in science, like the application of established codes of conduct, or the adaptation of existing regulation schemes to special features of nanoparticles. It seems helpful to reconsider, as an example, the implementation of the precautionary principle in the genetically modified organisms (GMOs) case: "The European Directive 2001/18 (superseding directive 90/220), concerning the deliberate release of GMOs into the environment, is the first piece of international legislation in which the precautionary principle is translated into a substantial precautionary framework. [...] In the framework of this directive, the precautionary principle is translated into a regulatory framework based on a so-called *case-by-case* and a *step-by-step* procedure. The case-by-case procedure facilitates a mandatory scientific evaluation of every single release of a GMO. The step-by-step procedure facilitates a progressive line of development of GMO's by evaluating the environmental impacts of releases in decreasing steps of physical/biological containment (from greenhouse experiments, to small scale and large scale field tests up to market approval). This procedural implementation of the precautionary principle implies an ongoing scientific evaluation and identification of possible risks" (Schomberg, 2005). This model could serve as a framework of reference also for the nanoparticle case. The application of the precautionary principle would, in this argumentation, imply "that there is a need for a cautious step-by-step diffusion of risk-related activities or technologies until more knowledge and experience is accumulated" (Haum et al., 2004, p. 6).

Against this background the first conclusion is that the existing measures of regulation *need not to be extended* for dealing with possible nanoparticle risks. There is a wide consensus that no really new regulatory measures are needed: "[...] the existing regulative framework is adequate to deal with the introduction of new substances in the nanotechnology sector" (Haum et al., 2004, p. 27). Treating nanoparticles as new chemicals seems to be the adequate risk management approach. Though there are particular points of criticism (for example towards the TSCA, see the section on Current Regulatory Situation; Wardak, 2003), these points may be taken into account by modifications and supplements. Special awareness about possible side-effects should be given to possible long-term effects of nanoparticles on health and environment. This special awareness, however, should also and always be given to the use of new chemicals and their impacts. Consequently, from the view of engineering ethics and ethics in economy, there is no principally new situation in the field of synthetic nanoparticles.

There are urgent tasks to be accomplished by science, especially to close the still large knowledge gaps. Within the framework presented, the main task is to improve the knowledge base by research in several respects, mainly by means of human and environmental toxicology. A second line of required activities consists of

---

[8] One of the most dramatic lessons to be learned from the asbestos story is exactly the lesson about the crucial necessity of such a systematic assessment.

systematically collecting all relevant knowledge and of conducting regular and comprehensive evaluation of the respective state of knowledge and what would follow from that state. A third field of activities should be directed towards contributing to public debate and to improving communication at the interface between science and society. Schmid et al. (2006) arrived at the following steps for further action, thereby meeting several of the recommendations made by other groups (Haum et al., 2004; Royal Society, 2004):

- Develop a nomenclature for nanoparticles and assign a new Chemical Abstract Service (CAS) registry number to engineered nanoparticles
- Group and classify nanomaterials with respect to categories of risk, toxicity, and proliferation
- Treat nanoparticles as new substances, and develop and approve tools for screening and testing
- Improve the knowledge base in toxicology
- Develop guidelines and standards for good practices in cautiously dealing with nanoparticles
- Avoid or minimise production and unintentional release of waste nanoparticles
- Establish comprehensive evaluation of the state of knowledge and its evaluation with respect to implications for risk management as well as for identifying knowledge gaps which should urgently be closed
- Create institutions to monitor nanotechnologies and the knowledge about possible risks
- Establish a permanent and open dialogue with the public and industry.

This step-by-step approach conforms to present knowledge and seems to be the adequate way of applying the precautionary principle in a pragmatic way. However, there is no guarantee of preventing all kinds of possible risks by applying these steps. To repeat an important lesson from the asbestos story, the present situation "no evidence of harm" must not be re-interpreted in the way of "evidence of no harm" (Gee & Greenberg, 2002).

Conclusions

Nanoparticles and nanomaterials, produced by a wide variety of physical, chemical and biological processes, are providing novel and radically different properties which can be used to develop new products for the marketplace (Schmid et al., 2006). Many of these materials are actually handled only in research laboratories. However, specific nanoparticles are already present in many areas of our life, and their number will considerably increase in the next years. Most of these nanomaterials are still not sufficiently investigated with regard to possible impacts on health and environment. Therefore, classical risk assessment is – due to knowledge gaps concerning hazards resulting from nanoparticles – not applicable. Quantitative measures of the probability of damage and of the extent of possible hazards are not yet available. Insofar it is not surprising that lively discussions about the applicability of the precautionary principle take place in the industrialised countries.

For the precautionary principle to be applied, there must be a reasonable concern about the possibility of hazards, based on scientific investigation and knowledge assessment. According to the state of the art in toxicological research on nanoparticles (Schmid et al., 2006, sect. 5.2), there is such a reasonable concern of the *possibility* of hazards. However, there is no "reasonable concern" for an *evidence* of hazards caused by nanoparticles which would legitimate hard measures (like a moratorium) according to the precautionary principle.

A lot can be done and is, in parts, on the way already: the gaps of knowledge with regard to risk assessment have to be filled up, the possible sources of exposure must be identified, and valid methods for the analytics of environmental compartments and biological samples have to be developed for the synthetic nanoparticles. New toxicological methods have to be established to measure the biological effects not only in relation to mass and number but also to the active surface area of nanoparticles; and the assessment of risks to human health, the environment, consumers and workers at all stages of the life cycle of nanotechnologies has to be integrated in research and development (Schmid et al., 2006).

With respect to the "reasonable concern for the possibility of hazards", specific caution in responsibly dealing with synthetic nanoparticles is required. Such particles should be handled analogously to *new* chemicals even in the case that the chemical composition is well-known beyond the nano character. Dealing with new nanoparticles is still based on a case-by-case approach because established nomenclature and classification schemes are not prepared to be applied to nanoparticles. This situation should be changed as soon as possible (Renn & Roco, 2006; Schmid et al., 2006).

Beyond risk management and regulation, public risk communication has to be observed carefully because irritations in this communication could have dramatic impact on public acceptance and political judgment. Public perception of risk has to be taken seriously even in case that it seems to be inadequate against the results of scientific risk assessment – taking it seriously, however, does not imply to simply follow it, but to start an open dialogue about risk perception, patterns and underlying reasons of perception, acceptability, acceptance conditions, trust-building measures, etc. Denying the existence of risks of nanotechnology often causes mistrust and suspicion instead of creating optimism. To speak frankly about possible risks increases the chance for a trusty relation between science and society: building trust in public debate needs an open debate about chances and risks, and often requires schemes for comparing different types and amounts of risk to be available (Grunwald, 2004; Renn & Roco, 2006). Transparency about the premises of different risk assessment exercises is urgently required. Knowledge about risks includes knowledge about the validity and the limits of that knowledge. Communicative and participative instruments of technology assessment (Decker & Ladikas, 2004) could help improving mutual understanding and public risk assessment.

Acknowledgement I am deeply indebted to Torsten Fleischer (ITAS). He commented on the draft version of this paper and contributed considerably to its improving.

References

Ball, P. (2003). Nanoethics and the Purpose of New Technologies. Lecture at the Royal Society for Arts, London, March 2003, http://www.whitebottom.com/philipball/docs/Nanoethics.doc [2.10.2006]

Coffrin, T., MacDonald, C. (2004). Ethical and Social Issues in Nanotechnology. Annotated Bibliography, http://www.ethicsweb.ca/nanotechnology/bibliography.html [2.10.2006]

Colvin, V. (2003). Responsible Nanotechnology: Looking Beyond the Good News. Centre for Biological and Environmental Nanotechnology at Rice University, http://www.eurekalert.org/ [2.10.2006]

Decker, M., Grunwald, A. (2001). Rational Technology Assessment as Interdisciplinary Research. In: Decker, M. (ed.): *Implementation and Limits of Interdisciplinarity in European Technology Assessment*. Berlin, pp. 33–60

Decker, M., Ladikas, M. (2004, eds.). *Bridges Between Science, Society and Policy. Technology Assessment – Methods and Impacts*. Berlin: Springer

Drexler, K.E. (1986). *Engines of Creation – The Coming Era of Nanotechnology*. Oxford: Oxford University Press

ETC Group (2003). The Big Down. Atomtech: Technologies Converging at the Nanoscale, http://www.etcgroup.org [2.10.2006]

Friends of the Earth (2006). Nanomaterials, Sunscreens, and Cosmetics: Small Ingredients, http://www.foe.org/camps/comm/nanotech/nanocosmetics.pdf, 11–19, 2006

Gannon, F. (2003). Nano-Nonsense. *EMBO Reports* 4(11), 1007

Gee, D., Greenberg, M. (2002). Asbestos: from 'Magic' to Malevolent Mineral. In: Harremoes, P., Gee, D., MacGarvin, M., Stirling, A., Keys, J., Wynne, B., Guedes Vaz, S. (eds.): *The Precautionary Principle in the 20th century. Late Lessons from Early Warnings*. London: Sage, pp. 49–63

Grunwald, A. (2000). Against Over-Estimating the Role of Ethics in Technology. *Science and Engineering Ethics* 6, 181–196

Grunwald, A. (2002). *Technikfolgenabschätzung – eine Einführung*. Berlin: Edition Sigma.

Grunwald, A. (2004). The Case of Nanobiotechnology. Towards a Prospective Risk Assessment. *EMBO Reports*, Special Issue 5(October 2004), 32–36

Grunwald, A. (2005). Nanotechnology – A New Field of Ethical Inquiry? *Science and Engineering Ethics* 11/2005(2), 187–201

Grunwald, A. (2007). Converging Technologies: Visions, Increased Contingencies of the Conditio Humana, and Search for Orientation. *Futures* 39(2007), 380–392

Harremoes, P., Gee, D., MacGarvin, M., Stirling, A., Keys, J., Wynne, B., Guedes Vaz, S. (2002, eds.). *The Precautionary Principle in the 20th century. Late Lessons from Early Warnings*. London: Sage

Haum, R., Petschow, U., Steinfeldt, M., von Gleich, A. (2004). Nanotechnology and Regulation Within the Framework of the Precautionary Principle. Schriftenreihe des IÖW 173/04, Berlin

Jonas, H. (1979/1984). *Das Prinzip Verantwortung*. Frankfurt/M.: Suhrkamp, English: *The Imperative of Responsibility* (1984)

Joy, B. (2000). Why the Future Does Not Need Us. *Wired Magazine* (April 2000), 238–263

Munich Re (2002). Nanotechnology – What Is in Store for Us? Münchener Rückversicherungs-Gesellschaft, München, http://www.munichre.com/publications/302-03534_en.pdf [12.11.2005]

Nanoforum (2004). Nanotechnology. Benefits, Risks, Ethical, Legal, and Social Aspects of Nanotechnology, http://www.nanoforum.org [2.10.2006]

Paschen, H., Coenen, C., Fleischer, T., Grünwald, R., Oertel, D., Revermann, C. (2004). *Nanotechnologie. Forschung und Anwendungen*. Berlin: Springer

Phoenix, C., Treder, M. (2003). Applying the Precautionary Principle to Nanotechnology. The Centre for Responsible Nanotechnology, http://www.crnano.org/Precautionary.pdf [2.10.2006]

Renn, O., Roco, M.C. (2006). Nanotechnology and the Need for Risk Governance. *Journal of Nanoparticle Research* 8(2), 153–191

Rescher, N. (1983). *Risk. A Philosophical Introduction to the Theory of Risk Evaluation and Management*. Lanham, MD: University Press of America

Royal Society (2004). Nanoscience and Nanotechnologies: Opportunities and Uncertainties. London: Royal Society

Schomberg, R. von (2005). The Precautionary Principle and Its Normative Challenges. In: Fisher, E., Jones, J., von Schomberg, R. (eds.): *The Precautionary Principle and Public Policy Decision Making*. Cheltenham, UK/Northampton, MA, Edward Elgar pp. 141–165

Schmid, G., Brune, H., Ernst, H., Grünwald, W., Grunwald, A., Hofmann, H., Janich, P., Krug, H., Mayohr, M., Rathgeber, W., Simon, B., Vogel, V., Wyrwa, D. (2006). *Nanotechnology – Perspectives and Assessment*. Berlin: Springer

Shrader-Frechette, K.S. (1991). *Risk and Rationality. Philosophical Foundations for Populist Reforms. Berkeley*, CA: University of California Press

Swiss Re (2004). Nanotechnologie. Kleine Teile – große Zukunft? Zürich: Risk Perception.

UK Gov, H.M. Government (2005). Response to the Royal Society and Royal Academy of Engineering Report, http://www.ost.gov.uk/policy/issues/nanotech_final.pdf [March 7, 2005]

VDI-TZ (2004). Industrial Application of Nanomaterials – Chances and Risks. Future Technologies Division of VDI-TZ (W. Luther, ed.), http://www.techportal.de.[2.10.2006]

Wardak, A. (2003). Nanotechnology & Regulation. A Case Study Using the Toxic Substance Control Act (TSCA). Woodrow Wilson International Center, Foresight and Governance Project, Paper 2003-6

Wood, S., Jones, R., Geldart, A. (2003). *The Social and Economic Challenges of Nanotechnoloy*. Swindon, UK: UK Economics and Social Research Council

Anticipating the Unknown: The Ethics of Nanotechnology

Joseph C. Pitt

Abstract It is argued that only a pragmatist ethics is sufficient for making ethical decisions in situations dealing with the unknown. The pragmatist ethics described here is one that emphasizes the role of the community when making descisions. Making a moral decision requires asking what the consequences of your actions are for the living the Good Life, where what the Good Life includes accounting for the actions and decisions of the other members of the community. I argue that it is crucial to making moral decisions to recognize that they are not made in a vacuum. Hence making decisions about implementing innovative technologies involving nanotechniques requires taking into account what others are doing and how their actions are likely to impact yours and the environment.

Keywords Pragmatism, ethics, the good life, law of unintended consequences, nanotechnology

The prospects of a fully exploited knowledge of how to manipulate the nano-world can be frightening. The possibilities imagined by both advocates and opponents of research and development in the nano-world range from the mundane, pants that can't be stained, to the horrific in military applications. The problem is what we should do *now* to ensure the benefits of nanotechnologies and to avoid the predicted horrors. This situation is no different in principle from the one we faced with the advent of techniques for splicing and recombining DNA in the early 1980s. How do we go about shaping the future when the genie is out of the bottle? Specifically, how should we think about the ethical and social consequences of developing various nanotechnologies?

In this paper, I propose that the only form of ethical reasoning capable of giving us some purchase on the set of issues future developments in nanotechnology present, comes from Pragmatism. After offering some suggestions as to why more traditional philosophical theories of ethics are inadequate, I will sketch a pragmatic

Virginia Tech

theory of ethics and show how it helps us deal with these issues. Fundamentally, pragmatic ethics forces us to a kind of wholism that is both naïve and yet, nevertheless illuminates the incomplete nature and limits of ethical situations considered in the abstract apart from their full cultural context. It is, in William James' sense, forward looking, rather than a search for first principles. And, following a lead from Charles Saunders Peirce, the founder of Pragmatism, it asks us to think about developing a conception of The Good Life that, in the long run, can receive universal endorsement.[1] In short, the proposal developed here asserts that concentrating on the particular merits or demerits of a particular action in some limited context or other is simply the wrong way to think about ethical issues. For, if ethics is concerned with The Good Life, as it is, is should be The Good Life for All.

My point of entry is fairly straightforward: in considering the consequences of various nano-scenarios, those scenarios have to be played out fully, thereby giving us a sense of the full scope of the impact of these technologies. Not only are there short term and long term consequences, but there are also local, regional and global consequences. It is only when we understand the full range of the impact of our actions that we can decide among our options. Anything else is merely the exercise of prejudice. But we cannot consider the full range of our actions merely in terms of physical effects. What these changes signify is to be understood against our values and our goals and the complexity of the real world situations in which we make our decisions, and it is in this sense that we make our Jamesean[2] decisions looking forward. That is, these decisions press against our current concept of The Good Life writ large pushing us to constantly rewrite it in the hope of what we do now mean for the future. I will elaborate on the notion of The Good Life below when I talk about the range of ethical theories open to us.

There are then three components to the view I wish to sketch here: a problem about ethics, a theory of rationality, and the concept of The Good Life.

I made a rather bold claim when in my opening paragraph I stated that *only* a pragmatic theory of ethics can handle the potential problems posed by nanotechnology. This claim has already been challenged by a colleague, Tom Staley, who says, naturally, that it is false. So, let me put this in perspective. There are roughly four major kinds of ethical theories in the western tradition of philosophy: utilitarian, deontological, virtue, and pragmatic. I am going to argue that of these four, the first three fail, leaving us with pragmatism.[3] But I will do more than offer a proof by elimination; I will offer a positive argument in defense of pragmatism.

It is tempting to cheat here. The cheating move I have in mind, but will not capitalize on, is this: only pragmatism allows us to learn from our mistakes while still being ethical, hence, since learning from our mistakes is the heart of being rational

---

[1] The concept of The Good Life, capitalized, is a regulative ideal – as such it is in constant motion – an evolving ideal that develops as we learn more about who we are, what we can do and where we are going as a world. As I will argue below, there is no one conception of The Good Life currently in hand, and there may never be one, but it seems to me to be a exemplary goal.

[2] See James (1907, chap. 2).

[3] It is curious that there has not been much of an effort to develop a full pragmatist theory of ethics. The most recent effort in that direction can be found in *Pragmatist Ethics for a Technological Culture*, edited by Keulartz et al. (2002).

and is the only successful theory of rationality, only pragmatism offers genuine solutions to ethical problems. *Q.E.D.* To develop this argument is merely to win by default – for the theory of rationality I am appealing to is not the only one available to us. I happen to think it is the best one and that it captures in the most fundamental way what we seek in a theory of rationality. But I cannot argue for that here, it is too big a job – so, with apologies for the appearance of arrogance, I refer you to *Thinking About Technology* where I develop the view in some detail.

I will not attack utilitarian, deontological and virtue ethics theories separately. For our purposes here I want to suggest that one argument holds equally well against all three: they fail to *guide* our actions, with an emphasis on "guide" and "actions". Each approach fails for different reasons, but the result is the same. Some reasons why they fail are: for utilitarians, it often happens that we cannot calculate the consequences accurately enough to determine what produces the greatest amount of happiness in either the short or long term, or, for that matter, what constitutes "happiness". For deontologists, G.E. Moore's open question looms as large today as when it was first posed. We may have arrived at a definition of "good" or "right", but is it really good or right? I frankly don't understand virtue ethics, maybe because I don't know what virtue is, but here is my best shot. With help from another colleague, William FitzPatrick, virtue ethics shifts the focus from principles to the character of individuals. So we ought to seek to do those things that improve our character. Here, the issue is similar to the one facing deontologists. Unless there is something else against which to measure whether this or that virtue actually improves character and such an improved character has independent merit, who can say?

But when all the philosophical arguments are concluded, the fact of the matter is that when people act, they do not consult deep and profound philosophical theories first – they do what they do and only when challenged do they seek justifications, and even in those cases it is rarely a philosophical principle they pull out of their bag of excuses. Finally, the approach taken by all these theories is a holdover from the Enlightenment concern with the individual as the prime object of concern. These views are on a par with epistemological theories that view the individual as the possessor of knowledge, all of which lead unavoidably to solipsism. The pragmatist move to the community is the genuine next step in understanding both epistemology and ethics.[4]

The Original Ethical Question was "what is the nature of The Good Life?" Somehow, after 2,500 years of philosophical perversion, that question has come to be wrongly understood as "Should I be a utilitarian or a deontologist, or a virtuist?" But there is more to living The Good Life than adhering to a set of rules that have been shown to be bankrupt for hundreds of years. And one reason the now classical ethical systems have failed to provide the guidance we seek is because their framers have forgotten that we are concerned with living The Good Life, which doesn't always translate into individuals deciding how to perform abstractly isolated moral actions. Not all actions are moral actions, and not all actions designed to contribute

---

[4] I owe this final articulation of how to understand the major pragmatist move to a positive response I felt I needed to formulate to a presentation by L.L. Bucciarelli at the Workshop on Philosophy and Engineering, October 31, 2007 in Delft, The Netherlands.

to The Good Life are moral actions. And those who would have us think so would have us replace an awareness of the genuine complexity of living with superficial formulaic solutions. In short, whether or not to go to a nude beach is not a question that can be decided by appeal to the principle of doing what is best for the greatest number, but it is a question about the kind of life you want to lead. And to be frank, whether or not you want an all over body tan has nothing to do with anyone else's well-being and don't let moral Fascists tell you otherwise. I may be personally disgusted by seeing you in the all together, but there may be nothing moral about my disgust, just the application of a perfectly good set of aesthetic principles!

So, if we are interested in living The Good Life, instead of fleeing from it, then we need to locate the domain of philosophical inquiry most appropriate to solving that problem. And there's the rub, for it is not as easy as one might think to locate the appropriate domain wherein lie the best principles to help you decide about and then live The Good Life. Today we tend to characterize philosophy as being composed of five areas, epistemology, metaphysics, ethics, logic and the history of philosophy. Unfortunately this way of capturing philosophy doesn't make sense to those of us who live the messy place we call the real world. If we see philosophy the traditional way, we end up with some notable leftovers, such as the philosophy of science, political philosophy, aesthetics and more.[5] Now those fascinated with neatness will try to stick the leftovers in one of the established areas. Thus, philosophy of science gets shoved under logic or epistemology, depending on your preferences. Political philosophy gets lumped with ethics, or under it, another mistake; and we are always left holding the questions of aesthetics in our hand looking for some sort of philosophical rug to sweep it under, hoping no one will catch us.

The mistake, of course, is to try to package neatly something as unruly and complex as philosophy. There is no obvious reason why philosophy should come readily dividable into those five packages. To see this, we need to understand what the goal of philosophy is. Those who adhere to the five box theory seemed to have forgotten what we are trying to achieve as philosophers. We know about the goal of physics, to understand the components and structure of the physical universe. Biology is concerned with the nature of life and living things. History with the nature of the past. But what about Philosophy? What is its goal?

According to Wilfrid Sellars,

> The aim of philosophy, abstractly formulated, is to understand how things in the broadest possible sense of the term hang together in the broadest possible sense of the term. Under 'things in the broadest possible sense' I include such radically different items as not only 'cabbages and kings', but numbers and duties, possibilities and finger snaps, aesthetic experience and death. To achieve success in philosophy would be, to use a contemporary turn of phrase, to 'know one's way around' with respect to all these things, not just in that unreflective way in which the centipede of the story knew its way around before it faced the question, 'how do I walk?', but in that reflective way which means that no intellectual holds are barred. (Sellars, 1963, p. 1)

If philosophy's goal is to see how it all hangs together, why should we assume that it all comes together under five simple headings or three or ten? We shouldn't

---

[5] For an elaboration of these ideas see my "Against the Perennial" (2003).

because it doesn't. The complexity of the world demands a philosophical method that accommodates that complexity. That doesn't mean we should abandon the five categories or three or ten, they have their function. It just means, among other things, that if we insist on them we will probably end with philosophy being irrelevant to the world in which we live since as new topics come up, like nanotechnology, the opportunity to address them is lost. However, if philosophy is to proceed as a dialogue through which we attempt to see how all hangs together, then we cannot dismiss the approaches of the past out of hand, for the present is a function of the past and to understand the present we need to deal with the past. It is for this reason that many of those with whom we have this dialogue known as philosophy are thinkers from the past. To do so requires understanding them, understanding how they expressed their ideas and how to interpret what they said. Further, this is good, for surely there is no better way to be completely misunderstood than to refuse to use accepted language, even as we to try to change the meanings of some of the terms. And many of the words we use as philosophers obtained their meanings from earlier philosophical systems. So we must begin with the legacy that has been bequeathed to us, the world as it is, the language as it is spoken, our history as it is written, and then proceed to analyze, correct, rearrange and put it together in a more coherent form as we learn more and more. The problem being, of course, that as we learn more and as relationships become more complex, the more we will need to keep analyzing, rearranging and resorting, continually trying to make it all make sense. But surely that is preferable to trying to live according to rules put forth in a story written centuries, nay millennia ago, in a world vastly different from our present one with players and arrangements now foreign to our understanding.

What I am after is this: the philosophical job is on-going; it never ends, because the complexity of the world is as much a function of what human beings do as anything else. And since we have this seemingly infinite capacity to mess up whatever appears to work, there remains the constant need to keep trying to rearrange things so that they fit together – where making things work in the world of ideas is to see how it all hangs together. This also means that no matter how pretty a picture we manage to create at any given time, we must accept the fact that it will shortly be out of focus because the world will not stop for temporary philosophical perfection.

The conclusion that follows from this discussion is that there is no absolutely preferable way of seeing how things hang together, since the bits and pieces we have arranged so nicely to justify doing what we want to do will soon be augmented by new pieces and players, or impoverished by the loss of old ones. The world will not stand still for us – we live in a sea of constant flux without any dry land on which to stand. And no matter where you go, everyone else will be trying to find some place firm enough on which to stand, frantically jockeying for position only to find the sand underfoot being undermined by yet another ocean current. We live in a world without absolutes, without constants, one that is constantly in change.

Well, things are even worse than that. Like my hero, David Hume, I feel it is important that we allow our philosophical ruminations to be carried to their logical end, to reduce us to melancholy, if you will, for only then will we allow a different control to take over.⁶ So for now let reason continue as our guide to complete and total dismay.

Let us assume that history is wrong and that our favorite ethical theories are not defeated, by that meaning that they have not been shown to be internally inconsistent, or based on false premises, or irrelevant. There remains, nevertheless, one fatal consideration which forces us to find them less than satisfactory as guides to living The Good Life and that is the one law of human existence that is incontrovertible, the Law of Unintended Consequences. Our best efforts notwithstanding, the best laid plans of men and mice will come acropper because of our in-principle inability to predict the future with complete accuracy. What do I mean?

The fact of the matter is this: when we act we set into motion events over which we cease to have complete control. When you consider the many people, i.e., the many actors in the world and how many actions we each initiate, then anticipating the results of the compounded effects of all these action on each other simply cannot be done. It is not just that there are too many variables. With a big enough computer and fine tuned programs, we could probably make accurate predictions, on one assumption. That assumption is that the world is deterministic. However, even if the physical world were governed by laws of nature, the *social* world is not governed by deterministic laws, and if you factor into the inconstancies of nature, such as tsunamis, the erratic actions of 6 billion people, is it any wonder that we cannot predict the future? The Law of Unintended Consequences says this: calculate as accurately as you will, the future will be different than you predict because you cannot factor in all the possible consequences of all the simultaneous new actions occurring at that moment and every other moment. No matter how hard we try to make things work out, something always seems to happen which spoils the result.⁷

Consider the following somewhat strained example, based on fictional events at my home university. You reserved rooms in Blacksburg's best hotel for graduation the day you came to Virginia Tech as a freshman. The week before graduation there is a horrible thunderstorm which blows the reservations program in the hotel computer. Furthermore, the hotel has recently been bought by a private local investor and there is no national backup. Further, the original owner destroyed all records after it sold the hotel because it needed the storage space for current matters. Net result: no hotel reservations for your parents, grandparents, or your Swedish great-grandmother who has flown in just for this occasion and who at age 102, after hearing of this disaster has a heart attack and expires. Unlikely? No. You did all the right things, you made early reservations; each year you called to make sure you still had them. You alerted your family to the need to make early plane reservations

---

⁶ See Hume (1739, pp. 263–274).

⁷ A number of year ago I was sent to a conference on Chaos Theory hosted by the U.S. Naval Academy at Annapolis by our adventuresome director of our Center for Programs in the Humanities, Wilfrid Jewkes. The ideas I was introduced to there concerned mainly predictions about the physical world. But as I read what I have written here I now realize how deeply I was impressed by those ideas and they apply to the human world multi-fold.

for great grandmother, you studied hard so you would graduate on time, etc. You did everything right and still lost. And, in the immortal words of Kurt Vonnegut, "so it goes".

So, not only are your favorite ethical theories worthless, because they are flawed and unworkable, no matter what you do, as this example is intended to show, there is a certain sense in which you are doomed to fail.

If I may be so bold, at this point I would like to make a suggestion. Since we are, in so many words, doomed to not do the right things, no matter how we try, then maybe we shouldn't be concerned about doing the right things, where "doing the right thing" means meeting the criteria of some abstract ethical theory. Instead, why not be concerned about the way in which you go about trying to do the right thing. I am talking about form, method, or, as I prefer, style.

In the world of horse showing, in particular hunter jumpers, there is a phrase I would like to appropriate: *Way of Going*. It refers to that component in the competition in which points are awarded for the way the horse and rider together make their way over the obstacles. For their way of going, their sense of each other, of their being a team working together, listening and responding to one another, they receive points, as well as for not knocking down any jumps.

In a world without absolutes, in a world in which we are constantly changing and responding to new demands made by nature and other people, what better model for us than to evaluate ourselves in terms of our way of going? Let us take a look at what is involved here. But so you are not mislead, I am not going to start with nice and neat definitions and deduce all the relevant consequences. That is exactly the approach I reject. What I will be doing now is trying to give a feel for the kind of view I am advocating.

First, we need to reevaluate the conclusion of the graduation story. You did nothing wrong. Several events transpired over which you had no control. In a world where that is understood, Great-grandmother might not have died just then. Instead you would all have crammed into your one room apartment and made do, being left with a marvelous experience to reflect back on over the years, and you would walk away with wonderful stories to tell *your* grandchildren about trying to get 127 people showered and dressed and out to graduation in the football stadium by 8:30 am, where it then rained.

Why would this Pollyanna ending result if I am right? Well, it sort of goes like this. The key concept in "Way of Going" is team-work. The horse and rider get more points the more they seem to act as one in accomplishing their task. Living The Good Life means working with your fellow human beings to create the best possible world. How you go about accomplishing this is not easy, nor is it obvious what these loaded concepts mean. Their meaning is to be worked out in times and places in which you will find yourselves. Furthermore, we need to recognize that sometimes some things work and sometimes they don't. Our job is not to decry the failure of ethical principles, but to find something that works at the time it is needed.

This is where pragmatism makes its entrance. The fundamental principle of the form of pragmatism I am advocating is the principle of rationality introduced earlier: learn from experience or Common Sense Pragmatism (CSP). It starts by recognizing that when faced with selecting among options, we do not make our choices in a vacuum. We come to any situation equipped with a set of values, goals, and background knowledge and with a lot of other actors equally well equipped. This is important: we must factor in the other factors from the beginning – individuals do not act in a void. When we make decisions, we rely on some subset, or maybe more, of what we bring to decision-making in general. But when we select an option, and act on it, that is not the end of the story. As I have been emphasizing, actions have consequences. Under CSP we are required to follow through and consider the consequences of what we did and then to reevaluate the assumptions, goals, values, and background knowledge we used to make the decision that led to the action that had those consequences. One of those assumptions concerns what the other people in the equation were expected to do. Once we have done that, we must make whatever appropriate adjustments we can to those assumptions, goals, values, and background knowledge to improve our ability to select the best option the next time. This is not to say that the option we selected had the expected consequences, but only that whatever the consequences, they ought to affect the set of factors we used to make that decision. If things turned out the way we hoped, then we can say that those things we employed in making that selection seem to be solid for the time being. If things did not turn out the way we expected, then we need to reconsider the importance of that goal, or the place of that value in our preference ranking, etc. In so doing, we are trying to make coherent those various factors that come into play when we make decisions and act, recognizing that not all of those values are moral or ethical values and concerns. In other words, we are trying to make it all hang together in a way that helps us to do the right thing which is to fashion and live The Good Life.

Common Sense Pragmatism is a method for making choices that lead to actions that have consequences for our conception of The Good Life. Why does this make CSP the only ethical theory that we can use in the face of the uncertainties raised by the possibilities of nanotechnology? As Anna Russell might note here, "Remember nanotechnology?" And how did a method, CSP, become an ethical theory?

Well, we also have a bit more than CSP. In our discussion so far we have noted that the world is a far more complicated and difficult place than even the most diehard cynic would have it. Not only can we not act assuming that everyone else can be held still. We are all acting and our collective actions have impacts on each of us and on the social and physical environments in which we live, and all of this ramifies and ramifies. It is truly amazing that we can make any plans at all that come out half way close to what we had hoped.

We also have the role of the concept of The Good Life to consider. It is clear that there is no one view of The Good Life that has priority or that even claims the allegiance of a majority of the world's population. Under those circumstances, how can we expect the concept of The Good Life to play any kind of significant role in ethical deliberations? I propose we take a page out of Peirce's book. According to Peirce, scientific enquiry occurs in the context of a community of investigators. What constitutes knowledge, on this account, is a function of the norms of the community as they evolve over time. Further,

Different minds may set out with the most antagonistic views, but the progress of investigation carries them by a force outside of themselves to one and the same conclusion. No modification of the point of view taken, no selection of other facts for study, no natural bent of mind even, can enable a man to escape the predestinate opinion. This great hope is embodied in the conception of truth and reality. *The opinion which is fated to be ultimately agreed to by all who investigate, is what we mean by truth, and the object represented in this opinion is the real.* (Peirce, 1955, p. 38)

Unlike Peirce, I would not propose that there is by necessity one perfect arrangement for humankind to live by. But I am convinced that in seeking for that one perfect arrangement, by measuring our actions in terms of how they contribute to or detract from contributing to that effort, is a very good way to determine how to act. But these deliberations should not take place *in foro interno*. There must be a dialogue with others making similar or even opposing decisions. It is only in the context of dialogue that we can triangulate our various visions of The Good Life and work to bring about a vision that harmonizes our differences. (I warned you at the beginning this would sound naïve.)

For a pragmatist, the meaning of our action is to be found in their consequences. When deciding on what action to take, we should consider the consequences, both for ourselves, and for others who will be affected by those actions. But I am not talking about just the few individuals I can imagine being affected by my actions. When I say that we should have The Good Life in mind, I am suggesting that you ask yourself this question: Will this action here and now contribute in the long run to my current view of The Good Life and how will it affect the various conceptions of The Good Life others hold? Further, if you are, and I suggest you must be, rational, then it would follow that you should reevaluate your conception of The Good Life in the light of the effects of your actions, and that this should be a constant activity. In this way, your and my conceptions of The Good Life will be a constantly evolving ideal, empirically informed, and this is the important pragmatist point, hopefully converging over time. It seems to me that there are some goods we can agree on from the start: a room over our heads, foods in our bellies, health care, safety, etc. What we see at work in the world are different efforts to achieve these goals – and it may be the case that for different peoples, different arrangements might make sense – but part of the living The Good Life is acknowledging that and factoring it into your conception of The Good Life.

So that is the general idea – be rational according to common sense pragmatism and seek The Good Life, knowing that it is always a changing ideal, one that we are reconstructing in light of what we have experienced and discovered, but one on which humankind can, with good intentions, most probably, in the long run, come to agreement. Thus, if you are committed to The Good Life in the long run, the ethical counterpart to Peirce's commitment to convergence on the truth is the normative injunction: *seek to bring about The Good Life, knowing ahead of time that not all, maybe none of your actions will ultimately contribute to that end in the long run.*

Now for nanotechnology. Lots of people are engaged in the design and manufacture of nanotechnologies. It seems to be everywhere. In our local paper, The Roanoke Times, a March 2004 Sunday special report, there was an article on a start-up firm in Wytheville, Virginia, making buckyball shaped nano structures based on a discovery by a Virginia Tech researcher. These devices are intended to be vehicles for carrying medications into the body. Is this something we should encourage or actively work to stop?

To begin with, let me put my general position on the table: there are no special ethical problems associated with nanotechnology, just as there are special problem unique to engineering ethics or bioethics, etc. In fact, I will be so bold as to suggest that those ethical theories that find the need to create ad hoc special arrangements for "new" problems have ipso facto demonstrated their inadequacy.

The reason I believe pragmatism provides the only viable framework for dealing with nanotechnology is that the perceived ethical problems associated with nanotechnology today derive primarily from a fear of the unknown. In short, a special case of deciding under uncertainty. Actually, it is fear not so much of the unknown as a fear based on speculation of the "what-if" sort. Thus, people worry about nanoparticles that could be sprayed like a gas in military contexts that would eat a person's insides – particles so small no filter could stop them from being inhaled. Other military applications include nano-bombs, nano-soldiers, nano-enhanced macro soldiers, nano-gases. To be frank, no ethical theory is going to deal with that. When it comes to military technologies, I am afraid that ethics is not the issue – beating the enemy no matter what is. I am perfectly aware that many struggle with issues like just war theory, torture, etc., but if is The Good Life we are after, I find it hard to image a vision in which these applications of nano knowledge have a legitimate place. However, it might be argued, if your vision of The Good Life acknowledges that it is a regulative ideal, one we are constantly struggling to obtain, but which seems to be a constantly receding horizon as more player enter the scene and with more options for novel action provided by novel technologies, then surely you need to be able to address factors that get in the way of achieving the final goal. That is, there will be many obstacles arising as we seek to bring about The Good Life. We cannot simply refuse to deal with them because they are irrational from the start. It also seems to me that dealing with the hard cases like military nano-options, may make it easier to deal with equally hard, but not so imminently threatening, like human enhancements.

We need to consider two separate cases of worry about nano-applications to military ends. The first concerns the worry about the potential use of Drexler's self-replicating nanomachines in a military context. Here the concern is that there will be no control on these devices once launched. That concern gains its credibility from accepting the premise that such machines are possible. There are serious objections to that possibility.

In a two page article in September 2001's *Scientific American* Nobel Laureate Richard Smalley argues that Drexler's monsters are not possible.

Smalley's account runs roughly like this:

1. Love is like chemistry: put two people together and there is a product that results.
2. Chemical reactions, however are actually a lot more complicated than just putting two atoms together and seeing what emerges. In the space (roughly 1 nm) of the atoms desired to react are many atoms 12 to 15 – engaged in a "three dimensional waltz".

Anticipating the Unknown: The Ethics of Nanotechnology

3. One nanobot would not be useful – generating even a tiny amount of a product would take a solitary nanobot millions of years.
4. "Making a mole of something – say 30 grams, or about one ounce – would require at least 6×10 to the 23rd bonds, one for each atom. At the frenzied rate of 10 to the 9th per second it would take this nanobot 6×10 to the 14th seconds – that is, 10 to the 13th minutes–, which is 6.9×10 to the ninth days, or 19 millions" which is pretty slow.
5. But, if the nanobot could replicate itself and then if the two could replicate themselves we could have an army of nanobots at our command – and then they could work together and increase the rate of production and maybe the world of plenty would be possible.
6. But – this not possible for two reasons, given the already mentioned small space in which atomic reactions occur:

    (a) Fat fingers and
    (b) Sticky fingers

7. Fat fingers: because the arms of a nanobot must itself be made of atoms, there is an irreducible size problem. "There just isn't enough room in the nanometer-size reaction region to accommodate all the fingers of the manipulators necessary to have complete control of the chemistry."
8. Sticky fingers "The atoms of the manipulator arms will adhere to the atom that is being moved. So it will be impossible to release this miniscule building block in precisely the right spot."
9. In conclusion Smalley returns to his love theme and also to his waltz theme –

    (a) Like the dance of love, chemistry is a waltz with its own step-slide-step in three-quarter time. Wishing that a waltz were a merengue – or that we could set down each atom in just the right place – doesn't make it so. (Smalley, 2001)

The second worry over military applications of nanotechnology concerns fears that we won't know how to overcome a nano-army. That, for example, if we let lose a nano-bug that eats our insides out, how will we turn these things off? That falls somewhat into the "what-if…" category. However, it also falls nicely into the purview of a pragmatic theory of ethics. Aside from Smalley's objections quoted above, if our concern is the construction of The Good Life, a state of affairs always in the future, then when faced with the potential of something like a nano-bug army, *we are obligated to develop a way to fix the problem*. There is no sense in arguing that we should resist the military use of nano-bugs – that genie is already out of the bottle. If it truly poses a threat, then if we are truly committed to The Good Life, we should also be committed to finding a way to defeat it. This is the general strategy. If we foresee a threat to The Good life, we are obligated to find a means for either neutralizing the threat or turning it to our advantage. And this a fundamentally different strategy from the ones employed by adherents to the other ethical theories discussed above. The other theories will commit you to a discussion of what the right thing ought to be – with the end result being: no nano. But, it is too late. Pragmatism is concerned with action. Given the current state of affairs, what should we *do!* In the immortal words of Eliza Doolittle – "Do something!"

The frustration with standard theories of ethics comes from their lack of contact with the way the world works. When the possibility of using recombinant DNA techniques to "fix" genetic disorders arose in the 1970s, it was already too late to say "we ought not to do that" – for the fact that we knew how to do it means that someone will. That is what is so silly about the current US administration's injunction against the development of new stem cell lines – not only is the research continuing elsewhere, but we in the United States will be negatively affected by the fact that our research is on hold while the rest of the world is figuring out how to cure cancer, regenerate failing organs, etc. That Governor Arnold Schwarzenegger saw this and authorized the state of California to expend funds to not only conduct such research, but to encourage it, thereby making California a stem cell research Mecca, proves the point.

A pragmatist ethics recognizes that we must deal with the world as it is. Ok, that said, how do we deal with the possibility of a developing technology that appears, by all accounts, to have potentially negative results as well as positive ones? For while stem cell research has clearly beneficial consequences, we need to ask the question of whether we should be repairing failing organs in an ever aging population. Overpopulation is already a problem, won't this just contribute to the problem? Well, consider the consequences. First, the argument that repairing failing organs using stem cells will contribute to overpopulation is of the same stripe as saving an 80 year old man who has had a heart attack, or using any medicine to keep people alive. Second, we need to ask, what is our conception of The Good Life? Do we want a world populated only by people who have never been ill? What about the 55 year old artist recognized by all as a genius who falls gravely ill and will likely die without medical attention? What about neonates who need that extra medical boost to get a start on life? Is our vision of The Good Life one in which no help is given? I hope not. Ok, we will allow stem cell research and find another way to deal with the overpopulation problem. But what about research that threatens to change the nature of being humans such as gene therapy and enhancements that offer to extend our potential? Well, what about them? Those who argue against using the best science we have to carve out the best possible future for humankind always amaze me. The only arguments for freezing us as we are derive from some commitment to a religious conception that privileges our current conception of what is a human. The fact of the matter is that we have been changing the species since we became a species. Sometimes by arranged marriages, sometimes by selecting a mate because he or she is sexually appealing – we can hardly argue that propagating the species has taken place according to a rational plan. What is our vision of The Good Life? Is it one in which human beings are kept from developing their fullest potentials?

That is the issue for a pragmatist: What should our future be like and what should we do to bring about that state of affairs?. There is no *a priori* answer to this question. The answer is what the evolving conception of The Good Life will reveal through argument and trial and error.

So, when we worry about the buckyball carriers noted above, our worry should be: What are the consequences of introducing these devices? These devices will be expelled from the human body – How will they affect the water supply, the soil, the air? Consider this parallel: antibiotics have decreasing efficacy because they are also being used to cure food animal diseases. We ingest beef treated with an antibiotic that we also use to cure an infection, but, because we are being constantly exposed to the antibiotic through the beef we consume, it ceases to have the effect it should when used to combat infection in humans.

There are numerous scenarios out there in which nanotechnologies are seen as the key to eternal youth or immortality – nano-devices will destroy cancers, delay aging, fight infections, etc. The question we face is not so much "is curing cancer a good thing?" as –"what effect on The Good Life is a population without cancer?" On the surface eliminating cancer is a good thing – in the short run – for me. But in the long run what is the effect? Likewise for wrinkle free pants – currently being manufactured. What if those pants turn out to be not only wrinkle free, but also non degradable? Landfills get filled. If the nano treatment the pants receive contributes to their remaining in good condition, good and useable for a much longer time, then what is the effect on the garment industry, and on the cotton growers, etc.? Cameras on cell phones seemed like a not so bad idea, even a cool gimmick – but today many gyms won't allow cell-phones in locker rooms because some men are taking pictures of their friends without any clothes on and sending them to their girlfriends or worse.

The problem with nano is that we don't know yet what it can do. Hence we need a way of thinking about it that not only has us considering the consequences of a specific type of device, but also the consequences in terms of their impact on our conception of how we want to live, i.e., The Good Life. But it is not enough to think about the consequences, the consequences must be weighed in the light of something else and I propose that the future well-being of *all* of humanity is the place to begin, recognizing that that future is constantly changing as we learn about more about the present. A real pragmatist ethics using an evolving conception of The Good Life demands we think hard and long about what our actions might entail and how to enhance the good effects and minimize or counter potentially negative effects as we continue in constant dialogue with as many other actors and visions as possible.

References

Hume, David (1739, 1888). *Treatise on Human Nature*, edited by L.A. Selby-Bigge. Oxford: Oxford University Press.
James, William (1907). *Lectures on Pragmatism*. New York: Longmans, Green & Co.
Keulartz, Josef, Korthaartz, Michel, Schermer, Maartje, Swierstra, Tsjalling, editors (2002). *Pragmatist Ethics for a Technological Culture*. Dordrecht, The Netherlands: Kluwer.
Peirce, Charles S. (1878, 1955). "How to Make Our Ideas Clear", in *Philosophical Writings of Peirce*, edited by J. Buchler. New York: Dover.
Pitt, Joseph C. (2000). *Thinking About Technology*. New York: Seven Bridges Press.

Pitt, Joseph (2003) Against the Perennial. *Techne; The Society for Philosophy and Technology Quarterly Journal (http://scholar.lib.vt.edu/ejournals/SPT/)* 7(2):

Sellars, Wilfrid (1963). "Philosophy and the Scientific Image of Man", in *Science, Perception, and Reality*. London: Routledge & Kegan Paul.

Smalley, Richard E. (2001) Of Chemistry, Love and Nanobots. *Scientific American Magazine* 285: 76–77.

Applications of Nanotechnology in the Biomedical Sciences: Small Materials, Big Impacts, and Unknown Consequences

Audy G. Whitman, Phelps J. Lambert, Ossie F. Dyson, and Shaw M. Akula*

Abstract Nanotechnology is at the forefront of a revolution in the biomedical sciences. It has the potential to give both researchers and doctors' abilities they would never have previously dreamt of, including everything from the capability to deliver engineered drugs to specific target tissues to filtering even the smallest harmful particles out of our water supply. With such increased power, however, also comes increased responsibility. Nanotechnologies have as much potential to do harm as they do good. For instance, nanotechnology could be an enormously effective tool in the hands of a bioterrorist. As such, it is critically important for mankind to fully appreciate the technology's awesome potential and the possible harm it may cause before this potential is realized. To this end, this review discusses not only the current and future applications of nanotechnology in the biomedical sciences, but also the incredibly important ethical ramifications of such applications.

Keywords Nanotechnology, ethics, nanite, nanate, medicine, review

Introduction

Nanotechnology takes advantage of the novel phenomena and properties (physical, chemical, and/or biological) that substances evince at the nanometer scale to create unique materials, devices, and systems (McNeil, 2005, pp. 585–595). Nanotechnologies can range from 10^{-7} to 10^{-9} m in size (1–100 nm). To put this

---

Department of Microbiology & Immunology, Brody School of Medicine at East Carolina University, Greenville, NC 27834, USA

*Corresponding author
Shaw M. Akula, Department of Microbiology & Immunology, Brody School of Medicine, East Carolina University, Greenville, NC 27834, USA
Email: akulas@ecu.edu

size into perspective, 1 nanometer (nm) is approximately the width of three to six atoms, or 10,000 times smaller than the width of a human hair (Hood, 2004, pp. A740–A749; Nel et al., 2006, pp. 622–627). As such, the manipulation of matter at such a small scale allows for the incredibly precise synthesis of molecules.

The concept of nanotechnology was first proposed by Richard Feynman in 1959. Since then, nanotechnology has become a reality, leading to numerous developments in many different scientific and engineering disciplines including biomedicine, plastics, energy, electronics, and aerospace industries, and has led to the development of many engineered nanoscale products, including fullerene derivatives, carbon nanotubes, and quantum dots (Hardman, 2006, pp. 165–172). Today, nanotechnology has gone a step further by incorporating recent developments in the physical sciences, biology, biotechnology, medicine, and molecular engineering into a common scientific field (Roco, 2003, pp. 337–346).

Nanotechnology is a hot commodity in today's scientific and economic world. Governments and organizations around the world are investing heavily in the future of nanotechnology, which is predicted to be a $1 trillion industry by 2012 (Roco, 2002; Hardman, 2006, pp. 165–172). The United States alone has over 23 federal agencies working in concert for the National Nanotechnology Initiative, which includes a multi-billion dollar R&D budget. Due to this enormous explosion of interest in the subject, this review strives not only to open the doors of this incredible technology to the uninitiated, but also to step back and reflect upon the ethical issues and responsibilities that come with the development and use of this nanotechnology so that our desire for progress does not outpace our moral sensibility.

Application of Nanotechnology in the Biomedical Sciences

In the biological sciences, researchers are applying developments in nanotechnology in both the laboratory and medicine. Many of these breakthroughs and applications are already in use in our everyday lives in everything from sunscreen to automobiles to the air we breathe (Friedrich, 2003, pp. 105–112; Stoffer, 2004, pp. 26; Fan & Lu, 2005, pp. 1561–1573; Zhang et al., 2005, pp. 46–49). Indeed, the possibilities for nanotechnology in biology seem limitless, with breakthroughs and new-found applications being discovered every day (Fig. 1).

Application in Research

Nanotechnologies hold great potential in research as nanosensors of biological conditions and processes. Scientists have proposed that in the future both nanomaterials, referred to as nanates, and nanomachines, referred to as nanites, can be synthesized and introduced into tissue to monitor disease states, biochemical and metabolic

Applications of Nanotechnology in the Biomedical Sciences

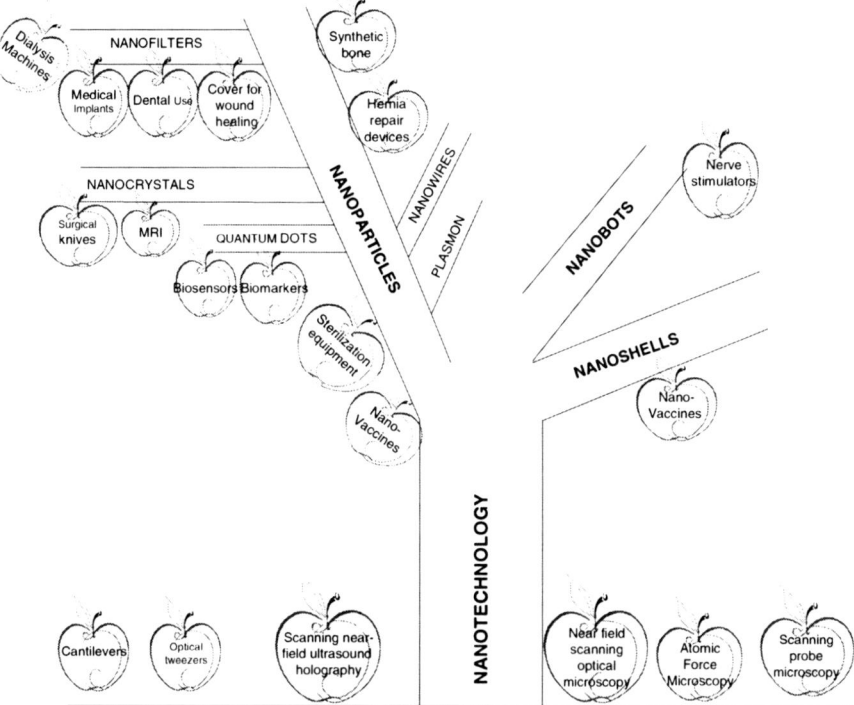

Fig. 1 Fruits of nanotechnology. The branches on the tree represent divergence from the origin of nanotechnology (Goose & Croquette, 2002, pp. 3314–3329; Lee et al., 2004, pp. 713–719; Alivisatos et al., 2005, pp. 55–76; Moghimi et al., 2005, pp. 311–330; Mukhopadhyay et al., 2005, pp. 2385–2388; Oh et al., 2005, pp. 4938–4943; Trache & Meininger, 2005, pp. 064023; Wabuyele et al., 2005, pp. 437–452; Xi et al., 2005, pp. 180–184; Zhang et al., 2005, pp. 933–946; Nidumolu et al., 2006, pp. 91–95; Pison et al., 2006, pp. 341–350; Portney & Ozkan, 2006, pp. 620–630; Ronco et al., 2006, pp. 48–50; Rodriquez et al., 2006, pp. 151–159; Smith et al., 2006, pp. 231–244; Uesaka, 2006, pp. 309–315). In addition, apples indicate various current medical uses or instruments by means of nanotechnology. Those that are attached to a branch specify the form of nanotechnology used to construct the product. On the other hand, the apples that are not attached indicate methods used to detect and/or manipulate nanoparticles

reactions, and other biological processes (Rodham & Olsen, 1997, pp. 103–136; Cote, 2001, pp. 15965–16045; Ng et al., 2001, pp. 375–379; Kasili et al., 2002, pp. 653–658; Mazzola, 2003, pp. 1137–1143; Mnyusiwalla et al., 2003, pp. R9–R13; Li et al., 2005, pp. 918–922). Biosensors of this order could provide researchers with a "snap-shot" or a "video-feed" of the biological processes going on in a cell at any time, and as such allow insight into cellular processes never before thought possible.

Nanotechnology, in combination with adenovirus vectors, has also been proposed as a possible multifunctional delivery vehicle (Glasgow et al., 2006). These delivery vehicles could transport components of interest into cells or cell organelles, opening up new possibilities in the world of research.

Nanotechnology has also found a significant niche in the world of research by providing new methods for separating biomolecules such as DNA, RNA, peptides, polysaccharides, and glycoproteins (McNeil, 2005, pp. 585–595; Baba, 2006a, pp. 271–277). As such, the technology has numerous potential applications, including SNP analysis, mutation analysis, haplotyping, DNA diagnosis, protein expression analysis, immunoassay, protein disease marker detection, point of care analysis, and other procedures (Baba, 2006a, pp. 271–277).

Lastly, nanotechnology has applications in the laboratory in the form of functional polymer coated nanoparticles and quantum dots that act both as biomarker detectors and in facilitating DNA separation (Nishiyama & Kataoka, 2006, pp. 199–205).

Applications in Medicine

Many of today's breakthroughs and applications of nanotechnology have made their way into the clinical setting. Developments in nanofabrication, nanocoating, and other molecular nanotechnologies have allowed for the incorporation of nanotechnology into medical applications, including preventative medicine. They promise to aid medical professionals in the detection, management, and healing of many injuries, illnesses, and diseases (Baba, 2006b, pp. 189–198).

Optimally, scientists and health professionals cure disease conditions. Realistically, most disease conditions are not cured, but at best managed. In an effort to increase the number of curable diseases, scientists and physicians have proposed various solutions which treat disease at the cellular level. These treatments include modification of cellular activities, incorporation of new materials into cells, targeting damaged and injured tissues for removal, replacing of damaged tissues with new healthy tissues or comparable materials, or repairing injured tissue. Some of the precise and interesting manners in which nanotechnology can be used in modern day medicine are as follows:

1. Diagnostics: Just as nanotechnology's potential to provide detailed insight into cellular processes and conditions has found an eager audience in research, so has it in medicine. Nanotechnologies can both be used to fix and characterize living materials in their native confirmations, such as with the photonic-induced protein immobilization method, as well as be used in the analysis of genome networks of diseased cells, such as through the use of quantum dots, photonic crystals, and DNA chips (Baba, 2006b, pp. 189–198; Neves-Petersen et al., 2006, pp. 343–351). These capabilities make nanotechnology highly suited for fixing and diagnosing diseased tissue, such as cancer (Baba, 2006b, pp. 189–198).
2. Drug Delivery: Nanomaterials can be used in medicine as vessels to deliver medications to specific tissues (Langer, 1998, pp. 5–10; Smith et al., 2003, pp. 5407–5412; Ferrari, 2005, pp. 161–171; Yih & Al-Fandi, 2006, pp. 1184–1190; Yoshikawa et al., 2006, pp. 247–252). The ability of these nanates to act as medicinal vectors can potentially allow scientists or medical workers to target specific cells and alter their cellular activities, be it creating balance/imbalance in chemical signals (hormones, cytokines, receptors, etc.) or altering genes within a specific cell

(Vasir et al., 2005, pp. 47–64). These nanates also have been reported to have the ability to cross previous impassable obstacles, such as the blood-brain barrier (Emerich, 2000, pp. 279–287). Due to their enormous potential to reach specific and hard to reach tissues, researchers and pharmaceutical companies the world over are increasingly employing nanotechnologies into the creation and production of drug delivery systems (Wickline & Lanza, 2003, pp. 1092–1095).

Going beyond simple drug delivery, some ingenious researchers have also looked into the possibility of using these tissue specific delivery vehicle nanates as potential "smart bombs" to target damaged and diseased tissues for apoptosis (Perkel, 2004, pp. 14). Diseases ranging from cancer to virus-infected cells have been proposed as potential targets of this medicinal strategy. Nanotechnology has thus far been applied to treat several diseases, including diabetes (Yamaguchi & Igarashi, 2006). A recent study has shown that the use of a nano-particle coated with calcium carbonate ($CaCO_3$) may contribute to the regeneration of beta cells *in vivo*, which may be of great help to those suffering from type 2 diabetes (Yamaguchi & Igarashi, 2006, pp. 295–300). Pharmaceutical companies have also explored the applications of nanotechnology into the treatment of insulin-dependent diabetes, examining the possibility of targeted and customized drug vehicles that monitor blood glucose levels and release insulin into the bloodstream when necessary (Cao & Lam, 2002, pp. 419–422; Thrall, 2004, pp. 315–318; Li et al., 2005, pp. 918–922).

Nanotechnology has also been proposed as a possible photodynamic therapy of cancer (Baba, 2006a, pp. 271–277). Cancerous cells are characterized by abnormal and uncontrolled cell division (Weinberg, 1996, pp. 62–70). Because of the unique intracellular and extracellular signaling events associated with cell division, cancer cells have the potential to be easily targeted by special receptors and materials on which cell division is dependent (Ferrari, 2005, pp. 161–171). By associating anti-cancer medications to compounds that will bind to cancer cell receptors, scientists can essentially create cancer cell specific drug regiments that will not affect healthy, normal cells (Ferrari, 2005, pp. 161–171). While these target specific compound-cancer medication complexes are still theoretical, they offer a powerful example of how health professionals and scientists are applying nanotechnology tools and concepts to treat prominent diseases such as cancer.

3. Corrective Medicine: Yet another field in which nanotechnology holds enormous promise is corrective/ regenerative medicine. For instance, one novel approach in the treatment and curing of disease that is currently being considered is the use of ATP motors. Scientists are attempting to synthesize biological motors in the laboratory which use ATP as a fuel (Montemagno et al., 1999, pp. 225–231). These attempts have created a buzz in the scientific community, in which many researchers have imagined the synthesis of biological motors and/or cellular organelles that could be inserted into diseased cells that lack function in their own motors/ organelles, thus curing that cell, and therefore the organism, of a particular disease condition.

Nanotechnology may also have applications in corrective surgery. Numerous sources have proposed the concept of nanosurgery, in which nanites are injected or inserted into a patient to perform a specific surgical procedure at the nano level (Mason, 2005, pp. 665; Panchapakesan, 2005, pp. 22–26; Ashammakhi, 2006, pp. 3–7). These nanites would be much more precise than current scalpel and laser techniques, and would be much less traumatic to target tissues and their associated or surrounding tissues, thereby contributing to the reduced risk of surgical complications and shortened recovery time.

Nanotechnology may also be used in the field of regenerative medicine (Yamato & Okano, 2006, pp. 335–341). Researchers have found that tissues grown in temperature responsive culture dishes can be harvested as a single cell sheet when done in conditions below 32ºC (Yamato & Okano, 2006, pp. 335–341). This finding opens up many possible applications in the field of tissue reconstruction, such as corneal regeneration (Yamato & Okano, 2006, pp. 335–341). Researchers have also recently found that soft and wet substances, such as hydrogels, may have important applications in articular cartilage replacement (Murosaki et al., 2006, pp. 206–214). These researchers report that hydrogels have desirable properties similar to actual articular cartilage, without many of the undesirable consequences of more traditional methods of articular cartilage replacement, which usually are restricted to hard and dry substances and are prone to friction, bacterial infection, and fracturing (Murosaki et al., 2006, pp. 206–214). In addition, private firms have also reported the successful production of synthetic bone that is identical to naturally occurring bone in both structure and composition (Emerich & Thanos, 2005, pp. 177–188). Nerve tissue has also been proposed as a tissue type that can benefit from nanotechnology (Vetter et al., 2004, pp. 896–904). Several researchers have proposed the use of nanomaterials to encourage nerve cell regeneration and division (Webster et al., 2004, pp. 48–54). Others have proposed the synthesis of entire organs in the lab for transplantation, thereby eliminating the problem of finding suitable donors.

4. Storage of Patient Medical Information: A lesser, but still significant contribution of nanotechnology to medicine may be in the form of small implanted biosensors and chips. These biosensors or chips could also store an individual's medical information, which would then be readily available to health professionals during disease or drug screening, health check-ups, or emergency situations in which the patient is unable to communicate (Rath & Tolles, 2003, pp. 1746–1759).

Benefit to Society and Environment

Nanotechnology also holds the promise of improving the state of our society. Counter-terrorism nanotechnologies involved in the detection of nanoscale concentrations of agents (biological, chemical, nuclear) have the potential to provide us with peace of mind against terrorist attack (Iqbal et al., 2000, pp. 549–578). By detecting molecules at such small concentrations, governments will have an improved capacity to protect their citizens from terrorist attacks.

Society might also benefit from nanotechnology applications in filtration and waste disposal. Small nanofibers have the capacity to filter nanoparticles out of the water supply, and therefore hold great potential in preventing the spread of waterborne infectious diseases throughout the world (Thrall, 2004, pp. 315–318; Van der Bruggen et al., 2004, pp. 1347–1353). Other nanotechnologies have demonstrated the ability to bind and encapsulate pollutants in our environment, which may be beneficial in future industrial accidents (Yuan, 2004, pp. 2661–2670). Waste decomposition is also an area of interest to nanotechnologists. Reports in the literature have described the need for nanites that can break down wastes and rearrange them into useful materials (Gilette, 1996, pp. 177–182).

Finally, nanotechnologies promise to reduce pollution and clean our environment via energy conservation. By making materials that can endure wear and tear for longer periods of time at greater stresses, and by synthesizing energy producing components that reduce friction and energy waste by fitting together better, nanotechnology can and will lower energy waste while improving our quality of life (Smith et al., 2000, pp. 10–12).

Ethical Issues of Applying Nanotechnology in the Biological Sciences

Being a relatively new field, there are no extended studies on the consequences of nanotechnology (Williams, 2005, pp. 9–10). In spite of the US government offering millions of dollars of funding for research into the ethical and societal implications of nanotechnology, few investigators have made use of these funding opportunities (National Nanotechnology Initiative; Mnyusiwalla et al., 2003, pp. R9–R13). However, a growing number of groups and organizations from around the world have called for an increase in the number of studies on the societal and ethical impact of nanotechnology. Reports from the literature indicate there are three primary groups of people vying for control of the future of nanotechnology and its applications: (1) Groups that do not concern themselves with the technology's moral, environmental, or health issues; (2) Groups that are overly cautious and/or apocalyptic about the application of nanotechnologies that lack a firm understanding of the technology and its present capabilities; and (3) Groups that advocate the incorporation of ethics into nanotechnological applications, realizing the importance of both nanotechnology science applications and the ethical issues involved in these applications.

A quick search of the literature, both scholarly and obtuse, reveals that nanotechnology is heralded by some as the future savior of mankind and the planet and feared by others as inevitably leading to an apocalyptic cataclysm beyond Oppenheimer's greatest nightmares (Gordijn, 2005, pp. 521–533). Many of these ethical concerns originate from some of the inherent qualities of nanotechnological products, including their unpredictability and the scale on which they operate. For instance, nanoproducts may not have the same properties on the nanoscale that they do on the macroscale. These properties include everything from a product's conductivity to its color, strength, and catalytic potential (Roco, 1999, pp. 1–16). As such, it is difficult to predict how a nanomaterial will act at the nano level based on its characteristics at the macro level. In addition, we are uncertain how these nanates will act as individual particles versus a conglomeration of thousands or millions of tiny particles (Roco, 1999, pp. 1–16). This uncertainty should create a certain degree of reserve regarding the degree and immediacy with which we will incorporate nanotechnology into our daily lives. Two of the major ethical concerns with respect to the use of nanotechnology are as follows:

1. Potential Threat to Personal Safety and Rights: Terrorism is a topic in which ethics and nanotechnology come to crossroads. In an ever changing world of destabilizing states, raw materials such as nuclear waste and weapons or biological agents are at an ever greater risk of falling into the hands of individuals or organizations that wish to force their views on others via the threat and/or act of physical violence. With nanotechnologies being almost undetectable, and with their potential being so great, what moral responsibilities do scientists and governments have in preventing these materials from falling into the hands of terrorists and becoming weapons of mass destruction? This is a dilemma that, left unaddressed, holds the potential to harm many innocent citizens.

Health risks are another factor that must be taken into account when considering the ethical issues of applying nanotechnologies (Williams, 2005, pp. 9–10). Scattered reports in the literature indicate that the toxicity due to nanotechnology is dependent on physicochemical and environmental factors (Hardman, 2006, pp. 165–172). Other literature reports the potential of nanates to freely enter and collect in the food chain and individual cells of organisms (Toensmeir, 2004, pp. 14–17; Matsudai & Hunt, 2005, pp. 923–927). The environmental and health hazards from this accumulation of nanates are unknown, but have serious potential consequences. As such, the long-term health consequences for the patient, physician, and environment have not been well explored.

Privacy may also be threatened in the future age of nanotechnology. The small, almost undetectable size of nanates and nanites could allow individuals to be monitored without their knowledge. Nanorecorders or nanocameras could invade the privacy of every human being on earth, whether their rights are protected by law or not. Additionally, information collected by nanotechnological medical applications on an individual's health and well-being, as well as their risks for disease, could fall into the wrong hands via inappropriate handling. Without a better handle on the implications of nanomonitoring, scientists and physicians could be opening a Pandora's Box from which there is no returning.

2. Potential Social Issues: One of the most pressing concerns raised by nanotechnology ethicists revolves around the increasing divide in the development of nanotechnology between the industrialized and many third world countries

(Baird & Vogt, 2005, pp. 101–107). While some argue that developing nations lack the resources and/or foresight to truly appreciate the impact nanotechnology will have on the lives of their citizens and the prosperity of their countries, most ethicists disagree and instead believe that once properly educated regarding its incredible potential, all countries will fully appreciate and embrace nanotechnology. In the current state of world affairs, however, nanotechnology is on track to become a trillion dollar industry by the end of the decade, and at their present course, developing nations will be missing out on most if not all of this economic opportunity (Featherstone & Specht, 2004, pp. 1–6).

Nanotechnology also has the potential to fan the flames of the debate over biological cloning (Seeman et al., 1994, pp. 1895–1903). Because nanotechnology typically fall in a realm smaller than the cellular level, the potential of nanotechnological cloning has been overlooked by governing bodies that regulate biological cloning methods. Many ethicists foresee this lack of oversight and governance as a potentially explosive issue if nanotechnology's potential in cloning is ever realized.

Another, perhaps more disturbing, dilemma is the event of nanites becoming self-aware. It has been proposed that at some point in the future, nanites could become a conscious entity (Lin, 2005, pp. 10). Would nanites have the ability to evolve into intelligent beings? And if so, could they develop capacities of thought and ability of function beyond our own? Would we be slave or master to these nanites? Would these nanites take it upon themselves to enslave or destroy us and life as we know it? These questions, while science fiction in the present, are stepping ever closer to science fact with every development we make in the world of nanotechnology. The fundamental questions of what makes an intelligent living being and our role in this world will need to be addressed as these technologies progress.

Incorporation of nanates and/or nanites into an organism such as a human also has ethical ramifications. As described previously, nanotechnologies have been proposed as potential replacements for damaged cellular components, cells, or tissues. Is a synthetic mitochondria or cell classified as a living material? At what point of incorporation would an organism cease to be living and be considered cybernetic or artificial? These are very basic questions that promise to be at the heart of the ethical argument in the future public's mind.

With such human-technology interactions foreseeable in the future, would nanotechnology have the potential to give individuals abilities which many would currently characterize as "God-like"? Such abilities could potentially be applied to everything from regulating or controlling thought via neural networks to genotypically altering an individual. As a society, how will we determine what capabilities should be permitted? Who should make these distinctions: the individual, social groups, or the government?

Confounding the ethical issues even further is the fact that all cultures and nations of the world do not share the same moral principles and values. Being that every nation, culture, and indeed individual has different educational backgrounds, family values, professional activities, social learning, religious beliefs, and needs; it is to be expected that each nation, culture, and individual will have different and possibly conflicting morals. Since ethics is the study of the general nature of morals, it is to be expected that differing entities will have different and sometimes conflicting ethical views. What may be acceptable in one nation or culture may be apprehensible to another. How we address the potential ethical conflicts between different cultures and nations in an increasingly interconnected, interdependent world will have significant implications on the future of nanotechnology.

What Does the Future Hold?

As demonstrated in this review, nanotechnology is a popular and new field encompassing many scientific disciplines. The applications of nanotechnology seem limitless, ranging from industry to engineering to disease diagnosis to tissue regeneration. Nanotechnology combines the best and brightest people from multiple fields of science and the most innovative technologies of our time to address some of the greatest needs facing humanity.

However, what is not known is the breadth and scope of the consequences the application of nanotechnology into our everyday lives will sew. As a newly emerging field, little to nothing is known on how these tiniest of materials will impact our health, environment, well-being, and very existence. As such, extreme caution should be taken when considering and partaking in the application of nanotechnology. Nanotechnology holds the promise of drastically changing and improving our quality of life, from cleaning the environment to helping us live longer, healthier lives. However, without proper precautions, nanotechnology may prove to become detrimental.

In short, researchers and medical workers should make the objectives of nanotechnology's applications clear to patients and society as a whole. This openness should include any questions, concerns, or uncertainties pertaining to these budding technologies. The only way nanotechnology can find a lasting place in the laboratory and in the clinical setting is if professionals charged with running and maintaining these facilities are open and honest about the applications, goals, uncertainties, limitations, and pitfalls of these said technologies. An open dialogue must be present between scientists, physicians, patients, and the public at large.

No science or technology is inherently good or evil. It is only mankind's application of science and technology that can be benevolent or malicious. As such, it is the responsibility of members of the bio-medical profession to apply developed and developing nanotechnologies in the best and most prudent way possible so as to not harm the individual and society as a whole. In turn, it is the duty of all people to make certain that bio-medical professionals do not abuse these responsibilities.

Nanotechnology is at the forefront of a modern renaissance in the scientific world. It has the power to change the very essence of matter, and as such has applications on an ever-expanding horizon. We, who have never before had the ability to realize such near divine power, must be more careful than ever before, to ensure that we use it to benefit mankind and not become mankind's destructor.

Acknowledgements SMA is funded by a Grant (R21EB006483) from NIH/NIBIB. We thank Dr. Adrian Reber (University of Georgia, Athens) and Huxley, A.M., for graciously accepting to proofread the manuscript.

References

Alivisatos, A. P., Gu, W., & Larabell, C. (2005). Quantum Dots as Cellular Probes. *Annual Review of Biomedical Engineering*, 7, 55–76.
Ashammakhi, N. (2006). Nanosize, Mega-impact, Potential for Medical Applications of Nanotechnology. *Journal of Craniofacial Surgery*, 17, 3–7.
Baba, Y. (2006a). Diagnosis of Diseases by Nanodevice. *Nippon Rinsho*, 64, 271–277.
Baba, Y. (2006b). Nanotechnology in Medicine. *Nippon Rinsho*, 64, 189–198.
Baird, D., & Vogt, T. (2005). Societal and Ethical Interactions with Nanotechnology ("SEIN")-An Introduction. *Nanotechnology Law and Business Journal*, XXN, 101–107.
Cao, Y., & Lam, L. (2002). Projections for Insulin Treatment for Diabetics. *Drugs Today (Barc)*, 38, 419–427.
Cote, G. L. (2001). Noninvasive and Minimally Invasive Optical Monitoring Technologies. *The Journal of Nutrition*, 131, 1596S–1604S.
Emerich, D. F. (2000). Recent Efforts to Overcome the Blood-brain Barrier for Drug Delivery. *Expert Opinion on Therapeutic Patents*, 10, 279–287.
Emerich, D. F., & Thanos, C. G. (2005). Nanomedicine. *Current Nanoscience*, 1, 177–188.
Fan, Z., & Lu, J. G. (2005). Zinc Oxide Nanostructures: Synthesis and Properties. *Journal of Nanoscience and Nanotechnology*, 5, 1561–1573.
Featherstone, D. J., & Specht, M. D. (2004). Nanotechnology Patents: A Snapshot of Nanotechnology Patenting Through an Analysis of 10 Top Nanotech Patents. *Intellectual Property and Technology Law Journal*, 16, 1–6.
Ferrari, M. (2005). Cancer Nanotechnology: Opportunities and Challenges. *Nature Reviews Cancer*, 5, 161–171.
Friedrich, H. E. (2003). Challenges of Materials Technology for Low Consumption Vehicle Concepts. *Advanced Engineering Materials*, 5, 105–112.
Gilette, S. L. (1996). Nanotechnology, Resources, and Pollution Control. *Nanotechnology*, 7, 177–182.
Glasgow, J. N., Everts, M., & Curiel, D. T. (2006). Transductional Targeting of Adenovirus Vectors for Gene Therapy. *Cancer Gene Therapy*, 13, 830–834.
Goose, C., & Croquette, V. (2002). Magnetic Tweezers: Micromanipulation and Force Measurements at the Molecular Level. *Biophysical Journal*, 82, 3314–3329.
Gordijn, B. (2005). Nanoethics: From Utopian Dreams and Apocalyptic Nightmares towards a more Balanced View. *Science and Engineering Ethics*, 11, 521–533.
Hardman, R. (2006). A Toxicological Review of Quantum Dots: Toxicity Depends on Physicochemical and Environmental Factors. *Environmental Health Perspectives*, 114, 165–172.
Hood, E. (2004). Nanotechnology: Looking as we Leap. *Environmental Health Perspectives*, 112, A740–A749.

Iqbal, S. S., May, M. W., Bruno, J. G., Bronk, B. V., Batt, C. A., Chambers, J. P. (2000). A review of Molecular Recognition Technologies for Detection of Biological Threat Agents. *Biosensors and Bioelectronics*, 15, 549–578.
Kasili, P. M., Cullum, B. M., Griffin, G. D., & Vo-Dinh, T. (2002). Nanosensor for in vivo Measurement of the Carcinogen Benzo [a] Pyrene in a Single Cell. *Journal of Nanoscience and Nanotechnology*, 2, 653–658.
Langer, R. (1998). Drug Delivery and Targeting. *Nature*, 392(6679 Suppl), 5–10.
Lee, J. H., Kang, M., Choung, S. J., Ogino, K., Miyata, S., Kim, M. S., Park, J. Y., & Kim, J. B. (2004). The Preparation of TiO2 Nanometer Photocatalyst Film by a Hydrothermal Method and its Sterilization Performance for Giardia Lamblia. *Water Research*, 38, 713–719.
Li, J., Wang, Y. B., Qiu, J. D., Sun, D. C., & Xia, X. H. (2005). Biocomposites of Covalently Linked Glucose Oxidase on Carbon Nanotubes for Glucose Biosensor. *Analytical and Bioanalytical Chemistry*, 383, 918–922.
Lin, P. (2005). Nanotechnology's Dilemmas. *The Scientist*, 19, 10.
Mason, D. S. (2005). The World According to Nanotechnology. *Journal of Chemical Education*, 82, 665.
Matsudai, M., & Hunt, G. (2005). Nanotechnology and Public Health. *Nippon Koshu Eisei Zasshi*, 52, 923–927.
Mazzola, L. (2003). Commercializing Nanotechnology. *Nature Biotechnology*, 21, 1137–1143.
McNeil, S. E. (2005). Nanotechnology for the Biologist. *Journal of Leukocyte Biology*, 78, 585–594.
Mnyusiwalla, A., Daar, A. S., & Singer, P. A. (2003). 'Mind the Gap': Science and Ethics in Nanotechnology. *Nanotechnology*, 14, R9–R13.
Moghimi, S. M., Hunter, A. C., & Murray, J. C. (2005). Nanomedicine: Current Status and Future Prospects. *The Faseb Journal*, 19, 311–330.
Montemagno, C., Bachand, G., Stelick, S., & Bachand, M. (1999). Constructing Biological Motor Powered Nanomechanical Devices. *Nanotechnology*, 10, 225–231.
Mukhopadhyay, R., Sumbayev, V. V., Lorentzen, M., Kjems, J., Andreasen, P. A., & Besenbacher, F. (2005). Cantilever Sensor for Nanomechanical Detection of Specific Protein Conformations. *Nano Letters*, 5, 2385–2388.
Murosaki, T., Gong, J. P., & Osada, Y. (2006). Creation of Artificial Cartilage by Nanotechnology. *Nippon Rinsho*, 64, 206–214.
Nel, A., Xia, T., Madler, L., & Li, N. (2006). Toxic Potential of Materials at the Nanolevel. *Science*, 311, 622–627.
Neves-Petersen, M. T., Snabe, T., Klitgaard, S., Duroux, M., & Petersen, S. B. (2006). Photonic Activation of Disulfide Bridges Achieves Oriented Protein Immobilization on Biosensor Surfaces. *Protein Science*, 15, 343–351.
Ng, H. T., Fang, A., Li, J., & Li, S. F. (2001). Flexible Carbon Nanotube Membrane Sensory System: a Generic Platform. *Journal of Nanoscience and Nanotechnology*, 1, 375–379.
Nidumolu, B. G., Urbina, M. C., Hormes, J., Kumar, C. S., & Monroe, W. T. (2006). Functionalization of Gold and Glass Surfaces with Magnetic Nanoparticles using Biomolecular Interactions. *Biotechnology Progress*, 22, 91–95.
Nishiyama, N., & Kataoka, K. (2006). Nano-engineering for Biomedical Applications. *Nippon Rinsho*, 64, 199–205.
Oh, S. H., Finones, R. R., Daraio, C., Chen, L. H., & Jin, S. (2005). Growth of Nano-scale Hydroxyapatite Using Chemically Treated Titanium Oxide Nanotubes. *Biomaterials*, 26, 4938–4943.
Panchapakesan, B. (2005). Nanotechnology: The Promise Tiny Technology Holds for Cancer Care. *Oncology Issues*, Sept/Oct, 22–26.
Perkel, J. M. (2004). The Ups and Downs of Nanobiotech. *The Scientist*, 18, 14.
Pison, U., Welte, T., Giersig, M., & Groneberg, D. A. (2006). Nanomedicine for Respiratory Diseases. *European Journal of Pharmacology*, 533, 341–350.
Portney, N. G., & Ozkan, M. (2006). Nano-oncology: Drug Delivery, Imaging, and Sensing. *Analytical and Bioanalytical Chemistry*, 384, 620–630.

Rath, B. B., & Tolles, W. M. (2003). Nanotechnology, a Stimulus for Innovation. *Current Science,* 85, 1746–1759.

Roco, M. C. (1999). Nanoparticles and Nanotechnology research. *Journal of Nanoparticle Research,* 1, 1–16.

Roco, M. C. (2002). Government Nanotechnology Funding: An International Outlook. National Nanotechnology Initiative. Senate of the United States (January 16, 2003). 21st century Nanotechnology Research and Development Act. (http://www.nano.gov/intpersp_roco.html)

Roco, M. C. (2003). Nanotechnology: Convergence with Modern Biology and Medicine. *Current Opinion in Biotechnology,* 14, 337–346.

Rodham, K. J., & Olsen Jr., D. R. (1997). Nanites: An Approach to Structure-based Monitoring. *ACM Transactions on Computer-Human Interaction,* 4, 103–136.

Rodriquez, B. J., Kalinin, S. V., Shin, J., Jesse, S., Grichko, V., Thundat, T., Baddorf, A. P., Gruverman, A. (2006). Electromechanical Imaging of Biomaterials by Scanning Probe Microscopy. *Journal of Structural Biology,* 153, 151–159.

Ronco, C., Breuer, B., & Bowry, S. K. (2006). Hemodialysis Membranes for High-volume Hemodialytic Therapies: The Application of Nanotechnology. *Hemodialysis International. International Symposium on Home Hemodialysis,* 10(Suppl 1), S48–S50.

Seeman, N. C., Zhang, Y., & Chen, J. (1994). DNA Nanoconstructions. *Journal of Vacuum Science and Technology A: Vacuum, Surfaces, and Films,* 12, 1895–1903.

Shekhawat, G. S., & Dravid, V. P. (2005). Nanoscale Imaging of Buried Structures via Scanning Near-field Ultrasound Holography. *Science,* 310, 89–92.

Smith, A. M., Dave, S., Nie, S., True, L., & Gao, X. (2006). Multicolor Quantum Dots For Molecular Diagnostics of Cancer. *Expert Review of Molecular Diagnostics,* 6, 231–244.

Smith, G., Davies, G., & Saxl, O. (2000). Counting up the Benefits of Nanotechnology. *Materials World,* 8, 10–12.

Smith, R. A. J., Porteous, C. M., Gane, A. M., & Murphy, M. P. (2003). Delivery of Bioactive Molecules to Mitochondria In Vivo. *Proceedings of the National Academy of Sciences of the United States of America,* 100, 5407–5412.

Stoffer, H. (2004). Nanotechnology: Big Changes in Small Science Await. *Automotive News,* Mar 8, 26HH.

Thrall, J. H. (2004). Nanotechnology and Medicine. *Radiology,* 230, 315–318.

Toensmeir, P. A. (2004). Nanotechnology Faces Scrutiny over Environment and Toxicity. *Plastics Engineering,* Nov 2, 14–17.

Trache, A., & Meininger, G. A. (2005). Atomic Force-multi-optical Imaging Integrated Microscope for Monitoring Molecular Dynamics in Live Cells. *Journal of Biomedical Optics,* 10, 064023.

Uesaka, M. (2006). Application of Nanotechnology to Hemodialysis Membrane. *Nippon Rinsho,* 64, 309–315.

Van der Bruggen, B., Koninckx, A., & Vandecasteele, C. (2004). Seperation of Monvalent and Divalent Ions From Aqueous Solution by Electrodialysis and Nanofiltration. *Water Research,* 38, 1347–1353.

Vasir, J. K., Reddy, M. K., & Labhasetwar, V. D. (2005). Nanosystems in Drug Targeting: Opportunities and Challenges. *Current Nanoscience,* 1, 47–64.

Vetter, R. J., William, J. C., Hetke, J. F., Nunamaker, E. A., & Kipke, D. R. (2004). Chronic Neural Recording Using Silicon-substrate Microelectrode Arrays Implanted in Cerebral Cortex. *IEEE Transactions on Biomedical Engineering,* 51, 896–904.

Wabuyele, M. B., Culha, M., Griffin, G. D., Viallet, P. M., & Vo-Dinh, T. (2005). Near-field scanning Optical Microscopy for Bioanalysis at Nanometer Resolution. *Methods in Molecular Biology,* 300, 437–452.

Webster, T. J., Waid, M. C., McKenzie, J. L., Price, R. L., & Ejiofor, J. U. (2004). Nano-biotechnology: Carbon Nanofibres As Improved Neural And Orthopaedic Implants. *Nanotechnology,* 15, 48–54.

Weinberg, R. A. (1996). How Cancer Arises. *Scientific American*, 275, 62–70.
Wickline, S. A., & Lanza, G. M. (2003). Nanotechnology for Molecular Imaging And Targeted Therapy. *Circulation*, 107, 1092–1095.
Williams, D. (2005). The Risks of Nanotechnology. *Medical Device Technology*, 16, 9–10.
Xi, J., Schmidt, J. J., & Montemagno, C. D. (2005). Self-assembled Microdevices Driven by Muscle. *Nature Materials*, 4, 180–184.
Yamaguchi, Y., & Igarashi, R. (2006). Nanotechnology for Therapy of Type 2 Diabetes. *Nippon Rinsho*, 64, 295–300.
Yamato, M., & Okano, T. (2006). Nanotechnology-based Regenerative Medicine—cell Sheet Engineering Utilizing Temperature-responsive Culture Dishes. *Nippon Rinsho*, 64, 335–341.
Yih, T. C., & Al-Fandi, M. (2006). Engineered Nanoparticles As Precise Drug Delivery Systems. *Journal of Cellular Biochemistry*, 97, 1184–1190.
Yoshikawa, T., Tsutsumi, Y., & Nakagawa, S. (2006). Development of Nanomedicine Using Intracellular DDS. *Nippon Rinsho*, 64, 247–252.
Yuan, G. (2004). Natural and Modified Nanomaterials As Sorbents Of Environmental Contaminants. *Journal of Environmental Science and Health, Part A: Toxic/Hazardous Substances and Environmental Engineering*, 39, 2661–2670.
Zhang, G. D., Xia, X. X., Li, X. C., & Hu, F. (2005). Strategies for Controlling Harmful Vehicle Emissions. *Huaqiao Daxue Xuebao Ziran Kexue Ban*, 28, 46–49.
Zhang, Y., Lim, C. T., Ramakrishna, S., & Huang, Z. M. (2005). Recent Development of Polymer Nanofibers for Biomedical and Biotechnological Applications. *Journal of Materials Science: Materials in Medicine*, 16, 933–946.

Glossary

Atomic Force Microscope: An instrument able to image surfaces to molecular accuracy by mechanically probing their surface contours. A kind of proximal probe.

Quantum Dots: Tiny crystals that glow when stimulated by ultraviolet (UV) light; an in vivo application of nanotechnology.

Nanate: A nanomaterial whose dimensions are measured in nanometers (in units of 10^{-9} m).

Nanite: Also known as nanobot, it is a nanotechnological robot nanomachine which is a mechanical or electromechanical device whose dimensions are measured in nanometers (in units of 10^{-9} m).

Nanotube: A one dimensional fullerene with a cylindrical shape.

Scanning Probe Methods: A method for imaging nanoscale features of surfaces by scanning a sensor (probe) over a surface. Near-field effects such as tunneling, van der Waals forces, local fields and more are serially detected at localized points on the surface and used to create an SPM image.

Public Policy and (Bio)Nanotechnology

Nanobiotechnology and Ethics: Converging Civil Society Discourses

Alexandra Plows[1] and Michael Reinsborough[2]

Abstract Nanobiotechnology as a "converged" technological platform (CT = Converging Technologies) is discussed in relation to discourse within civil society. The conflicts and ethical debates surrounding nanobiotechnology can be intuited from these larger discursive frames of reference. Complimenting Glimell and Fogelberg's (2003) research documenting an emergent epistemic culture amongst scientists researching and working on nanotechnologies, and more recent research on the multiple meanings of nanotechnology in the political economy (Wullweber, 2007), this paper traces an emergent ethnography of engaged actors within civil society as they develop discursive and mobilization repertoires. Whilst on occasion ambivalent about the combination of specific promises and risks in relation to nanobiotechnology, in general a broad critique of the politics of technology is emerging as a counter epistemology or "Master Frame" (Snow & Benford, 1992) amongst certain predisposed UK civil society groups. Converging Technologies provide the issue around which this broad critique is solidifying. Thus whilst many of the specific risks raised by nanobiotechnology (and other CT) are definitively new, many of the potential risks and grievances, have been raised before in relation to other issues of scientific and environmental controversy, often by the same actor groups. Thus convergence is a useful metaphor for appreciating that broader frame of reference from within which the emerging conflicts and ethical debates about nanobiotechnology are being situated.

If you go ten, fifteen years in the future, you're not going to be able to distinguish between what's nano technology, what's bio technology, what's information technology or what's genetic engineering. They're all going to be the same kind of technologies ... just employed in different ways and different places. ("Mike", technology watchdog campaigner, in interview January 2004)

Keywords Nanotechnology, biotechnology, converging technologies, civil society, risk discourse, social movements

---

[1]CESAGen, Cardiff University, plowsa@cardiff.ac.uk
[2]Queens University, Belfast, m.reinsborough@qub.ac.uk

Introduction

Nanobiotechnology is the convergence of existing and new biotechnology with the ability to manipulate matter at or near the molecular level.[1] This ability to manipulate matter on a scale of 100 nanometers (nm) or less is what constitutes the nanotechnology revolution occurring today, the potentially vast economic and social implications of which are yet to be fully understood (Royal Society, 2004). The most immediate way to understand the implications of nanobiotechnology for ethics is to consider the real life concerns of communities that are mobilizing within civil society. The conflicts and ethical debates surrounding nanotechnology will, almost by definition, emerge on the fault lines between different civil society actors, researchers and financial interests associated with nanobiotechnology, as well as (potentially) government regulators. These fault lines are all reflected within the concerns (as expressed discursively) of the communities mobilizing. This chapter will explore converging discourses regarding converging technologies.

Converging Technologies (CT) are already a familiar theme in the next generation of biotechnology, nanotechnology, pharmacogenomics and proteomics research and development. Nanobiotechnology[2] means that previously separate disciplines (IT, physics, chemistry, and biology) are merging and converging to create new applications and even new life forms through converged technological platforms. Schummer (2004), and Glimell and Fogelberg (2003, p. 43), note the predominance of interdisciplinarity as a core theme of nano-discourse. This technological and domain convergence is now so familiar a concept as to be given acronyms like GRAIN (Genetics, Robotics, Artificial Intelligence and Nanotech) and NBIC (Nano, Bio, Info, Cogno).[3] Convergence and its implications are becoming standard regulatory and policy concepts and strategies, especially within the EU[4] (Nordmann, 2004). For example:

> The rapidly mounting level of interdisciplinary activity in nanostructuring is truly exciting. The intersections between the various disciplines are where much of the novel activity resides, and this activity is growing in importance. (WTEC Panel, 1998, p. 5)

Current social science in this emergent arena has mostly focused on analysing core discourses of those involved in producing the science, such as Glimell and Fogelberg's (2003) report on scientific framings of nanotech, which seeks to define an "epistemic

---

[1] This definition is consistent with more general definitions of nanotechnologies (Royal Society, 2004, p. 5).

[2] Nanobiotechnology is used in this chapter as the prime example of CT; however it should be noted that the actors identified in case study work tend to talk about CT and nanotechnology more broadly, rather than specifically focussing on nanobiotechnology.

[3] "We talk about BANG and that's Bits Atoms Neurons Genes...." ("Mike", technology watchdog campaigner in interview January 2004).

[4] See also "Converging Technologies for a Diverse Europe" report of the "Foresighting the new technology wave" expert group (2004); dissemination conference programme available at http://216.239.59.104/search?q=cache:THyA8mWIwAQJ:europa.eu.int/comm/research/conferences/2004/ntw/pdf/programme_en.pdf++DG+EU+converging+technologies+ of+the+knowledge+society&hl=en

culture of technology" (Glimell & Fogelberg, 2003, p. 82) amongst scientific producers. More recently, a discourse analysis of nanotechnology within market and governance settings notes the broad meaning of the word nanotechnology makes it a useful label to particular "interests and strategies aiming at the reconstruction of industrialised states" (Wullweber, 2007). Nanotechnology is not a particular technology, but rather, it is an industrial strategy.[5] Nanotechnology is associated with the consolidation of the competition state within the knowledge economy (Wullweber, 2007). We can generalize from these studies to the more particular case of nanobiotechnology. However, to appreciate the conflicts and ethical debates emerging around nanobiotechnology, the types of conversations happening within civil society also need to be examined.

This chapter seeks to compliment previous research by identifying emergent patterns of engagement with CT by specific network clusters of (UK) civil society actors. Based on case study work, this chapter focuses primarily on networks comprised of genetic technology watchdogs, environmental and anti-globalization activists (including, as discussed below, New Luddites and New Chartists), feminists, and Disability Rights campaigners. Although the networks considered here are predominantly UK-based, there are important globalized trends; activists are networking across Europe and beyond. To use Glimell and Fogelberg's term, these actors are also developing an epistemic culture, in that they tend to be *predominantly* critical, and raise a variety of concerns about the implications of nanobiotechnology and other converged technologies. It might be possible to identify an emergent master frame (Snow & Benford, 1992) of the politics of technology. This master frame incorporates a broad range of issues-other frames- such as commodification, control and identity. These frames have as yet only been broadly traced through the more visible engagement of NGOs such as ETC group and GeneWatch with the policy process (Wood et al., 2004; Grove-White et al., 2000). Presumably these discourses are still in a relative state of latency and emergence (Melucci, 1996) amongst these predisposed actors (Evans et al., 2007; Welsh et al., 2007).

Whilst some potential risks relating to nanobiotechnology are of course completely new, such as a specific type of health or environmental risk, many broad themes are familiar territory. Tellingly, in relation to CT/nanobiotechnology, many of these same critical or oppositional frames are being raised by the same networks of actor groups, and social movements who previously mobilized over precursive single issues such as nuclear power and agricultural GM (Welsh, 2000; Nelkin, 1995; Purdue, 2000; Plows, 2004b; Welsh et al., 2007). Thus convergence is not only an important techno-scientific reality; it is also a useful metaphor for appreciating the ways in which core critical frames articulated by civil society have been raised before in relation to other issues of scientific controversy, and often by the same actor networks – this is what Nelkin (1995) terms discursive linkage (see also Bauer, 1995). Nanobiotechnology, as a key example of converging technology

---

[5] This point was originally made to us in a conversation with Jim Thomas, ETC campaigner.

(CT), is thus a site in which, for example, established (academic, activist) discourses on public engagement with science (Irwin & Michael, 2003; Wynne, 1996; CorporateWatch, 2005a; Fischer, 2000) and likewise established discourses on the impact of market forces upon science (Glasner & Rothman, 2004; CorporateWatch, 2005b, 2007) are once again becoming linked.

Thus, triggered by new technological developments, converging civil society discourses are producing an emergent[6] master frame (Snow & Benford, 1992) or epistemic culture (Glimell & Fogelberg, 2003). A critical perspective on the politics of technology is developing which incorporates questions about the relationship between science and society, i.e. in what type of society do we want to live, and by what norms and values should it be driven? This broader framing of specific single issues (for example, the development and potential use of nanobiotechnology) was a core feature of much qualitative data collated in the UK between 2003 and 2006.

> *Bigger questions around technology [are] playing on the sidelines, as they have in the nuclear energy debate, in the toxic chemicals debate...whereas talking about converging technologies... if you can find a way of opening that up then you are talking about the politics of technology. ...* ("Mike", technology watchdog campaigner in interview January 2004)

In 2003, an international event of prime movers expressing a range of concerns over medical bioscience asked the question:

> *How much do we focus on the technologies themselves, and how much on the social justice and global equity values that motivate our concerns?*[7]

A very significant issue thus raised by this emergent master frame, is whether it is socio-political relations more generally, or the specific (environmental, health, social, ethical) risks/implications of these new technologies, which should be the core grievance frame, and whether it is possible, or indeed even useful, to separate out the two, given that such multiple framings arising from a single issue are the norm in these milieu (see Table 2). This issue is also a case of chicken and egg, as campaigners argue that the technology both (re)produces, and is itself a product of, existing social inequalities and power structures. This conceptual problem has policy implications in terms of the ways public engagement around issues of medical bioscience (generally, as well as nanobiotech specifically) is sought and framed (Welsh et al., 2007).

This chapter will: (a) provide a scientific overview of nanobiotechnology, placing it within the context of Converging Technologies (CT); (b) provide a short ethnographic overview of emergent oppositional and critical UK civil society networks and groups; and (c) demonstrate core converging discourses amongst these

---

[6] Emergent in the context of engagement with nanobio. To reiterate, a core point in this paper is that such "bigger picture" debates have been expressed in many other settings over decades of environmental and related social movement activity (Plows, 2003, 2004b; Nelkin, 1995; Welsh, 2000).

[7] http://www.genetics-and-society.org/analysis/opposing/2003_berlin_report.html

networks. The conflicts and ethical debates surrounding nanobiotechnology can be seen emerging from these discourses.

Section 1: Scientific Overview: Nanobiotechnology as a Converging Technology

The convergence of different disciplines through nanobiotechnology opens up an uncharted territory of actual, unrealised, and in some cases possibly unrealisable applications and inventions. Utopian and dystopian futures are projected by various actors (Nordmann, 2004; Glimell & Fogelberg, 2003; Wood et al., 2004).[8] Nanotechnology itself involves changing the nature of elements by breaking them down to the atomic level, where the atomic particles of, for example, gold or steel, possess different qualities and attributes (ETC, 2003; Royal Society, 2004; Corporate Watch, 2005a, b, 2007). This intrinsic change at the atomic level is what invokes both benefit frames in terms of potential applications, and risk frames relating to health and environmental impacts, such as potential toxicity, and the unquantifiable impacts of such potential applications in the "lab without walls" (Szerszynski, 2005).[9] Research and Development has focused on how to grow these new nano-materials, through the converged technological platforms of IT, nanotech and molecular biology, where the replicatory function of DNA is being harnessed; this is one important aspect of nanobiotechnology.

> *Nanobiotechnology involves the integration of biological materials with synthetic materials to build new molecular structures or products....* (ETC, 2004)

The use of different disciplines together to create new functions, new elements, new life forms, and the sense in which atomic level manipulation of biological elements can be interpreted from different disciplinary perspectives, is triggering the interdisciplinarity discourse referred to by Glimmel and Fogelberg (2003) and by Wood et al. (2004).[10]

> *...Biology is the nanotechnology that works.* (Tom Knight, Senior research scientist, MIT's Computer Science and Artificial Intelligence Laboratory[11])

"Cut and paste" somatic gene therapy has been referred to as first generation nanotechnology; with claims that faulty genes could now be made to work by altering

---

[8] The potential futures for these rapidly emerging technologies are difficult enough for those with science and technology backgrounds to understand and predict, and it is certainly beyond the author to do more than identify core principles.

[9] And the "social risks" posed by the development of specific applications.

[10] *Cells themselves are very complex and efficient nano-machines, and chemists and biochemists have been working at the nanoscale for some time without using the nano label....* (Wood et al., 2004, p. 21).

[11] Quoted in ETC (2004) "Green Goo and Red Herrings".

the molecules which make up the specific proteins, or using computer modelling to design new protein molecules which are then introduced into the patient's body. Efforts to create such self-replicating building blocks out of individual nano atoms are now focusing on harnessing the reproducing mechanisms in DNA:

A nanotechnological dream machine is one that can replicate. (Nadrian C. Seeman[12])

These have a myriad of potential applications, such as the creation of nano tools and organisms – nanobots – or of new nano particles, for example for drug delivery (Wood et al., 2004; Glimell & Fogelberg, 2003; Roco & Bainbridge, 2002; ETC, 2003). The potential to create nano-organisms specifically designed to clean up environmental problems, such as oil spills, is a frequently cited example. Nano particles are also being used to split open DNA, acting as markers for DNA sequences, further accelerating the rapid advance of mapping genes and proteins and the genomes of entire species. Converging Technology (CT) nano applications are thus either *aiding* biotechnology applications through better genomic/proteomic mapping techniques, or *becoming* biotechnology applications themselves.[13] The new field of synthetic biology is working to artificially produce organisms from mappings of their genetic code.[14]

Section 2: Ethnographic Overview[15]

The following is a short overview of "prime movers" (McAdam, 1986) and "early risers" (Tarrow, 1998) amongst a specific civil society network cluster, who are expressing the convergent oppositional frames outlined in the introduction and detailed in the final section of this chapter. These actor networks, whilst resisting simplistic definitions, could be generally characterised as environmental, social justice, feminist and anti globalization movement actors. Disability Rights actors have also been traced and project work has sought out points of crossover amongst these different networks. Actors comprise both NGOs such as ETC[16] who tend to

---

[12] Cited in http://www.etcgroup.org/documents/comBANG2003.pdf

[13] *To take the step from "fabrication" to "manufacturing" is by no means something that will come easily. This is one important point at which "bio-nano" is expected to make a difference. Nanobiology holds, on the one hand the promise of providing new and more refined tools for the advancement of biotechnology, whereas on the other hand the "biogenic" strategies may be an effective way to produce nanostructures en mass....* (Glimell & Fogelberg, 2003, p. 89)

[14] The primary research in this new field comes from the J. Craig Venter Institute (http://www.jcvi.org/research/) Some of the response from Civil Society organisations is presented at http://wolbring.wordpress.com/category/synthetic-biology/ See also http://www.etcgroup.org/en/materials/publications.html?pub_id=631

[15] This chapter has been written from findings of ESRC funded research; Cesagen flagship project "The Emerging Politics of Human Genetic Technologies". http://www.cesagen.lancs.ac.uk/research/projects/newgentechs.htm

[16] http://www.etcgroup.org/

primarily engage with the policy and regulatory sphere,[17] and more radical, direct-action focused groups who generally do not. Critical frames in relation to nano CT are being identified by the UK radical environmental campaign and action networks who previously mobilized over agricultural GM; emerging from loose, biodegradable (Wall, 1999) activist networks like Earth First! and the Genetic Engineering Network.[18] These networks have links to established NGOs such as GeneWatch, but are generally autonomous, decentralised dis-organisations who tend to deliberately stay outside of the policy processes emphasizing direct action (protest, civil disobedience) as their primary, or often only, tactical repertoire.

Whilst labels tend often to obscure, rather than clearly define, the discourses and identity of these fluid, shifting, radical networks (Seel & Plows, 2000; Plows, 2003), amongst them it is possible to identify New Luddites, including anarcho-primitivists who are anti-technology almost *per se*, and New Chartists who are primarily concerned with enclosures and ownership issues (patenting) in relation to CT. All groups/networks have strong ties[19] links to environmental and anti-globalisation networks, and identify multiple forms of risk accruing to CT, whilst often specialising in particular frames/issues. Further, important ideological differences between NGO and more activist groups definitely exist, though it is vital to stress that all groups engaging with CT are explicitly articulating bigger picture frames incorporated by the politics of technology, as well as specific risk and grievance issues.

As the social movement literature would predict (Diani, 1992, 1995; Doherty et al., 2003), activists with pre-existing social bonds developed through previous cycles of contention, and new actors brought into these networks, are developing strategies to oppose the development and commercialisation of nanotechnology and CT. Whilst the NGOs tend to have a more formal and static identity, the more radical networks tend to develop multiple, micro identities and campaign brand names. These reflect the more informal, often biodegradable processes of social network interaction (Diani, 1995) which spawn specific direct action events. For example, on December 9 2004, THRONG (The Heavenly Righteous Opposed to Nanotech

---

[17] For example, ETC was invited to take part in public engagement which led in part to the findings of the Royal Society (2004) publication; ESSF has sought, since its inception, to find allies within the EU to discuss core issues of political economy and "market- led science", for example at the EU "Science and Society" conference in Berlin in March 2005, where the ESSF petition was launched.

[18] See Plows (2003, 2004b) for an overview of the eco movement in the UK. Current and previous work by the project team (Welsh, 2000; Welsh & Chesters, 2004, 2005; Doherty et al., 2003) has emphasised how in the UK "radical environmental movement", different currents of seemingly disparate "single issue" movements in fact emerge from a broad–based, occasionally latent, movement of "biodegradable networks" which have developed capacity through diffusing discursive and mobilization repertoires across "weak and strong ties" networks (Granovetter, 1973; Diani, 1992, 1995), over several decades of activist generations and cycles of contention (Tarrow, 1998; Doherty et al., 2003; Whittier, 1995).[18]

[19] The term "strong ties" means direct connections. Compare Granovetter, "The Strength of Weak Ties" (1973).

Greed) disrupted the "Nanotechnology-delivering business advantage" conference in Buckinghamshire, presenting a can of worms award to one of the conference participants, Harry Swan, formerly of Monsanto. The symbolic (Melucci, 1996) and playful nature of much UK direct action (Wall, 1999; Szerszynski, 2002) is evident; the THRONG activists dressed as angels to disrupt the conference, stating:

Where these nano fools rush in we angels fear to tread[20]

The identification of specific corporations' operations as a clear target for criticism and action, and the use of such action as a means of framing broader concerns about risk and uncertainty, echoes previous campaigns against GM crops which were focused on the operations of specific companies such as Bayer and Monsanto, with clear critical reference to globalization processes.[21] Such actions are likely to seed further mobilizations within the UK. Similar actions are taking place in other countries. A less symbolic direct action took place in Grenoble, France, where the site of the proposed Minatech nanotechnology research centre was occupied by activists on December 12, 2004.[22] Again this action is the visible peak of a more latent/nascent activist opposition, and there are network links between French and UK activists.

Risks and Benefits: Complexity and Ambivalence

A key finding of the "Emerging Politics of Human Genetic Technologies" project (from which this data is drawn) is that the engagement of different publics cannot be polarised into broad pro- and anti-positions on technologies, such as biotechnology, human genetics (Plows, 2004a; Evans et al., 2007). Having said that, in some contexts, clear battle lines are emerging. The social and political contexts in which the new technologies are being developed and introduced have catalysed many critical actors into taking an oppositional position. These critical actors are more likely to articulate, and mobilize over, risk frames, thus opposing nanobiotechnology and other CT (Table 1). At the same time possible benefits of the new technologies are also being highlighted by some of these critical civil society actors, such as better drug delivery or better solar power technology. For some actors (New Luddites), the threats of perceived risks mean that the technology is taboo per se; many actors also (re)emphasise an established critique of technological fixes for what are perceived as being socially created problems (such as oil spills). Other actors are much more ambivalent, acknowledging possible future promise medical

---

[20] From THRONG press release – quote by THRONG "angel" Sarah Phimms.

[21] Links, both network and discursive, between "single issue" campaigns and the broader "anti globalisation movement" are thus in evidence. For a full discussion on the network links between activist groups, see Plows (2004b).

[22] http://www.indymedia.org.uk/en/2004/12/302727.html

Table 1 Examples of UK critical actors predisposed to mobilise, or mobilising, over CT[23]

Type of actor network/group	NGOs (policy, public engagement)	Social movements and SMOs (informal networks using radical forms of engagement e.g. protest)
Environmental actors	Greenpeace Friends of the Earth[24]	UK Earth First! Genetic Engineering Network
Disability rights campaigners	Disability Awareness in Action	Direct Action Network
Technology and bioscience watchdogs	Genewatch ETC group EcoNexus Institute of Science In Society	e-lists such as Techne
Social justice, feminist, human rights, anti-globalisation, anti-corporate actors	The Corner House Liberty Corporate Watch	Loose, informal, fluid and biodegradable networks, campaigns and protest identities emergent from broader activist networks such as Dissent (see also environmental actors above) – including New Luddites and New Chartists, THRONG, FemACT

applications such as better treatments for disease. Many activists, tellingly, admit to being confused about nanotechnology[25]. The extent to which groups or individuals are completely opposed to CT per se, and the extent to which they are merely opposed to the technologies as they are likely to manifest under the current socio-economic paradigm (which primarily produces, they argue, the prioritisation of cures for the Western rich, and the medicalisation of syndromes for the worried well), is a complex point. This type of complexity and ambivalence is not easily summarised; risks and benefits are seen as tied together by many actor groups (Royal Society, 2004; Wood et al., 2004; Plows, 2004a; Evans et al., 2007). Thus many critical actors will cautiously also frame hopes and promise claims, and often refute the charge of being anti-science, as the quotation below highlights:

---

[23] It must be stressed that this is not a definitive table but rather presents some key examples. Further, these are by far from being the only critical groups and networks mobilising in this arena. The groups/networks in this table are representative of a particular "cluster" of actors.

[24] Only very recently have traditional large scale environmental organisations become concerned about CT. This new activity is evidence of emergent movement mobilization. See for example press release "Broad International Coalition Issues Urgent Call For Strong Oversight of Nanotechnology" July 31, 2007 (http://www.icta.org/press/release.cfm?news_id=26).

[25] Opinions expressed during activist workshop on nano and CT, Oxford, 2005. In part, this confusion is evidence of the mismatch between careful definitions of a specific technology (which would have been expected by communities previously engaged in GM crop politics) and the broad meaning of the word "nanotechnology" within market and governance settings as a label for a particular industrial strategy associated at the corporate level with acquiring competitive advantage and at the governance level with refashioning the "competition state". As such the confusion is evidence for Wullweber's thesis that "nanotechnology" is an "empty signifier" that serves "interests and strategies aimed at the reconstruction of industrialized states" (2007).

To cut short on all sorts of silly accusations that are frequently made to criticisms of technological development...let us state clearly that we are neither obscurantists nor «against science».

(From French anti nanotech action December 04 press release; English version, emphasis in original)[26]

Section 3: Convergent Discourses on Converging Technologies – The Politics of Technology

As well as emphasising completely new (health, environmental) risks or (much less often) promises associated with CT, the actors in this case study frame their concerns within a broader challenge to the epistemological and value claims of more traditional technology narratives, for example the preconception that technology equals progress and that there is only one correct direction of technical development. The critique also draws attention to the implications of an economic mandate driving the pace and nature of development (Mayer, 2002; Wood et al., 2004; Birch, 2006), and the perceived lack of citizen input into this process. Thus opposition to nanobiotechnology is currently being framed by prime movers in UK activist networks within the broader framework of the politics of technology. As discussed at the start of the paper, this developing epistemic culture (Glimell & Fogelberg, 2003) with its explicit master frame (Snow & Benford, 1992) is a significant theme in civil society engagement with CT, and demonstrates discursive convergence. Nanobiotechnology and other CT become spaces in which pre-existing discourses and critiques, across a broad range of issue frames, are further explicated and developed. As a key interviewee noted,

> *what we need to get to is ... the politics of new technologies ... that people can engage with, so that we're not dealing with nuclear power and then dealing with genetic engineering and then dealing with...those mutually just separate local little areas... [but] to have a real live politics of how new technologies impact on society, how society has some control over that.*
>
> ("Mike", technology watchdog campaigner in interview January 2004)

Many other actors, explicitly rejecting claims of a value- neutral, objective science, also identify emergent technologies as political, noting that

> *technology is political. Social forces shape technology...technology... becomes a social force that in return shapes society...The desires and interests (or ignorances) of those who control the design process are what shape technology.*[27]

The above quotes are examples of critiques which have been developed over time, by many different social movements and social networks, in multiple arenas – converging in the new arena of nano-related CT. This master frame or epistemic

---

[26] http://www.indymedia.org.uk/en/2004/12/302727.html

[27] From "Technology, Politics and Democracy": Green Action briefing for the European Peoples Global Action 2004 conference, held in Slovenia.

culture is, in the context of nanobiotechnology, concerned with broad themes about the relationship between science and society, political economy, and cultural value systems. Power, social control,[28] ownership, commodification, the impact of market mechanisms on the science and society relationship, the lack of public engagement in the agenda-setting process are all familiar themes in many previous scientific controversies (Wynne, 1996, 2006). Under these conditions, CT as a fait accompli is a core grievance (see for example Mayer, 2002; Corporate Watch, 2005a, 2007).[29] This final section uses preliminary analysis of ethnographic data to provide some examples of these core converged frames articulated by the critical and oppositional actors. While the examples are not definitive, they nevertheless serve to highlight the depth and breadth (Welsh et al., 2007) of convergent discourses of opposition, some of which are summarised in Table 2 below.

Such frames are highly complex, interlinked in many ways, and require more elaboration than this chapter, which aims simply to provide an overview of an emergent ethnographic field, can provide. Some of these key converged discourse frames are discussed in more detail in the following sub-sections.

Table 2 Converging discourses: emerging epistemic culture/master frame[30]

The politics of technology is a broad discourse, comprising multiple, cross-cutting frames and issues, some of which are included in the following list:

Critical perspectives on Market-led science, economic globalisation pushed forward by a Neoliberal agenda

Discourses on ownership, control, surveillance, enclosure (patents, biocolonialism, intellectual property regimes), commodification of knowledge

Concerns over environmental and health risks which are specifically identified; also indefinite risks and the threat of uncertainty more generally

Risk and concern discourses in relation to human nature, culture, identity, and values. Specific issues cited include therapy/enhancement, cures, eugenics, bio subjectivities, Disability Rights, feminism, and threats to indigenous peoples and their cultures

Critical perspectives on public engagement practises, power dynamics, the defining and privileging of specific forms of knowledge and expertise; as indicated above over the contested meanings of terms (enhancement) and their implicit values

Calls for sustainable development and social justice approaches to health and equity; critiques of medicalisation and biological reductionism

---

[28] Surveillance is another core risk/grievance frame in terms of opposition to nanotech applications (in particular military surveillance inventions such as nano "smart dust"– linking again to eco/health risks) – surveillance/privacy issues have also been a central grievance frame for opposition to DNA biobanks (see for example http://www.genewatch.org/HumanGen/GeneticResearch.htm)

[29] These are of course also familiar themes in STS and PUS literature – Bauer (1995), Fischer (2000), Wynne (1995, 1996).

[30] These frames continuously surface in a variety of ethnographic contexts (interviews, weblogs, workshops, press releases and so forth), are explicitly linked together by actors, and tend to consistently surface in relation to multiple "single issues" such as the development of specific technological applications or technological domains; here, the context is CT.

Market-Led Science

> *A lot of the body of the risk associated with genetic technology actually comes from the capitalist structure underlying the usage of that technology....*
> (Interview with "Alice", genetic sequencer, November 2003)

The impact of market mechanisms, a sustained critique of competitiveness and Neoliberalism forms a substantive part of the converged master frame or epistemic culture of activists challenging the politics of technology and articulating opposition to technologies of control. Activists (with strong ties links to anti–globalisation networks) mobilising over nanobiotechnology emphasise the unequal power relations which ensure that market mechanisms drive the pace, nature and applications of change in society. Highly critical of the impacts of such globalized processes, they also highlight the production (or reinforcement) of potentially harmful social and cultural norms that are intrinsic to the demand for specific products and services.

> *It's to do with the bigger picture which is the framework within which science is working... if the framework which science is working in [is] a neo liberal version of a market economy which is the rich get richer and the poor get poorer, you are going to get the one version of human genetics.*
> ("James" GM activist and journalist in interview January 2004)

Nanobiotechnology markets are developing rapidly already having first world applications particularly in cosmetics – face and suntan creams (Wood et al., 2004; Corporate Watch, 2007).[31] Emphasising the inter-connectedness of specific eco/health risks and the bigger picture, many actors feel that market mechanisms are more likely to produce medical risk scenarios than publicly funded science:

> *The rush to market means that we rush it before we really have a full understanding....*
> ("Mike", technology watchdog campaigner in interview January 2004)

Activists who frame these types of critiques have developed the capacity to do so though years of engagement with environmental and social justice issues. It is these types of networks which formed the backbone to the anti-globalization movement. It is significant that opposition to GM crops was explicitly framed in conjunction to critiques of emergent WTO policy and patenting strategies via the TRIP (Trade Related Intellectual Property) agreements (Plows, 2004b; Purdue, 2000). Thus, broader critiques of nanobiotechnology are set in the context of global opposition to the "economic competitiveness" and "biotech competitiveness" (COM, 2001; Salter & Jones, 2002) driving the Research and Development agenda (Birch, 2006) presenting

---

[31] *Cosmetics and* personal *products companies have been extremely active in using nanotechnology to improve their existing products and to develop new ones. The company L'Oreal famously holds more nanotechnology patents than many companies in high technology sectors (though again this is in part a matter of labeling). Cosmetics companies were among the first to get products that were labeled as being nano-enhanced to market. Shampoos and skin creams, containing nanoparticles with the ability to deliver the desired ingredient to where it is needed, for example deeper into the epidermis, are already on the market.* (Wood et al., 2004, p. 21)

society with a fait accompli (Mayer, 2002), i.e. the "juggernaut of modernity" (Giddens, 1990; Habermas, 2003; Bauman, 1994). While modernity is not reducible to the associated economic motives, certainly the links between modern technologies and the economic motives that drive development can be drawn. From many activists' perspectives, the economic imperatives driving the speed and directions of research mean that any resulting regulation or mitigation (governance) is simply "shutting the barn door" after the horses have already bolted.[32] Hence the market-led science frames are intrinsically connected to the public engagement process and calls from activists for a broader debate on the politics of technology. In this context, anti-nanotech activists were disappointed by the Royal Society's (2004) rejection of a moratorium on nanoproducts, and it would appear that the "realpolitik" of economic competitiveness as the primary driver in EU policy is not likely to be seriously challenged by policy makers, as the Royal Society report makes clear:

> *Nanoscience and nanotechnologies are evolving rapidly and the pressures of international competition will ensure that this will continue.*
> (Royal Society (July 2004) "Nanoscience and nanotechnologies: opportunities and uncertainties"; Summary and Recommendations, p. 7)

In direct response to this perceived lack of appropriate governance in the face of identified risks, the NGO Corporate Watch in its 2007 publication on nanotechnology calls for a moratorium on industrial nanotechnology, together with several other key UK NGOs.

Intellectual Property

Linked closely in with the above anti-globalization discourse, is opposition to patenting – key words here are the enclosure, ownership, commodification and privatisation of nature, life and the human. These are core frames raised in relation to patents on genes and plant, human or animal DNA sequences. These previous stances inform the immediate opposition to nanobiotechnology and the patenting of nano particles/organisms. Privatising nature, the commodification and ownership of life, and the risks associated with the uncertainty of molecular manipulation were the core grievance frames articulated by the THRONG during the angels against nanotech action. The enclosures metaphor used for over a decade by UK GM activists in the context of patenting is reflexively converging centuries-old grievances about elites owning the commons, linking past and future together. The enclosures frame grounds the nanobiotechnology debates about the patenting of genomic, proteomic and atomic information in the history of the Diggers and Chartists' fights for

---

[32] *Use of nanotechnology has already begun ... by the time we can easily notice its presence (including its environmental and civil society effects) it will be firmly established in the economy, difficult to uproot. By the time we can see it coming it will already be here.* (Green Action briefing, 2004)

common ownership of the land, of shared resources.[33] The abstract information codes (which New Chartists define as a shared resource rather than claimable territory or property) are made more real through this association with land. New Chartists is one way of understanding the perspectives of civil society actors who are framing critiques in relation to corporate enclosures and corporate power.

Risk and Uncertainty – Health and Environment

Civil society articulations of risk and unquantifiability have been core themes in many previous scientific controversies; hence the actor groups discussed throughout are predisposed to identify similar risk potential in relation to nanobiotechnology, most notably through their oppositional engagement with agricultural GM. Whilst the concept of risk is thus a converged frame, the latest nanobiotechnology developments would mean that these are often specifically new risks. Critical and oppositional actors identify that nanobiotechnology poses qualitatively new health and environmental risks of types not seen before. The potential toxicity of nanoparticles raises medical and environmental risks. Significantly, these are not only addressed by oppositional groups but are also raised within the regulatory and scientific sphere. Two examples are the (2004) Royal Society report[34] and a recent ESRC report (Wood et al., 2004). Despite their acknowledgement of uncertainty in the domain of risk and the emphasis on the need to uphold the precautionary principle, neither of these reports supported civil society calls for a moratorium on development or for a recall of nanoproducts already on the market[35]; despite the fact that only a handful of toxicology studies have been funded (ETC, 2003; Corporate Watch, 2007). This lends further weight to Mayers' (2002) warnings that nanotechnology is likely to trigger a public backlash which mirrors that of GM crops. The ecological and health risks accruing to nanobiotechnology specifically are explicitly addressed by what might be the leading NGO in the field, ETC group, who also identify specific risks of what they term "green goo":

> *Nanobiotech raises many potential concerns: will new life forms, especially those that are designed to function autonomously in the environment, open a Pandora's box of unforeseen and uncontrollable consequences? That's the specter of green goo.* (ETC, 2004)[36]

---

[33] Some potentially important "strange bedfellows" in the form of assemblages (convergences) of opposition to patenting are emerging (Plows, 2004a), the most interesting of which in terms of sustained critiques of market processes on science perhaps coming from the advocates of IT and molecular biology "open source" within the science community. See for example http://science.slashdot.org/science/05/01/18/2120255.shtml?tid=191&tid=155

[34] Royal Society report (2004).

[35] Anecdotally, campaigners have reported that over 300 products using nanotech are already on the market.

[36] ETC Group (2004).

As with crop GM, it is the uncertainty of as yet unimaginable impacts and interactions of nanobiotechnology out in the environment (e.g. clean-up nanobiotechnology organisms) which is the core environmental/health risk frame for actors. Nanobiotechnology also significantly converges ecological risk with medical risk, i.e. a nanobiotechnology organism or supervirus loose in the environment might well have both environmental and human health impacts; and nanobiotechnology also collapses the boundaries between human and non–human, life and non-life, with yet unconsidered massive philosophical implications. Whilst the atomic manipulation of particles and their potential to interact in unquantifiable ways with DNA and cultured GM viruses once loose in the environment *is* new, this is not a completely new risk area, as, for example, genetically engineered superviruses as vectors for "cut and paste" gene therapy have long been identified as potentially encapsulating a mix of eco and medical risk (Mae-Wan Ho, 2003).[37] Likewise military research on deliberately created superviruses for germ warfare (Altmann, 2004) is a previous and continuing concern issue. The risk of nanoparticle release and unquantifiable toxicity as discussed above is identified by the Royal Society (2004) nanotechnology report which again emphasises the need for the precautionary principle. The point that nano production processes are already in motion without standardised, legalised, safeguards is not addressed in the Royal Society report, but it is one of the core concerns of leading critical NGOs ETC group and Corporate Watch. Nor are ETC group and other actors' concerns about "green goo" perceived as legitimate within the few policy documents which currently chart any civil society responses at all (Royal Society, 2004; Wood et al., 2004), this despite uncertainty, risk and the precautionary principle being identified as core policy issues (Wood et al., 2004).

Specific medical risks are also being identified by oppositional actors, such as potential toxicity of nanoparticles in face and sun creams already on the market (Wood et al., 2004; Corporate Watch, 2007). There is some evidence, for example, that nano particles become stored in the liver, and are not flushed out of the body (ETC, 2003). There is a potential for adverse auto-immune responses to nano-scale interventions in the human body, similar to the risks highlighted through the case of Jesse Gelsinger - whose death halted many first generation somatic gene therapy trials when his body had a massive immune rejection response to the GM virus vector transmitting the replacement gene (Bowring, 2003).[38] It is the unquantifiability, the uncertainty, of risk in these circumstances which is triggering much of the criticism amongst oppositional actors (Mayer, 2002; ETC, 2003, 2004; CorporateWatch, 2005b).[39] In the light of the recent FMD (foot and mouth) outbreak in the UK in

---

[37] *SARS Virus Genetically* Engineered? http://www.i-sis.org.uk/SVGE.php

[38] It is worth emphasising that caution and risk, and unquantifiability of outcomes, are also often explicitly framed by scientists carrying out these types of research.

[39] Adverse reactions to the TGN1412 immunity drug by volunteers in the UK in March 2006 is another example of a similar type of risk and uncertainty controversy. See for example "*Volunteers never think it will happen to them*", p. 10, *The Daily Telegraph*, March 16, 2006.

late 2007,[40] one should also consider bio risks in the form of *accidental* release of nanobiotechnology organisms/applications from the lab, as well as their (potential) deliberate release in the future.

Knowledge and Expertise

Contested definitions of knowledge and expertise arose in relation to public engagement with science and technology during previous environmental and other controversies, such as nuclear power, climate change, and GM crops (Bauer, 1995; Wynne, 1996, 2006; Fischer, 2000; Welsh, 2000; Purdue, 2000; Horlick-Jones et al., 2007). These tensions included calls for better public engagement in the context of an epistemological debate on what science is for, the nature of knowledge, and the value systems which inform science. At the core of the debate are disputes about who has discursive legitimacy to discuss, let alone to set the terms of the debate in the first place (Grove-White et al., 2000; Horlick-Jones et al., 2007; Wynne, 2006). While these types of debates are well detailed in Science and Technology Studies (STS) literature, the critiques are heard most stridently from the grassroots groups framing challenges to the specific issue in the first place (Mayer, 2002; Plows, 2003; Plows & Boddington, 2006; Welsh et al., 2007).

Linked closely with the challenging of specific forms of knowledge is the controversy over who counts as experts in the debate (Fischer, 2000; Irwin & Michael, 2003; Wynne, 1996, 2006; Kerr et al., 1998a, 1998b; Collins & Evans, 2002; Evans & Plows, 2005, 2007; Plows & Boddington, 2006). Challenging the discursive legitimacy of certain types of knowledge, and certain types of expertise, has also formed a core part of radical environmental activist opposition to different issues over decades. This is in fact one of the core reasons why radical groups choose not to engage with the policy process, but instead take direct action of various types, stating that the regulatory process is a fait accompli, as Research and Development agendas have already been set by vested interest in the form of powerful lobbyists, and by experts; a process from which they are excluded (Doherty et al., 2003; Plows, 2003; Seel & Plows, 2000). It is now acknowledged that local and lay expertise, embodied expertise, expertise organically acquired through personal involvement, should be acknowledged within the policy making process (Evans & Plows, 2007; Wynne 1995, 1996; Grove-White et al., 2004). What is seen as relevant or useful knowledge, as expertise is also a complex subject.[41] These are well established debates within academic and civil society domains are converging once again via the nano frame.

---

[40] http://www.defra.gov.uk/footandmouth/latest-situation/index.htm

[41] Although some commentators appear to continue to privilege scientific and medical accounts (see in particular Collins & Evans, 2002), where the most relevant forms of knowledge and expertise intrinsic to the issue may well be at the socio-political/ cultural level.

Human "Nature" and Identity

Changing the atomic nature of our own bodies, as nanobiotechnology is potentially capable of doing, also re-invokes, or converges, an established discourse on challenges to human identity and human nature. What might be termed socio-cultural risk frames have been raised in relation to gene therapy and interventions at the genetic level (Habermas, 2003; Fukuyama, 2002; Bowring, 2003). Civil society groups have been framing such opposition at the grassroots for some time:

> *And it worries me particularly that the definition of what is normal is this cultural thing, and... human nature, we don't really know what it is to be human... and yet that's being... decided for us. And the technology to enforce that decision is being put in place to improve profits at the end of the day.*
> ("Mike", technology watchdog campaigner in interview January 2004)

Such clashes over meanings and definitions, provide contextualised examples of debates over knowledge and expertise discussed above, not to mention ethical stakes. Such definitions (enhancement, normal functioning and so on) are quite clearly normative and value-laden, and it is often the case that the civil society prime movers discussed here have difficulty getting this fact across; that there are even knowledge gaps about the very existence of their knowledge claims. Evidence of more discursive clusters or convergence around these identity and human nature concerns are, interestingly, in some cases well framed by some within the scientific community in relation to opposition to reproductive cloning. Identity and human nature risks from nanobiotechnology cures or treatments are also accompanied by another familiar frame – what Habermas (2003) like many civil society actors, identifies as "market-led eugenics". Many activists articulate a concern that the push to maximise profit predictably drives research and development into a socio-cultural context which perpetuates and even exacerbates dominant and problematic ideals of enhancement and normality.[42] The text below is taken from the press release of the 2004 French anti-nanotech protest discussed earlier.

> *In the name of medicine or progress, are being launched research programs, aiming at body manipulations, cerebral control, and human beings standardization, whose practical uses are closer to eugenistic and Orwellian nightmares, and do not deal with any kind of promotion of... social links and diversity.*[43]

The potential for technologies to exacerbate existing socio-cultural divides and inequalities are already familiar debates for disability rights, feminists, social justice and some ecology activists in the context of "red", or medical, genomics. They are also well documented in academic literature (Kerr & Shakespeare, 2002), where the cures for disease frame is heavily problematised. Similarly, feminist and disability rights actors are articulating concerns about transhumanist/enhancement

---

[42] See *"Nano, Bio, Info, Cogno, Synthetic bio, NBICS: Where I post what I find interesting in regards to NBICS and social implications"*. http://wolbring.wordpress.com/

[43] http://indymedia.org.uk/en/2004/12/302727.html

discourses and accompanying possible nanobiotechnology applications. Feminist opposition to sexual stereotyping is likely to be a recurring theme in opposition to developments in cosmetic enhancement. This is yet another example of converging discourses.

The cautious support from some oppositional actors regarding the potential for diseases to be cured[44] is usually accompanied by ambivalence about the use of drugs and interventions, particularly in when and why they might be used. Many actors such as patient groups frame such applications, straightforwardly as an unproblematic potential to find cures or treatments, i.e. avoiding suffering or early death. Familiar social and cultural critiques are also being raised about what it means to be cured, and "where to draw the line" (Kerr et al., 1998b). Disability Rights (DR) perspectives on difference, diversity, the blurred boundaries between abnormal/normal, and therapy and enhancement are core (Wolbring, 2006)[45]; DR campaigners and others indicating that the push for cures could translate as a form of "eugenics by the back door". This debate is well under way in the case of genetic embryonic screening[46] (see also the 2004 Human Genetics Commission consultation on PGD and PND[47]). It is likely to re-emerge again in the context of nanobiotechnology and its many possible screening applications, if the future promise (Brown & Michael, 2003) of the technologies manifests. Calls for sustainable development, social justice approaches to health and equity, critiques of medicalisation and of biological reductionism are all part of the cautious approach that oppositional actors bring to their ambivalent support for some types of medical innovation (Plows & Boddington, 2006).

Conclusion

This chapter has presented an ethnographic narrative of identifiable UK groups and networks which are relatively understudied, showing their converging discursive frames in relation to nanobiotechnology/CT. Preliminary lines of conflict and likely ethical concerns regarding nanobiotechnology can be discerned in these converging discussions, which are now sometimes described as the politics of technology. While some of this discourse is not any more specific to nanobiotechnology than it

---

[44] It should be emphasised that patient and patient advocacy groups such as "Seriously Ill for Medical Research" and Genetic Interest Group are highly pro "cures for disease" and do not problematise this frame in ways that disability rights groups do.

[45] See Aldred et al. (2003) for a highly controversial account of where the acceptable limits of genetic screening should be.

[46] These issues were the focus of an anti-eugenics protest event in September 2004; see http://www.indymedia.org.uk/en/2004/10/298454.html. Whilst it would be useful to give more detail about these and other protests mentioned in this chapter, constraints of space mean that a broad ethnographic overview, rather than a detailed focus, is possible.

[47] http://www.hgc.gov.uk/choosingthefuture/index.htm

is to other CT, nanobiotechnology is likely to be one of the early concerns of Civil Society actors because it can be seen as a direct continuation of recent and continuing conflict around biotechnology. This interlocking set of discursive frameworks, now being described as the politics of technology, is the way that these actors are approaching nanobiotechnology. They indicate the actual ethical conflicts present and latent within the developing field of nanobiotechnology. The stakes are nascent, multiple, not to say messy; it is crucial to trace these emergent, shifting network patterns and key discourses as they shape up within the public sphere, and to be open to the many different directions towards which such discourses point. In addition to the predictable concerns, nanobiotechnology is likely to produce specific conflicts that have not been present in previous new technology debates. Nonetheless, the importance of convergence as a metaphor should be reiterated here. Whilst some of the risk frames highlighted here are definitively new, broader debates on, for example, the impact of globalization and the market on science and society; knowledge, power, control, in relationship to the nature of expertise and the necessity for better public engagement, are not new frames at all, but are increasingly louder, and perhaps better heard, critiques being articulated once again, often by the usual suspects in a new arena (CT). The clarity of the repetition has been partially responsible for the current emphasis on the necessity for "upstream public engagement" (Willis & Wilsdon, 2004) and "horizon scanning"[48] in emerging areas of potential scientific controversy.

This paper has outlined an emergent epistemic culture or master frame amongst certain actors critical of nanobiotechnology/CT, and detailed how these technologies are triggering frame convergences. Familiar discourses which have been expressed in relation to many different issues (see Nelkin's, 1995) concept of "discursive linkage") are now, through the CT nano lens, being explicitly framed as a broader critique of the politics of technology. As the social movement literature would suggest (Melucci, 1996; Tarrow, 1998), the prime movers mobilising over CT articulating predominantly critical and oppositional frames, are generally from predisposed actor groups who were already mobilising over other related issues, such as agricultural GM, feminism, and Disability Rights.

Just as technological convergences are triggering convergences within science and industry (Glimell & Fogelberg, 2003), they are set to trigger more convergences within and between different civil society actor groups (Evans et al., 2007; Welsh et al., 2007). However, it is likely that the more radical oppositional actor groups predominantly focused on in this chapter will tend to mobilize through the strong ties networks established within their own communities, taking confrontational direct action as their primary strategy for achieving change, given that they feel they have relatively little incentive to engage with the policy process as it currently stands. However, other civil society actors, such as NGOs, are engaging with the regulatory sphere, as evidenced in this chapter.

Whilst this chapter focused on data drawn from case study work in oppositional/critical radical activist networks, it should be reiterated that other pre-existing civil

---

[48] http://www.hgc.gov.uk/subgroups/horizon_scanning.htm

society campaign groups such as patient groups, faith groups, animal rights activists and pro lifers also existed as campaign networks before the advent of genomics, pharmacogenomics, and nanobiotechnology, and they have incorporated these new technologies into their existing epistemologies and mobilisation repertoires. Thus "progress, hope and cure" frames, for example, are core components of many civil society groups mobilising broadly in support of genomic technologies (Brown & Michael, 2003; Rose & Novas, 2004; Plows, 2004a; Plows & Boddington, 2006). Similarly the ways scientists, industry and regulators interact over, and set, Research and Development agendas and legislative frameworks and talk up, or talk down, risks/benefits associated with scientific development, are established patterns of interaction, involving many established actors (Glimell & Fogelberg, 2003). Thus the phenomenon of frame convergence in relation to CT/nanobiotechnology is not exclusive to the usual suspects highlighted here; many actor groups have developed capacity, meaning that multiple convergences of predisposed, well resourced, actor groups in this arena are likely and are in fact occurring.

It must be reiterated that most "anti" technology networks and groups are not, despite receiving this label, opposed to technology per se. Many activists critical about the potential risks associated with CT and bioscience more generally, can also identify potential benefits such as better drug delivery, although health, cures, treatments and disease are themselves highly complex and contested fields (Plows & Boddington, 2006). Again, the socio-political/economic conditions under which CT science and products are being developed, and demands for civil society control (as opposed to corporate control) over the direction of development, are perhaps the core frames. The specific risks accruing to CT are generally expressed in relation to critiques of the continuing EU push for biotech competitiveness, and opposition to economic globalization and capitalism. Many risk frames articulated by radical and critical actors are also being framed by conventional and supportive ones, perhaps in particular the growing critique of patents (see for example Royal Society, 2003). It will be interesting to see whether more attention is paid, in the regulatory sphere, to the policy potential for strange bedfellow discursive crossover between these different groups where such ambivalence exists. However given their critical stance on how science and technology agendas are set, many activists will continue to feel that the policy process as it stands is "too little, too late".[49]

---

[49] "The left-wing think-tank Demos has… produced a report, See Through Science, that calls on industry, government and scientists to involve concerned groups in shaping research into problematic subjects such as nanotech at a much earlier stage than commercialisation. The problem with this argument is that for a development to be seen as "concerning"…usually happens well after industry and government have set their targets…" Corporate Watch, Newsletter 22, p. 8, February/March 2005.

References

Aldred, M., Crawford P., Savarirayan R., & Savulescu J. (2003) 'It's only teeth - are there limits to genetic testing?' *Clinical Genetics* 63 (5), 333–339.
Altmann, J. (2004) Military Uses of Nanotechnology: Perspectives and Concerns. *Security Dialogue* 35 (1): 61–79.
Bauer, M. (ed) (1995) *Resistance to New Technology: Nuclear Power, Information Technology and Biotechnology*. Cambridge: Cambridge University Press.
Bauer & Gaskell (eds) (2003) *Biotechnology: The Makings of a Global Controversy*. Cambridge: Cambridge University Press.
Bauman, Z. (1994) *Postmodern Ethics*. London: Blackwell.
Bey, H. (1991) *TAZ*. Brooklyn, NY: Autonomedia.
Bijker, W.E. (1987) The Social Construction of Bakelite: Towards a Theory of Invention, in Bijker, W.E., Hughes, T.P., & Pinch, T.J. (eds), *The Social Construction of Technological Systems: New Directions in the Sociology and History of Technology* (pp. 159–187). Cambridge, MA: MIT Press.
Birch, K. (2006) The Neoliberal Underpinnings of the Bioeconomy: The Ideological Discourses and Practices of Economic Competitiveness. *Genomics, Society and Policy* 2 (3): 1–15.
Brown, N. & Michael, M. (2003) A Sociology of Expectations: Retrospecting Prospects and Prospecting Retro-spects, *Technology Analysis and Strategic Development* 15 (1): 3–18.
Bowring, F. (2003) *Science, Seeds & Cyborgs: Biotechnology and the Appropriation of Life*. London: Verso.
Bucchi (2004) Can Genetics Help Us Rethink Communication? Public Communication of Science as a 'Double Helix'. *New Genetics and Society* 23 (3) (December): 269–283.
CMP Científica (2002) *Nanotech: The Tiny Revolution*. http://www.cmp-cientifica.com/
Collins, H.M. & Evans, R.J. (2002) The Third Wave of Science Studies: Studies of Expertise and Experience. *Social Studies of Sciences* 32 (2) (April): 235–296.
Commission of the European Communities (CEC) (2001) *Proposal for a Decision of the European Parliament and the Council Concerning the Multiannual Framework Programme 2002–2006 of the European Community for Research, Technological Development and Demonstration*. Brussels, 21.2.2001, COM (2001) 94 final.
Corporate Watch (2005a) *How Many Anti-Nano Angels Can Dance on the Head of a Pin?*. Corporate Watch newsletter no 22, February/March 2005, www.corporatewatch.org
Corporate Watch (2005b) *Nanotechnology: What It Is and How Corporations Are Using It*. Corporate Technologies briefing no.1, www.corporatewatch.org
Corporate Watch (2007) *Nanomaterials: Undersized, Unregulated & Already Here*. www.corporatewatch.org
Department of Trade and Industry/Office of Science and Technology (2002) *New Dimensions for Manufacturing: A UK Strategy for Nanotechnology* www.nano.gov/agency_rpts.htm
Diani, M. (1992) Analysing Social Movement Networks, in Diani, M. & Eyerman, R. (eds), *Studying Collective Action*, (pp 107–125) Newbury Park, CA: Sage
Diani, M. (1995) *Green Networks: A Structural Analysis of the Italian Environmental Movement*. Edinburgh: Edinburgh University Press.
Doherty, Plows, & Wall (2003) Studying Local Activist Communities Over Time: Direct Action in Manchester, Oxford and North Wales 1970–2001. Paper given at *European Sociological Association*, Social Movements stream, Murcia September 23–27, 2003.
Drexler, K.E. (1986) *Engines of Creation: The Coming Era of Nanotechnology*. New York: Doubleday.
ETC (2003) The Little BANG Theory. http://www.etcgroup.org/documents/comBANG2003.pdf
ETC Group Communiqué (May/June 2004) Issue #85 *Nanotech News in Living Colour: An Update on White Papers, Red Flags, Green Goo (and Red Herrings)*.
Evans & Plows (2005) *Discriminating Citizens: Making Judgements About Science*. Cardiff University Working Paper number 69. http://www.cardiff.ac.uk/schoolsanddivisions/academicschools/socsi/publications/ workingpaperseries/numeric-61-70.html

Evans, R., & Plows, A. (2007) Listening Without Prejudice? Re-Discovering the Value of the Disinterested Citizen. *Social Studies of Science* 37 (6): 827–854.
Evans, R., Plows, A., & Welsh, I. (2007) Towards an Anatomy of Public Engagement with Medical Genetics, in Atkinson, P. & Glasner, P. (eds), *New Genetics, New Identities* (pp.139–157). London: Routledge.
Fischer, F. (2000) *Citzens, Experts and the Environment: The Politics of Local Knowledge.* Durham, NC/London: Duke University Press.
Foresight Institute (2000) *Foresight Guidelines on Molecular Nanotechnology,* www.foresight.org/guidelines
Fukuyama (2002) *Our Posthuman Future: Consequences of the Biotechnology Revolution.* New York: Farrar, Straus, and Giroux.
Genewatch (2000) *Briefing 11: Privatising Knowledge, Patenting Genes: The Race to Control Genetic Information.* http://www.genewatch.org/Patenting/briefs.htm#Brief11
Giddens, A. (1990) *The Consequences of Modernity.* Cambridge: Polity Press in association with Basil Blackwell, Oxford.
Glasner, P. (2002) Beyond the Genome: Reconstituting the New Genetics. *New Genetics and Society* 21 (3): 267–277.
Glasner, P. & Rothman, H. (2004) *Splicing Life? The New Genetics and Society.* Aldershot, Hants, England/Burlington, VT: Ashgate.
Glimell, H. & Fogelberg, H. (2003) *Bringing Visibility to the Invisible: Towards A Social Understanding of Nanotechnology.* Göteborg: Göteborg University. http://www.sts.gu.se/publications/STS_report_6.pdf
Granovetter, M.S. (1973) The Strength of Weak Ties. *American Journal of Sociology* 78 (6): 1360–1380.
Green Action (2004) *Technology, Politics and Democracy.* Green Action briefing for the European Peoples Global Action 2004 conference, held in Serbia. http://www.mallusk.net/grassrootsgathering/greenaction.html
Grove-White, R., Macnaghten, P. & Wynne, B. (2000) *Wising Up: The public and New Technologies.* Lancaster: IEPPP, Lancaster University.
Habermas, J. (2003) *The Future of Human Nature.* Oxford: Polity Press.
Hammersley, M. (ed) (1993) *Social Research: Philosophy, Politics and Practise.* London: Sage.
http://angelsagainstnanotech.blogspot.com
http://www.demos.co.uk/media/pressreleases/pressreleases2004/pressreleasepaddlingupstream/
http://earthfirst.org.uk
http://www.etcgroup.org/
http://www.genewatch.org/HumanGen/GeneticResearch.htm
http://www.hgc.gov.uk/choosingthefuture/index.htm
http://www.hgc.gov.uk/subgroups/horizon_scanning.htm
http://www.i-sis.org.uk/fluidGenome.php
http://www.i-sis.org.uk/SVGE.php
http://indymedia.org.uk/en/2004/12/302727.html
http://www.genetics-and-society.org/analysis/opposing/2003_berlin_report.html
www.nano.gov/agency_rpts.htm
http://www.nanotechweb.org/articles/society/3/6/1/1
http://www.royalsoc.ac.uk/document.asp?tip = 0&id = 1374
http://science.slashdot.org/science/05/01/18/2120255.shtml?tid = 191&tid = 155
http://www.sts.gu.se/publications/STS_report_6.pdf
http://www.sustainabletechnologies.ac.uk/Project%20pages/site/brief10.htm
www.wired.com/wired/current.html
http://wolbring.wordpress.com/
Horlick-Jones, T., Walls, J., Rowe, G., Pidgeon, N., Poortinga, W., Murdock, G., & O'Riordan, T. (2007) *The GM Debate: Risk, Politics and Public Engagement.* Taylor & Francis, London: Routledge
Irwin, A., & Michael, M. (2003) *Science, Theory and Public Knowledge.* Maidenhead, UK: Oxford University Press.

Joy, B. (2000) Why the Future Doesn't Need Us. *Wired*: 8.04, www.wired.com/wired/current.html

Kerr, A. Cunningham-Burley, S., & Amos, A. (1998a) The New Genetics and Health: Mobilizing Lay Expertise. *Public Understanding of Science* 7: 41–60.

Kerr, A. Cunningham-Burley, S., & Amos, A. (1998b) 'Drawing the Line: An Analysis of Lay People's Discussion About the New Genetics', *Public Understanding of Science* 7: 113–133.

Kerr, A., & Shakespeare, T. (2002) *Genetic Politics – From Eugenics to Genome*. Cheltenham: New Clarion Press.

Knorr-Cetina, K. (1999) *Epistemic Cultures: How the Sciences Make Knowledge*. Cambridge, MA: Harvard University Press.

Mae-Wan Ho (2003) *Living with the Fluid Genome*. http://www.i-sis.org.uk/fluidGenome.php

Mayer, S. (2002) From Genetic Modification to Nanotechnology: The Dangers of "Sound Science", in Gilland, T. (ed), *Science: Can We Trust the Experts?* (pp. 1–15). London: Hodder & Stoughton.

McAdam, D. (1986) Recruitment to High-Risk Activism: The Case of Freedom Summer. *American Journal of Sociology* 92: 64–90.

Melucci, A. (1996) *Challenging Codes: Collective Action in the Information Age*. New York: Cambridge: Cambridge University Press.

Mnyusiwalla, A. et al. (2003) "Mind the Gap": Science and Ethics in Nanotechnology. *Nanotechnology* 14 (3): R9–R13. www.iop.org/EJ

MulCahy (1993) Rhetorics of Hope and Fear in the Great Embryo Debate. *Social Studies of Science* 23 (4), 721–742.

Nelkin, D. (1995) Forms of Intrusion: Comparing Resistance to Information Technology and Biotechnology in the USA, in Bauer, M. (ed), *Resistance to New Technology: Nuclear Power, Information Technology and Biotechnology* (pp. 379–90). Cambridge: Cambridge University Press.

Nordmann, A. (2004) *Converging Technologies – Shaping the Future of European Societies*. Report to the European Commission.

Plows, A. (2003) *Praxis and Practice: The 'What, How and Why' of the UK Environmental Direct Action (EDA) Movement in the 1990's*. Unpublished Ph.D. thesis, University of Wales, Bangor. http://www.iol.ie/~mazzoldi/toolsforchange/papers.html#counter

Plows (2004a) Mapping the Emergent Complexity of 'Social Movement Society Engagement with Human Genetic Technologies: Implications for Social Movement Theory. Paper prepared for: *Meetings of Research Committee 47 of the International Sociological Association (ISA) in conjunction with Centre d'Analyse et d'Intervention Sociologique (CADIS)* at EHESS, Paris, June 11–12, 2004.

Plows (2004b) Activist Networks in the UK: Mapping the Build-Up to the Anti-Globalization Movement', in Carter, J. & Morland, D. (eds), *Anti-Capitalist Britain*. Cheltenham: New Clarion Press (pp 95–113).

Plows, A. & Boddington, P. (2006) 'Troubles with Biocitizenship?', in *Genomics, Society and Policy* 2 (3): 115–135.

Purdue, D. (2000) *Anti-GentiX: The Emergence of the Anti-GM Movement*. Aldershot, UK, Avebury.

Roco, M.C. & Bainbridge, W.S. (2002) *Converging Technologies for Improving Human Performance: Nanotechnology, Biotechnology, Information Technology and Cognitive Science*. NSF/DOC-sponsored report.

Rolison, D.R. (2002) Nanobiotechnology and Its Societal Implications, in *Nanotechnology: Revolutionary Opportunities & Societal Implications*. EC-NSF 3rd Joint Workshop on Nanotechnology, Lecce, Italy, 31 January–1 February 2002.

Rose, N. & Novas, C. (2005) Biological Citizenship. In *Global Assemblages: Technology, Politics, and Ethics as Anthropological Problems*. Edited by Ong, A.; Collier, S., Oxford: Blackwell, (pp. 439–463).

Routledge, P. (2003) Convergence Space: Process Geographies of Grassroots Globalization Networks. *Transactions of the Institute of British Geographers* 28 (3): 333–349.

Royal Society (2003) *Keeping Science Open: The Effects of Intellectual Property Policy on the Conduct of Science*. http://www.royalsoc.ac.uk/document.asp?tip = 0&id = 1374

Royal Society report (2004) *Nanoscience and Nanotechnologies: Opportunities and Uncertainties*. www.royalsoc.ac.uk

Salter, B. & Jones, M. (2002) Regulating Human Genetics: The Changing Politics of Biotechnology Governance in the European Union. *Health, Risk and Society* 4 (3): 325–340.

Schummer, J. (2004) Multidisciplinarity, Interdisciplinarity, and Patterns of Research Collaboration in Nanoscience and Nanotechnology. *Scientometrics* 59 (2004): 425–465.

Seel, B. & Plows, A. (2000) Coming live and direct: strategies of Earth First! in Seel, B, Paterson, M and Doherty, B (Eds), *Direct Action in British Environmentalism*, pp. (London: Routledge).

Sexton, S. (2001) If Cloning Is the Answer, What Was the Question? Power and Decision-Making in the Geneticization of Health. *International Journal of Sustainable Development 2001* 4 (4): 407–433.

Snow, D. & Benford, R. (1992) Master Frames and Cycles of Protest, in Morris, A.D. & Mueller, C. (eds), *Frontiers in Social Movement Theory* (p. 137). London/New Haven, CT: Yale University Press.

Sweeney, A.E., Seal, S., & Vaidyanathan, P. (2003) The Promises and Perils of Nanoscience and Nanotechnology: Exploring Emerging Social and Ethical Issues. *Bulletin of Science, Technology & Society* 23 (4): 236–245.

Szerszynski, B. (2002) Ritual Action in Environmental Protest Events. *Theory, Culture and Society* 19 (3) (2002): 51–69.

Szerszynski, B. (2005) Beating the Unbound: Political Theatre in the Laboratory Without Walls in Stewart, in Nigel, Giannachi, & Gabriella (eds), *Performing Nature: Explorations in Ecology and the Arts*. Bern: Peter Lang, pp. 181–197.

Tarrow, S. (1998 2nd ed) *Power in Movement: Social Movements, Collective Action, and Politics*. New York: Cambridge University Press.

Wall, D. (1999) *Earth First! and the Anti-roads Movement*. London: Routledge.

Welsh, I. (2000) *Mobilising Modernity: The Nuclear Moment*. London: Routledge.

Welsh, I. & Chesters, G. (2002) Reflexive Framing and Ecology of Action: Engaging with the Movement 'For Humanity Against Neoliberalism'. Paper Given at *XV World Congress of Sociology: Session Globalization and the Environment*.

Welsh, I & Chesters G. (2004) "The Rebel Colours of S26: Social Movement: 'Frame-work' during the Prague IMF/WB protests", *Sociological Review*, 52(3): 314–335.

Welsh, I & Chesters G. (2005) Complexity and Social Movement: Process and Emergence in Planetary Action Systems, *Theory Culture and Society*, 25(5):187–211.

Welsh, I., Plows, A., & Evans, R. (2007) Human Rights and Genomics: Science, Genomics and Social Movements at the 2004 London Social Forum *New Genetics and Society* 26 (2) (August 2007): 123–135.

Whittier, N. (1995) *Feminist Generations*. Philadelphia, PA: Temple University Press.

Willis, R. & Wilsdon, J. (2004) *See Through Science: Why Public Engagement Needs To Move Upstream*, London: Demos.

Wolbring, G. (2006) The Unenhanced Underclass, in Miller, P. & Wilsdon, J. (eds), *Better Humans? The Politics of Human Enhancement and Life Extension* (pp. 122–128). London: Demos. http://www.demos.co.uk/publications/betterhumanscollection

World Health Organization (2004) Priorities for Research to Take Forward the Health Equity Policy Agenda. Report from the *WHO Task Force on Health System Research Priorities for Equity in Health*, pdf doc 47.

Wood, S., Richard, J., & Geldart, A. (2004) *The Social and Economic Challenges of Nanotechnology*. ESRC report. http://www.esrc.ac.uk/esrccontent/DownloadDocs/Nanotechnology.pdf

WTEC Panel (1998) *Nanostructure Science and Technology: A Worldwide Study*. National Science and Technology Council, Committee on Technology, Interagency Working Group on Nanoscience, Engineering and Technology.

Wullweber, J. (2007) Enclosure of Knowledge and the Emergence of a Global Nano-Divide. Paper presented at International Studies Association 48th Annual Convention.

Wynne, B. (1995) Public Understanding of Science, in Jasanoff, S., Markle, J.C., Peterson, J.C., & Pinch, T. (eds), *Handbook of Science and Technology Studies* (pp. 361–388). London: Sage.

Wynne, B. (1996) May the Sheep Safely Graze? A Reflexive View of the Expert-Lay Knowledge Divide, in Lash, S. Szerszynski, B., & Wynne, B. (eds), *Risk, Environment & Modernity: Towards a New Ecology* (pp. 27–83). London: Sage.

Wynne, B. (2006) Public Engagement as a Means of Restoring Public Trust in Science- Hitting the Notes, but Missing the Music?, *Community Genetics* 9: 211–220.

Allotropes of Fieldwork in Nanotechnology

Christopher Kelty

Abstract This article discusses the distinctive contributions that the discipline of anthropology (in particular, socio-cultural anthropology) might make to the study of nanotechnology. It focuses on recent research conducted by anthropologists on the subject of nanotechnology, human health and the environment at the Center for Biological and Environmental Nanotechnology (CBEN) at Rice University. The chapter plays on the chemical concept of "allotropes" as a way of understanding three variations on the method of anthropological fieldwork. These allotropes of fieldwork include a focus on site, in this case CBEN; a focus on method, especially the role of observations, participation and objectivity; and a focus on substance, the subject matter of nano-science and technology, in this case water and its relationship to nano-materials. The argument of the paper is that all of these are necessary for effective ethnographic work and they can focus attention on the human practices that shape research, concepts and results in nanotechnology. It further argues that such practices go unnoticed, or are deliberately downplayed, by some nano-scientists and engineers, and so the contribution of anthropology can be to highlight certain critical projects or potential alternative futures not otherwise visible.

Keywords Anthropology, fieldwork, environment, human health, methodology

> Here's an example that hit me the other day at an NSF conference on directions for potable water treatment. So in the past, you would notice the water supply getting shorter and shorter and you'd say, well, we're going to treat dirtier and dirtier water. We'll take all this stuff out of it, we'll take the salts out of the ocean, we'll take all the crud out of wastewater and reuse the wastewater and so we're going to take stuff out of it and we're going to clean it up. And the nano vision of this is, you know, let's just make water. Let's just construct it from oxygen and hydrogen. I mean, that would be a very nano thing to do.
>
> (Interview with Mark Wiesner, April 2004)

This chapter addresses two issues: first, it lays out some of the distinctive contributions that the discipline of anthropology (in particular, socio-cultural anthropology) might make to the study of nanotechnology and second, it touches on recent research conducted by anthropologists on the subject of nanotechnology, human health and the environment at the Center for Biological and Environmental Nanotechnology (CBEN)

Rice University

at Rice University. The chapter plays on the chemical concept of "allotropes" as a way of understanding how variations on the focus and meaning of *ethnographic fieldwork* amongst scientists and engineers should be used to reveal aspects of nanotechnology that are not rendered accessible in any other way. Each allotrope examines ethnographic fieldwork from a different organizing perspective. The first allotrope considers the *site*, in this case CBEN, as that which organizes research; the second allotrope considers *method* as an organizing force; and the third allotrope looks at *substance* or subject matter as the organizing force. The argument of the paper is that all of these are necessary for effective ethnographic work to reveal something that is not already known by the actors, or obvious from a review of the literature. Effective anthropological fieldwork in areas of emerging science and technology can focus attention on the human practices that shape research, concepts and results, but go unnoticed, or are deliberately downplayed by scientists and engineers, and ultimately contribute to certain kinds of critical projects that emerge from within science.

Allotropes of Fieldwork

Among social science disciplines, it is fair to say that none is as ecumenical, pluralistic, and even internally contradictory as that of anthropology. Few other disciplines are at home researching everything from archaeological origins of complex society to cosmopolitanism in modern Greece; from primate genetics to sexual behavior among Indian trans-gendered prostitutes. The range of work brings with it a concomitant range of methodological tools and approaches, ranging from symbolic analysis to archaeological investigation to DNA sequencing and comparison. A key component present across many of these methods in anthropology, however, is *fieldwork*. Fieldwork encompasses participant-observation, interview and dialogue, collaboration and critique. It is as central, if not more, to the definition of anthropology as any theoretical concern with the human (*anthropos*).

Although this chapter consists primarily of socio-cultural anthropology with an emphasis on elites and contemporary complex societies, one might well imagine the various ways in which many of the sub-fields of anthropology could become interested in aspects of nanotechnology. Medical anthropologists (and especially bio-medical anthropologists studying contemporary high-tech medicine) might be interested in the emerging therapies for cancer that make use of gold nano-shells and nano-rods. Linguistic anthropologists interested in creolization or pidginzation should find a fascinating project for study in the attempt to forge new nomenclatures and standards for nano-scale particles and materials. Archaeologists who theorize about tool use might see something of value in the claims for "human enhancement" so often promised by nanotechnologists (see Stone, n.d.).

Fieldwork in anthropology thus has a variety of "allotropes" – to play on one of the key concepts in a basic understanding of nanotechnology. Allotropes are the diverse shapes a single element can take, depending on the arrangement of the bonds between the atoms. The resulting materials can have vastly different thermal, electrical and structural properties. Carbon, for example, has diamond, graphite, and buckminsterfullerenes (buckyballs) among its allotropes. If fieldwork is the "carbon" of anthropology, then its arrangement and structure can yield similarly diverse results, depending on the tools and techniques used – depending on whether questions are being answered or hypotheses generated, and on the length, locale, and style of interaction that is imagined (Marcus & Fischer, 1986; Marcus, 1995, 1998; Gupta & Ferguson, 1997, 2002). Without pushing the simile too far, these allotropes of fieldwork can in turn have very different properties; some fieldwork projects are extremely flexible and robust – engaged in a kind of studied curiosity without explicit questions. Others are rigidly defined, seeking "qualitative" answers through survey, interview or dialogue. The version of socio-cultural anthropology presented here is primarily focused on fieldwork that starts *in medias res* – especially among things that appear to be novel, surprising or emergent, such as nanotechnology. In this setting, one might organize one's fieldwork into one of three possible allotropes: site, method and substance.

The most obvious defining factor of fieldwork can be that of the **site**. Site can be a classical geographic locale – a village, a street corner, a café, a lab – but more often today it is defined with respect to some kind of conceptual object as well – state surveillance and policing, economic development or identity politics, for instance. Objects that travel are also common reference points – sacred objects, commodities and artworks; or more often, money, policies and legal documents. Site allows for a way of keeping track of connections and social actions that make up a particular problem area. However, it is also the case that there are frequently issues that impinge on one site, but are not easily visible within it – the most common such situation is the effects of global institutions, or global financial capitalism on the lives and work of local populations in specific local contexts. If site alone defines an anthropological project, then the demand for novelty in method and or substance is high, since there is little analytic depth in simply choosing a particular site.

A second mode of defining fieldwork is that of **method**. At a very basic level, method concerns the what, where, who, when, and why of field research. In closely bounded settings (a village, a bar or a lab) these are easier to define than in the changed global conditions within which most anthropologists must now operate – alongside experts and fellow social scientists whose goals are different in kind from those of the socio-cultural anthropologist. For instance, the manner in which an anthropologist looks at "the economy" is quite different from how an economist might; similarly, the mental life of individuals can be approached much differently by the symbolic anthropologist than the manner in which a psychologist or cognitive scientist might approach it; furthermore, the fact that these experts – economists and cognitive scientists, for instance – are the principal actors defining what economies and brains are, makes the work of the anthropologist doubly challenging. Method, in this context, consists of questioning the foundations not only of the discipline of anthropology, but those of other disciplines as well. This immodest approach is risky, needless to say, and usually requires a significant investment in time and learning to carry out. It also suggests that anthropology often, though not always, will take a position of critique vis-à-vis neighboring disciplines. If method is the defining focus of an anthropological project, sites and subjects are in danger of proliferating wildly.

Finally, **substance** can also define the structure of fieldwork. Substance suggests not simply a tangible substance, but the "subject matter" of a study. The "politics of oil and energy" for instance, might define a set of sites and methods for investigation;

"human rights law" similarly might imply a number of possible sites and methods for achieving an anthropological understanding of the changing substance and definition of what counts as human and what as rights. By defining a topic in terms of substance, an anthropologist can avoid some of the traps of a too narrowly defined site-based project or too strictly defined a methodology, by being more flexible with respect to possible sites, people and forms of investigations. By the same token, a project defined only in terms of substance often faces the fact that other researchers in other disciplines have already staked out particular substances – and the question returns: what is distinctive about the anthropologist's observations?

In the remainder of this paper, I present each of the allotropes with respect to a project in the anthropology of nanotechnology. First, a description of the specific site within which the bulk of our research is carried out, and a detailed description of why the site is both distinctive and partially representative of emerging nanotechnology; second, a series of methodological approaches that can help orient the epistemological questions about the goal of an anthropology of nanotechnology; and third a focus on substance – in this case, water, and its relevance to nanotechnology – that has emerged by virtue of the combination of site and method.

Allotrope: Site

Rice University's Center for Biological and Environmental Nanotechnology (CBEN) was funded by the National Science Foundation in 2001, as part of the National Nanotechnology Initiative. Rice's presence in Nanotechnology has been significant given its small size (less than 4,000 graduate and undergraduate students and around 500 full-time faculty). This has been due in large part to the work of two committed Rice faculty; the scientific patron, recently deceased Nobel laureate Richard Smalley, and the political patron, former science adviser to Clinton, Neal Lane. Both men have been extremely well respected on campus, and have had signi-ficant influence over the direction of university vision and funding over the last 15 years.

The Center for Biological and Environmental Nanotechnology, however, is an unusual center for Rice to have – given the fact that Smalley, and many other campus scientists work primarily on the chemistry and engineering of fullerenes (especially buckminsterfullerenes, for which Smalley, Robert Curl and Harold Kroto were awarded the Nobel prize in 1996, and Single Walled Carbon Nanotubes, discovered in 1991). One might expect a center devoted to fullerene chemistry or the chemical engineering of nanotubes, or a center aimed specifically at contributions to the homegrown energy industry – not a center explicitly identified with the environment and biology.

CBEN was an idea hatched by chemist Vicki Colvin and Environmental Engineer Mark Wiesner and it includes chemists, physicists, environmental and civil engineers, chemical engineers and lately, two anthropologists. The center is one of 14 that were initially funded by NSF, modeled on their Science and Technology Center program, but focused in areas of interest in Nanotechnology. Of the 14 initial centers, and the handful of others recently funded, CBEN stands out as the only science and engineering center with any emphasis on human and environmental issues (see Box 1: NNI NSE Centers).

Box 1 List of nanotechnology NSF centers and networks: NSECs, NCLT, NISE, NNIN, NCN (http://www.nsf.gov/crssprgm/nano/info/centers.jsp)

	Prop. ID	Institution	PI name	Proposal title	NSF award (estimation at the award) 5 years	NSF funds obligated in FY 2005
1	0117795	Columbia Univ.	James Yadley jy307@columbia.edu 212-854-3265	Center for Electron Transport in Molecular Nanostructures	$10,800,000	$2,150,000
2	0117770	Cornell Univ.	Robert Buhrman rab8@cornell.edu 607-255-2103	Center for Nanoscale Systems	$11,600,000	$2,668,000
3	0117792	Rensselaer Polytech Inst.	Richard Siegel rwsiegel@rpi.edu 518-276-8846	Center for Directed Assembly of Nanostructures	$10,000,000	$2,000,000
4	0117795	Harvard Univ.	Robert Westervelt nsec@deas.harvard.edu 617-496-3275	Science for Nanoscale Systems and their Device Applications	$10,800,000	$2,358,000
5	0118025	Northwestern Univ.	Chad Mirkin c-mirkin@northwestern.edu 847-491-5784	Institute for Nanotechnology	$11,100,000	$2,450,000
6	0118007	Rice Univ.	Vicky Colvin colvin@ruf.rice.edu 713-348-8212	Center for Biological and Environmental Nanotechnology	$10,500,000	$2,474,861
7	0327077	UC Los Angeles	Xiang Zhang xiang@seas.ucla.edu 310-206-7699	Center for Scalable and Integrated Nanomanufacturing	$17,658,208	$2,827,493

(continued)

Box 1 (continued)

	Prop. ID	Institution	PI name	Proposal title	NSF award (estimation at the award) 5 years	NSF funds obligated in FY 2005
8	0328162	Univ. of Illinois Urbana Champaign	Placid Ferreira pferreir@uiuc.edu 217-333-0639	Center for Nanoscale Chem-Electr-Mechanical Manufacturing	$12,530,745	$2,516,749
14	0425914	UC Berkeley	Alex Zettl azettl@physics.berkeley.edu 608-265-8171	Center for Integrated Nanomechanical Systems	$11,910,000	$2,400,000
11	0425826	Northeastern Univ.	Ahmed Busnaina busnaina@coe.neu.edu 617-373-2992	Center for High Rate nanomanufacturing	$12,376,000	$2,450,000
10	0425626	Ohio State Univ.	Ly James Lee leelj@che.eng.ohio-state.edu 614-292-2408	Center for Affordable Nanoengineering	$12,923,000	$2,573,000
9	0425780	Univ. of Pennsylvania	Dawn Bonnell bonnell@sol1.lrsm.upenn.edu 215-898-6231	Center for Molecular Function at the Nanoscale	$11,426,000	$2,250,000
13	0425897	Stanford Univ.	Kathryn Moler kmoler@stanford.edu 650-723-6804	Center for Probing the Nanoscale	$7,459,709	$1,467,000

12	0425880	Univ. of Wisconsin	Paul Nealey nealey@engr.wisc.edu 608-262-5434	Center for Templated Synthesis and Assembly at the Nanoscale.	$13,365,000	$2,610,000
15a	0531194	ASU	David Guston david.guston@asu.edu (480)727-8787	Nanotechnology in Society Network_1	$6,155,000	$1,335,000
15b	0531184	UCSB	Bruce Bimber bimber@polsci.ucsb.edu (805)893-3860	Nanotechnology in Society Network_2	$4,970,000	$1,050,000
15c	0531160	USC	Davis Baird db@sc.edu (803)777-4166	Nanotechnology in Society Network_3	$1,475,000	$275,000
15d	0531146	Harvard Univ.	Richard Freeman freeman@nber.org (617)868-3900	Nanotechnology in Society Network_4	$1,625,000	$325,000
16	In review			Hierarchical Nanomanufacturing		$36,180,103

Centers from the Nanoscale Science and Engineering Education solicitation

1	0540658	Northwestern Univ.	Robert Chang r-chang@northwestern.edu (847)491-3598	Nanotechnology Center for Lerning and Teaching (NCLT)	$15,000,000	$80,001
2	0532536	Museum of Science	Lawrence Bell lbell@mos.org (617)589-0282	Nanoscale Informal Science Eduction (NISE)	$20,000,000	$4,500,000
						$4,580,001

(continued)

Box 1 (continued)

Prop. ID	Institution	PI name	Proposal title	NSF award (estimation at the award) 5 years	NSF funds obligated in FY 2005
NSF Networks and centers that complement the NSECs					
1 0335765	Cornell University-Endowed	Sandip Tiwari st222@cornell.edu 607-254-6254	National Nanotechnology Infrastructure Network (NNIN)	$35,000,000	$7,000,000
2 0228390	Purdue Univ.	Mark Lundstrom lundstro@purdue.edu 765-494-3515	Network for Computational Nanotechnology (NCN)	$2,848,333	2,848,333
3 9876771	Cornell University-Endowed	Barbara Baird bab13@cornell.edu (607)255-4095	STC: The Nanobiotechnology Center (ENG)	$20,000,000	$3,986,814
					$13,835,147

There are 15 Nanoscale Science and Engineering Centers (NSECs): Six awards were made in FY 2001, two new awards in FY 2003, six new awards in FY 2004, one new award in FY 2005. Each award for 5 years, renewable for another 5 years.

The trade-offs involved in making such a center happen are not insignificant, they will be explored in more detail in the third section.

CBEN, and the other centers like it, are NSF Centers whose mission is more complex than simply the performance of research. Unlike a simple grant for a research project, the center model assumes that research projects need to be *precipitated* in some way – and that rather than the NSF in DC attempting to second-guess the wide range of possible research directions in the area of science and technology, they might "outsource" some of that activity to focused centers. The centers, thus, are charged with promoting – and funding – research in their areas, locally, in the disciplinary sense, and usually in the regional sense (though many are also "networked" across many universities).

It is important to note, therefore, that both the NSF and CBEN are involved in the same kinds of "meta-theoretical" or "meta-research" attempts to define what constitutes research in nanotechnology. For many people who have been involved in nanotechnology, the period between 2000 and 2005 appears to have been one in which a large amount of money was available, resulting in a concomitant re-definition of a very wide variety of research as "nanotechnology". Nowotny et al. (2001) for instance, suggest that nanotechnology is a classic form of "weakly contextualized" science – precipitated in large part by political goals that involve ensuring national competitiveness in nanotechnology. The implication being that a very large number of people who claim to do nanotechnology are simply continuing research directions that long predate the NNI, and may have nothing to do with the presumptive core vision of nanotechnology.

While it may be the case that many scientists have done just such a thing, it does not, therefore, mean that there is no content to nanotechnology, and the existence of multiple centers (in addition to trade and industry associations, conferences, publications, and other standard organs of scientific practice) should be understood as an active transformation – possibly even *creation* – of the definition of nanotechnology. Indeed, prior to the funding of these centers, the most common definition of nanotechnology came from K. Eric Drexler and the Foresight Institute and their notion of bottom-up molecular manufacturing. Drexler and those who followed his vision were indeed very specific about what they suggested nanotechnology would encompass, and what it would not, citing various authorities along the way to bolster their case (Regis, 1996; Toumey, 2005).

Thus the creation of 14 diverse centers essentially inverted this situation by considerably broadening the various definitions of nanotechnology that circulate; this transformation was perceived by Drexler as the dilution and destruction of the vision of molecular manufacturing (Drexler, 2004). The success involved in creating the NNI was the failure in his eyes of focusing scientific energy on a specific problem. But for many new participants, this broadening and refocusing of the possible definitions of nanotechnology has been a welcome change, especially those made uncomfortable by the more radical and utopian fears and desires of Drexler and followers (Baum, 2003). Meanwhile, at least some philosophers and ethicists have busied themselves with the question of the distinctiveness of nanotechnology (Khushf, 2004; Dupuy, 2004a, b).

The fact remains, however, that most of the NSF Centers in nanotechnology are devoted not only to scientific and technical work but to what might be called the supporting activity of defining nanotechnology, disseminating ideas about it, educating various publics and stakeholders, communicating with the media, and in general, trying to do more than simply promote research amongst scientists and engineers, but to build a constituency of support for the center's activities far beyond the laboratory. Indeed, the continued existence of these centers after the initial 5 years, will hinge on their importance to the University, or to funding agencies or corporate interests beyond the NSF.

How then, has CBEN been defining its activities? Because of the close relationship between the chemical industry and the chemical engineers on the Rice faculty, CBEN's self-presentation starts this way: "[CBEN] aims to shape nanoscience into a discipline with the relevance, triumphs, and vitality of a modern day polymer science." The reference point is strategic – it appeals to chemical engineers, and it sets the boundaries within which it is possible to imagine a definition of nanotechnology as a kind of materials science. The relationship that materials science has to issues of human health and environmental safety.

But CBEN is also aimed at something much more general, in terms of the theoretical disposition of its research – they refer to this as the "wet/dry interface":

> Water, the most abundant solvent present on Earth, is of unique importance as the medium of life. The Center's research activities explore this interface between nanomaterials and aqueous systems at multiple length scales, including interactions with solvents, biomolecules, cells, whole-organisms, and the environment. These explorations form the basis for understanding the natural interactions that nanomaterials will experience outside the laboratory, and also serves as foundational knowledge for designing biomolecular/nanomaterial interactions, solving bioengineering problems with nanoscale materials, and constructing nanoscale materials useful in solving environmental engineering problems.
> http://cben.rice.edu/about.cfm?doc_id = 4998

This definition of CBEN's research distinguishes it from other centers and other research sites by defining it in terms of the interaction of environment and materials, with specific reference to the central role of water, which would be obvious to chemists and chemical engineers. However, there is also a very strategic trade-off hidden in this description. On the one hand, scientists interested in promoting nanotechnology perceive a risk that emphasizing the negative effects of nanomaterials might have grave consequences on future funding. As a result, many scientists, including Richard Smalley, have actively sought to downplay such research, and even in some cases, actively deter funding and research into the hazard and exposure risks of nanomaterials. By the same token, there is a perception among an equal number of nanoscientists (and corporate representatives as well) that ignoring such potential dangers is precisely what has led to major problems in the past (for instance, in DDT, paraquat, GM foods, or asbestos). Hence there are also a number of scientists vocally calling for more research, not less, on safety, toxicity and hazard/exposure risks.

The bargain that was struck in CBEN – a necessary bargain in order to get Richard Smalley involved – is clear in the definition of the research above. CBEN funds projects that research and test nanomaterials for biological and environmental use – and it does not simply study the risks that they may or may not present. Despite how it might appear, this trade-off is not disingenuous. The two principals primarily responsible for creating CBEN have conducted research that fits this profile precisely: Vicki Colvin, a chemist, has in fact conducted toxicity studies even though this is not her primary research focus (Colvin, 2003), and Mark Wiesner, an environmental engineer, has created new nanomaterials for use in water filtration and fuel cells (Wiesner et al., 2003). Hence CBEN has involved itself in both the nanotech-for-environment and the dangers-to-environment sides of this definition. It may be, in fact, that it was this bargain that gave CBEN its distinctiveness, a bargain by which toxicity and hazard studies were included insofar as they take a positive role in the promotion of the responsible creation of new materials.

Like all the NSF Nanotechnology centers, CBEN's mission is broader than only research. CBEN also has as part of its mission, various forms of outreach, education, and public relations work. CBEN's attempt to manage the image of nanotechnology is part of its "meta-research" activities; part of the activity of defining what nanotech is, who it benefits and who it might threaten. Most of the work of outreach, education and public relations was motivated implicitly, if not explicitly, by what researchers in the Public Understanding of Science call the "deficit model" of scientific literacy – that the public, whoever they are lacking some quantity of scientific information, without which they are unable to properly assess the work and meaning of modern science, and may even, in some cases (as when they are manipulated by popular novels like Michael Crichton's *Prey* or fear-mongering Drexlerian scenarios of grey-goo) become dangerous to the future life and funding of nanotechnology as a research science (Sturgis & Allum, 2004). Education, research and public relations are therefore necessary to stem this tide of dangerous, false-thinking publics by increasing the quantity of scientific literacy in the world. CBEN is by no means alone in adopting this attitude, even if it has been widely critiqued. The result is that a significant focus of the small amount of social science research undertaken or promoted by CBEN is on the perception of the risks of nanotechnology – usually conceived of primarily as a consumer product, and not, despite the sophistication of CBEN's environmental focus, a systemic issue of production and planning. This approach raises the question of whether or not anthropologists would be more at home pursuing such research, defined in large part by the directors of CBEN, or more comfortable offering criticism of the questions – and thereby risk being ignored. We return to this issue in the next section.

Within CBEN, and in many other circles of the nanotech world, the most dangerous publics (in terms of potential impact on the future of nanotechnology research) tend to be environmental advocates and activists, ranging from groups such as Environmental Defense and the National Resources Defense Council to the most learned and deeply critical, the ETC Group. As a result of attempting to understand the dangers posed by publics – especially these groups – CBEN directors formed a kind of institutional bud-growth: another center called the International Council on Nanotechnology (ICON). ICON is an institution co-funded and run by industry partners, CBEN, and various environmental advocacy stakeholder groups.

It is intended to be as impartial as possible, serving only to facilitate the interests of the participating stakeholders. The corporations involved are those who simultaneously have some of the biggest goals for developing nanotechnology, such as the cosmetic company Loreal and chemical company DuPont – but they are also those with the deepest concerns about the management of the risk and hazard of new nanotech, like the Swiss Re reinsurance corporation, whose experience with the banning of GM foods in Europe made them hyper-sensitive to the problems of ignoring potential risks.

ICON was inaugurated in summer of 2005. The immediate need for ICON arose out of a desire to meet the potential publics halfway – but since "the public" as such has no secretary to call, and no email address, ICON effectively narrows down the definition of public to the stakeholders perceived to be most important to – or critical of – nanotechnology research. Coordinating various groups – university, industry (including not just manufacture, but as the case of Swiss Re makes clear, other kinds of corporate observers), environmental advocacy, activist and social movement groups – required some kind of "independent" organization. As of this writing, ICON has almost achieved that independence, but even given its forthright attempts to meet these publics halfway could not convince, for instance, Environmental Defense or ETC Group to join as official members. Nonetheless, the very existence of this kind of institution suggests that the configuration of science and society is changing – and that CBEN (and ICON) represents one of the most concerted experiments in identifying participants, risks, and potential problems well in advance of any real dissemination of nanomaterials or consumer goods associated with them, without giving up the core desire of scientists and engineers to pursue the discovery, synthesis and promotion of new materials and new technologies.

Allotrope: Method

Broadly speaking, one of the key reasons anthropologists (and other social scientists) might be interested in nanoscience and nanotechnology research is in order to test theories about the historically changing relationship between social and governmental institutions and scientific research. Pioneering work in the history of science, such as Shapin and Schaffer's *Leviathan and the Air Pump* has suggested that the interpenetration of the two is deeper and more complex than is usually assumed (Shapin & Schaffer, 1989). Shapin and Schaffer detail how Hobbes political philosophy was also a theory of nature and natural process, and that Robert Boyles' experiments with an air pump within the brand new Royal Society were also political claims about relationship of knowledge to sovereignty.

Today, the relationship between forms of scientific knowledge and the social order they relate to is, if anything, vastly over-theorized. One can choose from theories of reflexive modernization (Beck et al., 1994; Lash & Urry, 1994; Beck, 1992) actor network theory (Callon, 1986; Law, 1987; Latour, 1987), mode 1/mode 2 theory (Gibbons et al., 1994; Nowotny et al., 2001), triple helix (Etzkowitz & Leydesdorff,

1995, 1996) or co-production (Jasanoff, 2004, 2005) among just some of the more well known. Social scientists have an abundance of theories to draw on in order to explain the changing nature of the relationship in science and society in the recent past. But by the same token, this embarrassment of theoretical riches is accompanied by a relative poverty of methodological innovation. What sites, methods, techniques, questions, observations or participations should the interested anthropological observer be engaging in vis-à-vis nanotechnology? How does one go about asking questions, performing research that helps sort out these various theories, refine them or extend them? To whom does one talk, and about what exactly?

Anthropologists who are interested in these questions about the changing relation of science and society, and the meaning more generally of those changes for social theory and for human action, must devise ways to conduct empirical research given the tools and methods of ethnographic fieldwork. Broadly speaking, the goal of anthropological fieldwork is not so much to test, in a statistical fashion, whether these theories are correct; rather it is to go into the field *with these theories* and try to discover where they lead, and where they need to be refined or abandoned. That is to say, anthropologists generally are not interested in approaching the question of the impacts of science on society or vice versa independent of the theory (as a practice of statistical confirmation or disproval), but with theory-in-hand, as a practice of situated, embodied re-thinking of these theories – and this is what ethnographic fieldwork provides.

As this article suggests, one might begin this practice simply by landing in a specific site – in this case CBEN – and through participation and observation, come to some kind of understanding of the role and meaning of the actions of individuals engaged in this site. But this is too simple – CBEN as a site is both too specific (it does not "represent" nanotechnology) and too diverse and heterogeneous (there are dozens of labs and hundreds of projects under its purview). It is necessary to make some choices about what kinds of participation and what kinds of observation will help distinguish the essential from the accidental, or the unique and interesting from the ephemeral and quotidian. At this point, some reflection on the methodological conundrums of anthropology is useful in order to understand why some people in CBEN and some activities may be more appropriate than others for observation.

To begin with, however, it should be made clear that there are (at least) two distinctive kinds of activities that happen in CBEN, which it is necessary to describe in more detail. First, there are of course a number of different scientific experiments running at any moment – experiments expected to produce results related to nanotechnology, biology and environmental engineering and specifically to the use of nanomaterials for biological and environmental purposes or for hazard, exposure, or risk (including perception of risk). These experiments range from long-term investigations that aim at understanding the fundamental properties of new nanomaterials like buckyballs and quantum dots to relatively short-term experiments aimed at filling in gaps in knowledge to creative forms of analyzing existing data in order to answer open questions. Usually this activity includes senior and junior scientists, post-docs, grad students and undergraduates, technicians and staff. Often they occur between departments, and include faculty from diverse fields (always on the grants, occasionally in the lab meetings, sometimes on the published papers). Senior scientists, post docs and grad students do varying amounts of paper-writing and travel to conferences to promote and share results. Oversight from CBEN or the university is at a minimum, though it claims credit for the results (http://cben.rice.edu/research.cfm?doc_id = 5012).

Second, there is the production of ideas *about* these experiments. CBEN actively promotes the direction of, sifting of, discussion and promotion of ideas, big and small, about the activities of CBEN and nanotechnology. Some are ideas about applications and implications of research results, some are about the clarification and dissemination of results, some are ideas that themselves need to be researched and confirmed or disconfirmed. In short, it is a kind of free form, intensive hypothesis-generation activity (and hence, quite similar to the normal pursuits of the anthropologist, albeit in a different substantive context). It can also include ideas about the creation of standards, rules, voluntary ethics, and objectives for promoting these kinds of experiments; creation of institutions to do same (e.g. ICON); promotion, outreach, media management, reports and conferences, meetings, conversation and reading, etc. All of this is conducted in the context of a heavy emphasis on the definition of nanotechnology as a field, and in CBEN's case, a field that includes research on human health and environment.

There is of course a spectrum between these two kinds of activities. The dominant mindset of most participants in CBEN and CBEN-like institutions is a techno-methodological one, in which all questions (even those about objectives and goals) are best answered through some kind of scientific method, not through philosophical reflection or democratic deliberation. If members of CBEN suggest that, for instance, it is important to know whether the public knows anything about nanotechnology, this hypothesis is seen as something requiring (and worthy of) research funding in order to answer definitively whether or not the public has any knowledge of nanotechnology. It cannot stand as an assertion based on anecdote or observation, regardless of how extensive an individual's experience might be. However, as a result of this mindset, there emerges a very small, and very privileged sphere, within which it is in fact possible for certain members to speak with consequence about "nanotechnology" without needing to find some kind of demonstration or proof of an idea. This activity therefore looks more like the second, the production of ideas about nanotechnology, than the first, the experimentation and testing of issues related to nanotechnology. Obviously certain realms (such as the potential benefits of nanotechnology or issues perceived as social science questions) are easier to speak about without demonstration or proof than others.

Anthropologists interested in making sense of these two activities are confronted with a problem: these actors themselves appear to be interested in answering questions about the social, organizational, and cultural aspects of nanotechnology – what it is, what it could be, how it is related to people, institutions, societies, markets, and how the cultural authority of science can be made more responsive, responsible or ethically attuned. That is to say, a central preoccupation of at least the second form of activities engaged in at CBEN concern precisely the relationship of science and scientific knowledge to social order. These are questions anthropologists ask regularly, across the various subfields of the discipline, but they are here asked not by anthropologists or other observers (e.g. philosophy or sociology of science or science studies) but by the actors themselves, the putative objects of our analysis. And as I have suggested, they too find at least partial solace in the wealth of existing theorizations of a changing relationship between science and society, which help them make sense (to themselves) of their own actions and goals. Most scientists do this kind of "informal" research without recognizing that it is a form of research that could be conducted more "formally" by anthropologists, or science studies scholars.

For many anthropologists, this surprising recognition has become a theoretical and practical problem – most starkly evidenced by the question "what exactly to anthropologists *add* to this practice?" The remainder of this section reviews four modes of answering this question, four proposals for thinking about where, why and how to do fieldwork in emergent fields like nanotechnology.

First off, there is the by now classic genre of the laboratory study, made popular in the late 1970s and early 1980s (Latour & Woolgar, 1979; Latour, 1987; Traweek, 1988; Collins, 1985; Pickering, 1992). Philosophers, sociologists and anthropologists, for various reasons have found themselves working amongst scientists, attempting to answer questions about their practice and about the vagaries of a sociology, or anthropology, of (scientific) knowledge (Fleck, 1981; Mannheim, 1955). In this frame, only the first of the two kinds of activities that occur in CBEN is explicitly observed – that of the scientific experiment itself. The anthropological "result" of such studies can vary depending on approach, from claims about social construction, to a focus on meaning and metaphor, to a concern with tacit knowledge, to a focus on lab infrastructure. A few studies in this area, especially that by Latour and Woolgar, recognize that both kinds of activities (the pursuit of experiments and the creation of ideas) occur in particular labs and institutions – especially in the distinction between the senior lab scientist (whose primary work consists of travel, grant-getting, idea-generating and credit-building) and the junior level scientists, postdocs, grads students and technicians who perform the daily work of science.

Second, Holmes and Marcus (2005) have recently proposed the notion of "para-ethnography" as a way of capturing the fact that many of the activities that people (such as those involved in CBEN) pursue are strangely similar to, and happen in the same times and spaces as those that anthropologists pursue (Holmes and Marcus study experts at the Federal Reserve and European Central Bank, and for them, high-profile persons like Alan Greenspan are "para-ethnographers" par excellence). That is to say, outside of the laboratory experiment, experts are involved in conferences and meetings, institution building, grant getting and organizing, conversation and reading, batting ideas around, focus groups and outreach, anecdotes and stories, public speaking, media relations, policy studies and promotion, testimony to congress – all things that anthropologists have often found themselves doing as part of their research. The "para-ethnographer" is the anthropologist's uncanny double, when the activities they engage in include investigation, reading, interview, survey, taking notes, offering critique and rethinking, etc.

CBEN researchers, when they engage in the second kind of activities listed above (and this is especially true of the three main directors of the center), might be said to be doing something like fieldwork – but for putatively different purposes. It is precisely those people within CBEN granted the ability to reason about what nanotechnology is or could be (without needing to find or fund scientific demonstrations or proofs to do so), who would be considered the para-ethnographers in this framework. The implication of such an approach is strictly methodological – it suggests "where ethnography might literally go in fieldwork" by seeking out the sites of the anecdotal, the deliberative, or the non-technocratic forms of reasoning that nonetheless remain most powerful within highly rationalized and technocratic discourses (Holmes & Marcus, 2003, p. 241).

In the case of CBEN, the ability to pronounce on nanotechnology takes a curious form: the directors are usually extremely careful when they speak about the risks of toxicity and exposure – and stick very closely to the available scientific literature. However, when they discuss the definitions, possibilities and potential benefits of nanotech, they are much freer in their discussion and explanation. The dominant language of risk assessment is an obvious outcome of this disparity – but risk assessment applied only to the potential costs, not the benefits of nanotechnology. For the anthropologist, or science studies scholar, the interest is precisely in identifying where these gaps are productive: the places where nanotechnologists can effectively speak about nanotechnology in an anecdotal frame, and where they are forced to adopt a technical-rational one. Holmes and Marcus caution against becoming too embroiled in the technical-rational discussions, because they fear that by doing so, the work of anthropology will be rendered inaccessible and irrelevant to anthropology itself. They suggest that the core practice of offering critique and discussion relevant across the diverse anthropological fields is the first priority. They nonetheless recognize that it is the informants (such as those at CBEN or in nanotechnology generally) who will be most likely to understand and profit from a creative critique or rethinking of the relationship of science and society, or of the growth of institutions and ideas related to emergent sciences.

A third and similar approach to the para-ethnographic might be that advocated by Paul Rabinow, apropos of Niklas Luhmann: that of the "second order" observer of "first order" observers of society (Rabinow, 2003). This approach intends to lend some distance to both kinds of activities CBEN carries out, and to forestall some of the more immediate forms of criticism that anthropology or science studies might be tempted to immediately engage. Instead, it suggests a kind of provisional functionalism – an observational mode that is concerned with the goals and objectives of institutions such as CBEN or ICON, both manifest and latent, but which does not seek a strictly Weberian typology – precisely because second order observers are always someone else's first-order observers. What is interesting here is the comparison of such an approach applied to the first and second kinds of activities listed above. In the case of scientific experiments (which can also include social science research) the observation of the world is conducted in a strictly rationalized scientific manner (even if there are gaps and leakages of all kinds throughout the work, it is still organized within a clearly defined methodology and classification). The activity of the second kind, however, is much more mysterious, ad-hoc, uncoordinated, and unstructured. It is more than likely that the actors themselves do not know exactly what kinds of activities they engage in. All the more so in the field of "nanotechnology" where the wealth of complex interactions, the constantly changing goals and demands, and the tricky incompatibilities of disciplines and histories intertwine. It might be the case that CBEN is observing itself more carefully than anything soundly external to it. What "second order observation" seeks to add to this series of methodological possibilities, however, is the provisional nature of this distancing. Anthropologists need not relinquish the ability to speak with authority about nanotechnology and society, only to take a provisional stance of "adjacency" by which the perspectives – ostensively differently informed and more capable of critical analysis – can be compared with those of the 1st order observers (who may also be anthropologists or sociologists) of nanotechnology and society.

A fourth approach might be the more general field of an "anthropology of intellectuals" in which the old Weberian question of vocations are asked anew, or the more familiar sociology of knowledge/critique of ideology traditions are brought to bear here (Weber, 2004). A twist on this tradition is captured by Thomas Osborne (2004) in the notion of a "mediator" – that is, a kind of intellectual who is neither a "public intellectual" (e.g. talking head, policy advocate or activist) nor strictly speaking a scientist engaged in pure or basic research, but an individual interested in moving ideas from one sphere to another, or in creating the right environment (in terms of media, opinion, and scientific data) for certain ideas (Osborne, 2004). Here the activities of CBEN (of the second sort) might be made analogous to the activities of think tanks (such as the Brookings Institute or the Heritage foundation), even though their explicit goals are rarely to influence government policy or law-making. The promotion, direction, management, public facilitation of research ideas (if not results) is what "sets the stage" for the kinds of questions that get asked in labs and field studies. Seen in this light, the activity of CBEN is largely directed at making CBEN's definition of nanotechnology into something that drives research questions and experiments beyond what it funds itself. To a large extent, ICON can be seen as a way to promote this activity in an even more independent "stake-holder"-oriented manner.

These four methodological approaches are not mutually exclusive – but they do give an indication of one of the trickier aspects of anthropological research amongst scientists and experts. Are anthropologists capable of observing and participating in projects such as CBEN with sufficient detail and depth without sacrificing the ability to offer independent critique and re-thinking? How does one avoid, on the one hand, becoming co-opted into ("going native" in an older idiom) the project of such centers and research initiatives, and on the other, avoid becoming completely irrelevant through the willful attempt to maintain distance? In what ways can the practices and research of anthropologists, which are often deliberately unconventional in style and critical in philosophical terms, become part of the practice of organizations and centers where experts consider issues of the relationship of science and society?

Allotrope: Substance

Methodological questions such as those raised above are most often a concern because of the clear disjunction between the subject matter of anthropology and that of its observed expert or elite subjects. Nanotechnology, for instance, appears on the surface to have almost no conceivable overlap with the putative subjects of study in anthropology. Particles of carbon that are less than 200 nm in size seem at first blush to be as far as possible from, for instance, issues of consanguinity in kinship or symbolic meaning in a religious ritual. In large part this is true, but it is nonetheless quite easy to identify some issues of technical and scientific detail that are related more or less directly to those of anthropological concern. In the case of CBEN, this issue is made all the easier by virtue of its explicit focus on the environment and human biology. Toxicity, environmental hazard and the potential for remediation or prevention through nanotechnology are all issues that have eminently social and human dimensions, and might well connect easily with the practices of anthropologists working in medicine and the environment.

For some scientists in CBEN, these same questions are also salient – but the normal organization of scientific and engineering research and funding does not facilitate their asking. An institute like CBEN is a rarity therefore, largely because the kinds of disciplinary questions that normally seen so central are subordinated to a set of concerns about shared objects: in this case human biology and the environment. This does not, however, mean that the work of scientists and engineers will naturally pursue these objects in some synthetic fashion – hence the need for both scientific work, and the work of producing ideas about it that CBEN directors participate in. Take for example, CBEN co-founder and environmental engineer Mark Wiesner.

Wiesner is an unusual scientist. A graduate of John's Hopkins Geography and Environmental Engineering department, Wiesner came to Nanotechnology accidentally (in much the same way the author did, largely because there was such a heavy emphasis on nanotechnology on the Rice campus), but it has created for him a completely novel approach, unusual in his own discipline and practice, as the opening epigraph illustrates. Wiesner's approach to environmental engineering has been transformed by nanotechnology, and in particular by the worldview in which nature is seen as inherently engineerable, from the bottom up. The idea of making pure water from scratch, as opposed to cleaning it of toxins and other harmful materials, is a nano-inspired mode of thinking – not a standard approach amongst environmental engineers.

However, for Wiesner, nanotechnology is not necessarily a revolutionary interdisciplinary science, but it is a field within which questions that have plagued environmental engineers and their precursors become exciting and novel once again.

> Q: Does nanotechnology in your lab provides the sense that you're at the cutting edge of something, you're the leading edge, you're in the process of discovery?
>
> A: Mmmm, I would say that it's not nanotechnology per se that has been a big stimulus in the group. I'd say that right now the most exciting sort of, you know, fast changing stuff that's going on has to do with understanding the environmental uses and properties of a new material, which is the fullerenes. All the other stuff is sort of interesting, but, um, it's been around in different forms. And, although we have some ideas about what we

think is unique... I still think that right now it's just, well, nobody's ever looked at, you know, what a fullerene can absorb or how it interacts with water or why it structures itself in this way with other fullerenes.

Q: So that's the exciting part is these new materials

A: [Yeah] And people see it as the material that's gonna have important economic uses. And so, you say, well this is important stuff to work on because we're not— it's not like plastics that are everywhere right now—but it could be. And so, this really is going to be a material that we need to understand, you know we need to...if we know something about it now it's gonna make a difference for the future.

Getting to the point where he could study the interaction of fullerenes and water, or think about the properties, hazards or uses of fullerenes, however, has not been an easy task. Wiesner's involvement in CBEN was largely serendipitous: he was the one environmental engineer at Rice who seemed like the sort of person who might be interested in writing a grant to the NSF – and so was invited to do so by Vicki Colvin. And though the process was difficult, his participation and his insistence on studying the potentially dangerous aspects of nano-materials proved crucial to the funding of CBEN, as he explains:

When CBEN was finally funded, in December 2001, this was really, I believe, the first time anyone had articulated publicly the issue of not only how you can use nanotechnologies to do good things for the environment, but also what are the implications of these nanomaterials for the environment? ... Another interesting aspect of that whole process was that ... the EPA at the time was funding things like how can you do nanotechnology to clean up the environment? or how can you use nanotechnology to treat drinking water? But the implications thing hadn't come through, and so when we were writing a proposal, Vicki's original idea was that she wanted me to group together people that were going to do technologies along that line, where we would be using nanotechnology for good [laughter] And every time I'd write the evil part [laughter]... It'd keep getting kicked back, and I'd say well we didn't have enough room, I'll cross that out. And it was really difficult to keep [the issue of the implications of nanotechnology] in the proposal. But it finally went through, and by talking with one of the people that was actually on the review panel, who's now a faculty member in this department, I came to understand actually even at that time when we went in for the visit, that that aspect of it was really one of the key things—along with having a Nobel prize winner—that really set us apart, and one of the key things that got anything funded. So, it was very controversial to have it accepted, and sort of an afterthought, but it turned out to be really important in getting it funded. And the controversy didn't stop after it was funded. I mean, it's continued to be something that people don't really know what to do with it.

The trade-off visible in CBEN's self definition – as an entity that can pursue research both on the potential dangers of nanotechnology and the potential uses of nanotechnology "for good" stem from the confrontation of one set of research concerns, represented by Wiesner (the interest in the environmental properties of fullerenes, such as their mobility in water) and those of another, represented by chemists like Richard Smalley (research into the synthetic and creative possibilities of fullerenes). This confrontation results in the peculiar and unique definition of nanotechnology promoted by CBEN. For Wiesner, such research is in fact fundamental research – not the kind of thing that is subsequent to research on new applications, materials or possibilities, as he makes clear:

One of the comments that came back on the proposal review for the Center, was "this is all too premature; this is all premature, we've got no nanochemistry industry, we have no idea what this stuff is. Why would you want to study this now?" And, the response to that is if you don't do it now a) when are you going to do it? And b) anything you do learn now, it's like a spaceship heading toward the moon, you know if you alter the trajectory just a little bit at the beginning it has huge impact down the line. And so, it's a very powerful time, even if you're not getting the full picture, to try to gather any information you can.

Perhaps ironically, even though Wiesner has fought to make the study of hazard and toxicity a core aspect of basic research in nanotechnology, his own research has not focused solely on this issue. Indeed, Wieser is one of the researchers who has most diligently pursued projects that fulfill the other side of the bargain – the desire to create novel nanomaterials that can be used in environmental engineering. Prior to the creation of CBEN, Wiesner had long been involved in collaboration with Andrew Barron on the creation of new kinds of nano-materials that could be used as membranes for filtration (specifically alumoxanes and ferroxanes). Wiesner's earlier work, before arriving at Rice had focused primarily on membrane science, but not with nano-materials, so Wiesner's collaboration with Barron allowed the two to explore areas of mutual interest that made use of new techniques and a new interest in nanoscale materials. For Barron, understanding the chemistry and synthesis of alumoxanes proved a challenging basic chemistry problem, while for Wiesner the potential use of such materials in order to filter fine particles from water proved a novel re-imagination of what environmental engineering could achieve in the area of Nanotechnology.

The approach of creating a membrane through the synthesis of nanomaterials requires something like a "nano worldview": rather than take existing materials and break them down into component parts and smash them together to form a membrane, Wiesner and Barron effectively created a material from scratch – a material whose properties and behaviors were more accurately theoretically understood and which could be explored using the visualization tools of nanotechnology (STM/AFM, etc.), as well as the more conventional testing of materials conducted by environmental engineers. Furthermore, because the materials are so well understood, they led the two scientists to imagine other membrane-like uses, such as the creation of alumoxane and ferroxane fuel cells – something seemingly far afield from the concerns with filtration of water.

Conclusion

But how are water filtration and fuel cells also anthropological problems? How can fieldwork amongst scientists and CBEN directors be understood critically as a practice that may have wider implications for an anthropological or social theoretical understanding of science and society?

From one angle, the success of the research and technologies pursued by CBEN rests primarily on decisions that will be made elsewhere and by other people about the value of the environment, or the value of human safety – Wiesner and Barron's work is understood to be a demonstration of certain technical possibilities that are intended to influence those decisions, whether promoted directly or not. Human safety, in terms of water quality, is demonstrably possible, as are cleaner forms of energy production with fuel cells (which also have the production of pure water as a by-product). From this angle, the changing cultural authority and power of science are obvious subjects of anthropological interest. To whom are these inventions and discoveries meaningful? To whom are they promoted and for what purposes? Who is expected to understand the potential of such work, and what forms of rational response are expected, either by the scientists themselves or by decision-makers or planners?

Wiesner himself, and many other scientists are acutely aware of the fact that merely demonstrating the possibility of human safety or energy efficiency are rarely sufficient to change any real practices in the world. Hence, an institution like CBEN does more than promote basic research – it can generate and promote ideas about this research that might set the stage for real changes. Whether or not it is more or less successful for doing so is one strong reason to observe it in action, as it makes the attempt; perhaps it means participating with these "para-ethnographers" as they go about investigating the values and practices that will help promote or deter what are understood to be better, safer, cleaner, or more feasible technical solutions; perhaps it means adopting a "second-order" point of view, in order to get at what these first order observers miss.

Yet seen from another angle, the creation of nanotechnology might simply serve the incessant demand for novel technologies, regardless of their effect on human health and the environment – that is, solely in order to satisfy certain economic demands in which growth, productivity, national or regional development, or international stature are tied to constant scientific and technical breakthroughs. In this light, it does not matter what kind of values the work of scientists and engineers are intended to serve – rather they are only valuable insofar as they can be translated into the terms of productivity and growth demanded by financial markets and businesspeople. CBEN might simply be a more efficient way to try out the maximum number of different routes, in a rush to identify profitable technologies. Even so, the critical approach of the anthropologist who can observe the constant interactions in such a setting can offer a critical view on this kind of relationship.

CBEN, and the scientists funded by it thus face a kind of puzzle concerning the cultural authority of science – and it is a puzzle that in some ways mirrors the puzzle faced by anthropologists who wish to study elites, experts or high-tech organizations and practices. It is a puzzle about the nature of the relationship between a critical scientific project – one in which the directions and values that underlie the practices are implicitly focused on issues of improving human health and safety through the investigation of alternatives to existing technologies, materials and chemicals.

CBEN offers a framework within which those pursuits are granted a provisional legitimacy, and a wider cultural authority than any single scientist might possess. And yet, in order to do this, it is necessary to risk co-optation, to risk creating new nanomaterials that may or may not be safe, may or may not be understood, and may or may not be responsibly used once discovered. Such scientific pursuits cannot be conducted solely in the voice of critical, defensive or precautionary research, that is to say, CBEN cannot simply be an institution that researches the potential risks and dangers of nanotechnology, but must also participate in the search for the novel and the undefined. This is not simply because CBEN and scientists serve other masters – whether the NSF or corporate dollars – but because this is in fact where the action is. The understanding of new materials and the invention of new uses and application is what gives the vocation of nanoscience and nanotechnology much of its momentum. Without this action, it becomes the mere bureaucratic accumulation of facts, devoid of the calling Weber so clearly identified as an essential component of the scientific mind. To ask scientists to occupy only the position of a caution is to prevent them from defining, contesting or arguing for the values that attach to new discoveries, new applications, or new materials, and to deny them the satisfaction of knowing that science can be used for good, as well as for evil.

By formalizing the insights that anthropology brings, and indeed, by making the "para-ethnographic" work of participants visible and tangible, anthropological fieldwork can help make good on this critical promise, if it is willing to take a parallel risk of participation. It is a common theme in the explanation of nanotechnology to insist that it is involved not only in an attempt to understand nature but to offer some kind of control as well; the same might be true of anthropology in this instance, it is not only an attempt to understand the social relations of science and implications of science, but to find novel ways of controlling them as well. Investigating the properties of the allotropes of fieldwork is a contribution to this endeavor.

References

Baum, R. (2003). Drexler and Smalley Make the Case For and Against 'Molecular Assemblers'. *Chemical and Engineering News*, 81(48), 37–42, available at http://pubs.acs.org/cen/coverstory/8148/8148counterpoint.html

Beck, U. (1992). *Risk Society: Towards a New Modernity*. Trans. Mark Ritter. London: Sage.

Beck, U., A. Giddens, & S. Lash (1994). *Reflexive Modernization – Politics, Tradition and Aesthetics in the Modern Social Order*. Cambridge: Polity Press.

Callon, M. (1986). Some Elements of a Sociology of Translation: Domestication of the Scallops and the Fishermen of St Brieuc Bay, in J. Law (ed.), *Power, Action and Belief: A New Sociology of Knowledge*. London: Routledge & Kegan Paul.

Collins, H. M. (1985). *Changing Order: Replication and Induction in Scientific Practice*. Beverley Hills, CA/London: Sage. [Second Edition, with a new Afterword, Chicago, IL: University of Chicago Press, 1992.]

Colvin, V. (2003). The Potential Environmental Impact of Engineered Nanomaterials. *Nature Biotech*, 21(10), 1166–1170.

Drexler, K. E. (2004). Feynman to Funding. *Bulletin of Science, Technology and Society*, 24(21), 21–27.

Dupuy, J.-P. (2004a). L'irréalité de l'avenir et l'impuissance de l'éthique. Cas des nanotechnologies. In G. Nivat (éd.), *Les limites de l'humain. 39èmes Rencontres Internationales de Genève*, L'Age d'Homme, Genève, pp. 115–138.

Dupuy, J.-P. (2004b). Pour une évaluation normative du programme nanotechnologique. *Annales des Mines* (February 2004), 27–32.

Etzkowitz, H. & L. Leydesdorff (1995). Emergence of a Triple Helix of University-Industry-Government Relations. *Science and Public Policy*, 23, 279–286.
Etzkowitz, H. & L. Leydesdorff (eds.) (1996). Universities and the Global Knowledge Economy: A Triple Helix of University-Industry-Government Relations. Book of Abstracts. Amsterdam: University of Amsterdam, 169 pp.
Fleck, L. (1981). *Genesis and Development of a Scientific Fact*. Chicago, IL: University of Chicago Press.
Gibbons, P., C. Limoges, H. Nowotny, S. Schwartzmann, P. Scott (1994). *The New Production of Knowledge: The Dynamics of Science and Research in Contemporary Societies*. Thousand Oaks, CA: Sage.
Gupta, A. & J. Ferguson (1997). *Culture, Power, Place: Explorations in Critical Anthropology*. Durham, NC: Duke University Press.
Gupta, A. & J. Ferguson (2002). *Anthropological Locations: Boundaries and Grounds of a Field Science*. Berkeley, CA: University of California Press.
Holmes, D. & G. Marcus (2005). Cultures of Expertise and the Management of Globalization: Toward the Re-Functioning of Ethnography, in A. Ong & S. J. Collier (eds.), *Global Assemblages: Technology, Politics, and Ethics As Anthropological Problems*. Oxford: Blackwell, pp. 235–252.
Jasanoff, S. (ed.) (2004). The Idiom of Co-Production. *States of Knowledge: The Co-production of Science and the Social Order*. Routledge: London.
Jasanoff, S. (2005). *Designs on Nature: Science and Democracy in Europe and the United States*. Princeton, NJ: Princeton University Press.
Khushf, G. (2004). A Hierarchical Architecture for Nano-scale Science and Technology: Taking Stock of the Claims About Science Made by Advocates of NBIC Convergence. In D. Baird, A. Nordmann, & J. Schummer (eds.), *Discovering the Nanoscale*. Amsterdam: IOS Press, pp. 21–33.
Lash, S. & J. Urry (1994). *Economies of Signs and Space*. London: Sage.
Latour, B. (1987). *Science in Action: How to Follow Scientists and Engineers Through Society*. Cambridge, MA: Harvard University Press.
Latour, B. & S. Woolgar (1979). *Laboratory Life: The Social Construction of Scientific Facts*. Los Angeles, CA/London: Sage.
Law, J. (1987). Technology and Heterogeneous Engineering: The Case of Portuguese Expansion, in W. E. Bijker, T. P. Hughes, & T. J. Pinch (eds.), *The Social Construction of Technological Systems: New Directions in the Sociology and History of Technology*. Cambridge, MA: MIT Press.
Mannheim, K. (1955). *Ideology and Utopia: An Introduction to the SOCIOLOGY (740) of Knowledge*. New York: Harvest Books.
Marcus, G. (1995). Ethnography in/of the World System: The Emergence of Multi-Sited Ethnography. *Annual Review of Anthropology*, 24, 95–117.
Marcus, G. (1998). *Ethnography Through Thick and Thin*. Princeton, NJ: Princeton University Press.
Marcus, G. & M. Fischer (1986). *Anthropology as Cultural Critique: An Experimental Moment in the Human Sciences*. Chicago, IL: University of Chicago Press.
Nowotny, H., P. Scott, et al. (2001). *Rethinking Science: Knowledge and the Public in an Age of Uncertainty*. Cambridge: Polity Press.
Osborne, T. (2004). On Mediators: Intellectuals and the Ideas Trade in the Knowledge Society. *Economy and Society*, 33(4) (November): 430–447.
Pickering, A. (1992). *Science as Practice and Culture*. Chicago, IL: University Of Chicago Press.
Rabinow, P. (2003). *Anthropos Today: Reflections on Modern Equipment*. Princeton, NJ: Princeton University Press.
Regis, E. (1996). *Nano: The Emerging Science of Nanotechnology*. Boston, MA: Back Bay Books.
Shapin, S. & S. Schaffer (1989). *Leviathan and the Air-Pump*. Princeton, NJ: Princeton University Press.
Stone, J. (n.d.) Anthropology and the Human Dimensions of Nanotechnology.... Manuscript, presented at the (2005). American Anthropological Association, on file with author.
Sturgis, P. & N. Allum (2004) Science in Society: Re-evaluating the Deficit Model of Public Attitudes. *Public Understanding of Science*, 13(1), 55–74.

Toumey, C. (2005). Apostolic Succession. *Engineering and Science*, 68(1, 2), 16–23, available at pr.caltech.edu/periodicals/EandS/articles/LXVIII1_2/Feynman.pdf

Traweek, S. (1988). *Beamtimes and Lifetimes: The World of High Energy Physicists*. Cambridge, MA: Harvard University Press. Paperback edition published in 1992 and reprinted in 1995.

Weber, M. (2004). *The Vocation Lectures: Science as a Vocation, Politics as a Vocation*. Edited and with introduction by D. Owen and T.B. Strong Tr. R. Livingstone. Indianapolis, IN: Hackett publishing [1917].

Wiesner, M., with M. M. Cortalezzi, J. Rose, G. F. Wells, J. Y. Bottero, A. R. Barron (2003). "Ceramic Membranes Derived from Ferroxane Nanoparticles: A New Route for the Fabrication of Iron Oxide Ultrafiltration Membranes." *Membrane Science*, 227, 207–217.

Interviews

Interview with Mark Wiesner, April 12, 2004, Rice University.

Law, Regulation and the Medical Use of Nanotechnology

Kenneth A. DeVille

Abstract While protection of society and individuals is essential, it is also important that legal and regulatory measures applied to nanotechnologies in medicine are rational and do not unnecessarily blunt the proliferation of valuable modalities. In most respects the existing U.S. legal and regulatory structure is flexible and sufficiently adaptive to accommodate the challenges posed by the use nanotechnology in medicine. In product liability and medical malpractice in particular, currently existing theories and approaches should apply to this revolutionary technology without fundamental revision. The regulatory and approval structure, too, should be mostly serviceable as well. However, the potential of biomedical nanotechnologies to have a wide range of post-approval, unintended and un-discoverable consequences pose a special challenge. As a result, an integrated, national post-approval registry tracking the use of nanotechnologies should be created. Because different nanosubstances used for different purposes may have similar ill-effects, it is important that the registry utilize a centralized repository in which all medical uses of nanotechnologies can be analyzed across uses and technologies and uncover pitfalls and limitations in the shortest possible period.

Keywords FDA; regulation; product liability, medical malpractice; post-approval registries

Nanotechnology is the label attached to a constellation of techniques that allow scientists and engineers to arrange and manipulate organic and inorganic matter on an atom-by-atom basis to construct devices or compounds that can perform specific functions. Nanotechnology manipulates substances at the scale of 100 nm (1/100,000,000 mm) or less and can rely on computer software to create nano-sized "molecular constructors" that can build unlimited copies the desired substance or mechanical tool using easily available elemental raw material (Masci, 2004). Awide range of potential medical applications are immediately apparent. Nanotechnology promises to provide the ability to create substances and devices with specified tasks with a size that allows free movement anywhere in the human body to detect, prevent or eradicate disease. Some medical usages are already in development. Scores of others challenge the imagination.

Brody School of Medicine, East Carolina University

The excitement surrounding nanotechnology applications in medicine, or "nanomedicine," is not surprising. Currently identified medical uses for nano-sized devices and substances alone promise extraordinary benefits in life extension, relief of human suffering and economics. Future and more fantastic uses are possible. Despite these observations, Western society is renowned for embracing new technologies without sufficient consideration of the potential consequences of the implementation (Jacobson, 2005). This practice is aggravated by the fact that medical technologies are invariably accompanied by unintended consequences and unforeseen limitations, many which pose substantial problems and even substantial harm for individuals or society. Therefore despite the genuine promise of nanomedicine, it is vital to forecast whether current legal and regulatory safeguards are sufficiently flexible and responsive to prevent, remedy and compensate for the foreseeable as well as for the unforeseeable consequences of this revolutionary technology. At the same time, unfounded or excessive fears can alter how nanotechnology is addressed in legal and regulatory contexts. Distorted legal and regulatory agendas can delay the development, use and proliferation of new and beneficial modalities and deny society the benefits of a promising technology.

After identifying a number of prospective medical applications of nanotechnology, this essay will consider whether its proliferation will require fundamental changes in existing drug and device regulation or common law liability doctrine and procedure. While most areas of the currently existing legal system are adaptive enough to accommodate the likely challenges of nanomedicine, some additional regulatory safeguards will be required to ensure that unintended and unforeseeable consequences of this new and powerful technology do not cause unnecessary and irreparable damage to individuals or society. While these few, new regulatory requirements will add to the cost of the technology, they have the advantage allowing proliferation and use of exciting and potentially beneficial modalities as long as their safety and efficacy have demonstrated by currently existing standards.

Medical Uses of Nanotechnology

A brief summary of the range of projected medical uses of nanotechnology affirms the excitement that accompanies its development. The use of silver nanoparticle-based coatings for infection control on anesthesia catheters is already under development as are brush-shaped gold and carbon nano-brushes that can be used to coat minute internal structures with beneficial substances or to clean unwanted deposits from veins and arteries (FDA News, 2006; Bourne, 2005; Pease, 2006). Nanodevices might supplement other current larger "micro devices" to serve as biosensors to detect such things as trace quantities of bacteria, disease signatures, and abnormal vascular pressure. In addition, nano-technology is expected to provide a wide array of ways to delivery drugs to affected cells. For example, thousands of injectable and coated nanospheres might circulate through the body, bond to affected cancer or other diseased cells, and release directed drugs in response to a radio or infrared transmission from the operator (Merkle, 2006). In this way drugs are protected in transit, healthy tissues are protected from the toxic drug, and the affected cells receive a premium dose of the medication. Nanosensors and nanosphere-drug delivery systems might be combined in order to monitor continuously blood glucose and be programmed to signal the release of the precise amount of medication at the most appropriate moment. Using carbon-based materials, nano-sized molecular constructors might produce biocompatible materials in vivo to repair bone and other human tissues, for example, nano-fiber mesh for the buttressing of heart or other tissue. Nanomolecules may eventually serve as a supplemental means of carrying oxygen to various parts of the body in patients with compromised circulation (Flattum, 2006). Nanodevices might be used for various forms of detoxification or clot removal. For example, specifically produced nanospheres containing a specified receptor would be introduced into the body. The nanospheres would circulate allowing toxins to bind to the spheres. After circulation, "an extracorporeal magnetic filter is tapped into a suitable artery…this filter magnetically holds the nanospheres carrying the toxins" while, unlike devices such as dialysis, the clean blood is never removed from the body (Sidawi, 2004).

The Risks of New Technologies

While the medical community sometimes errs in its delay in acceptance of a beneficial technology – the use of computer-assisted prescription programs for example – a more common phenomenon drives clinicians, patients and institutions to adopt medical innovations too soon, or before the technologies' limitations have been fully identified (DeVille, 2000; Kacmar, 1997). The so-called "technological imperative" is a short-hand description of the set of cultural factors in advanced countries, especially the US, that encourage the early and widespread use of new technologies, especially in medical care (Igelhart, 2001; Rothman, 1997). A number of motivating factors contribute to this tendency including physician and patient preference, a sincere desire to improve the quality of care, professional and institutional competition, and economic dynamics that frequently favor the adoption of newer technology. Whatever the predominant explanation, Western culture more frequently favors the rapid proliferation of medical technology to the slow, measured and deliberate use of innovations in health care (Grimes, 1993, p. 3030). Given the range and degree of potential benefit from nanomedical applications, and the anticipated financial windfall associated with those applications, nanotechnology is likely to follow a similar course and proliferate rapidly into everyday clinical use soon after development. Rapid proliferation and use raises the obvious specter of hidden and undiscovered risks as the technology is used more frequently on more individuals.

Despite the risks sometimes associated with the so-called technological imperative, it is important to acknowledge that there are two types of errors that can be made regarding technology, law and regulation. The first category of error, of course, is the premature adoption and use of a new modality. The second is the suppression and developmental delay of a medical technology through excessive and unnecessary legal and regulatory oversight. In the first case, individuals are harmed as unanticipated risks come to fruition during widespread use. In the second case, individuals and society are harmed because an unjustifiably risk-adverse stance in law and regulation limits access to beneficial options (Engelhardt & Jotterand, 2004). Both categories of error cause harm. It is a difficult balancing act but one that must be sensibly navigated when discussing the interaction of law, regulation and technology.

The FDA and Nanotechnolgy

In the U.S. drugs and devices must receive approval by the Food and Drug Administration (FDA) before they can be employed by the general medical community. The approval process requires formal scrutiny through an application process and the evaluation of required data before approval is granted by the FDA. For purposes of FDA jurisdiction and review, a "drug" is defined as an article intended for the use in diagnosis, cure, mitigation, treatment, or prevention of disease, or an article, other than food, intended to affect the structure or function of the body (U.S.C., 1988a). The definition of a "medical device" is similar except that it does not achieve its primary purpose through chemical action or metabolization within the body (U.S.C., 1988b). Another way of stating the distinction is that drugs employ "chemical action" while devices employ "mechanical action" to accomplish their respective tasks (Fiedler & Reynolds, 1994). Some medical products are deemed "combination products" because they are drug-device combinations. Combination products are ultimately treated as drug or device depending on the primary mode of operation of the remedy. Medical nanoapplications will almost certainly be classified as a drug, device or a "combination product" and, unlike medical and surgical "procedures" will be subject to the FDA approval process before they can be used on the general patient population.

The determination of whether a medical nano-product is classified as a "drug" or "device" is an illustration of the type of regulatory challenge that lies ahead with nanotechnology. To restate, under FDA classification a "drug" is product that relies on chemical action; a "device" relies on mechanical action. But, as one observer has explained, "The tricky part is that at the atomic level, it becomes virtually impossible to separate 'mechanical' from 'chemical' or 'electrical' effects." For example, some nanoremedies rely on enzymes which employ both chemical and mechanical methods to affect molecular change. A "nanobot," in the form of an ultra-fine brush, might be designed to remove plaque from an artery. Because the ultra-fine brush appears to chips away the plaque by mechanical force, the product may appear to be a "device" for purposes of FDA regulation. But more analysis might be required. If the plaque is being removed in the form of individual cholesterol molecules or calcium atoms the essential mechanism of the product might be more ambiguous. The plan for some such ultra-fine brushes is create them from chemically "sticky" materials to attract and hold the molecules that make up the plaque. In this configuration, the nano-plaquebuster product appears to function more like a drug than a device because it relies on chemical action to remove the unwanted substances (Miller, 2003; Fiedler & Reynolds, 1994).

Consider another example. A pharmaceutical company develops an injectable nanosphere coated with a thin layer of gold for the treatment of cancerous tumors. The nanosphere core contains properties that allow it to identify and bond to cancerous tumor cells. When infrared light produced by a medical laser is aimed at the tumor area the light passes through healthy tissue but superheats the gold-coated nanospheres destroying cancerous tissue with no damages to healthy tissue (Kahn, 2006, p. 105). Consider further that the developer/manufacturer packages the design, software and manufacturing equipment into freestanding "nanofactories" that can produce these nanoparticles. A developer might decide to offer these "nanofactories" for sale to medical centers with sufficient facilities and engineering resources to produce their own nanoshells for clinical use. On one hand, the nanoshells themselves might be considered a drug because they use their detection, infiltration and bonding capabilities to target cancerous cells. On the other hand, one might argue that the therapeutic function of the nanospheres is mechanical in that it relies on convection heat to incinerate cancer cells. In the latter case the technology might be considered a device. As rejoinder, one could legitimately suggest that incineration at the cellular level is as a much a "chemical" reaction as it is mechanical – thus making the product a drug and not a device. The method of delivery of the product to the medical center for use might raise similar questions. If the manufacturer provided the medical center with nanospheres ready-for-use on cancer patients, then the product might be viewed as a drug depending on one's ultimate conclusion in the previously recounted dialectic discussion. However, could the packaged design, software and manufacturing equipment be considered a "device" because it is a machine designed to produce substances? These are intellectually challenging questions which will require perceptive judgments by regulators, but once resolved, will probably allow medical nanotechnology to be reviewed under the FDA's existing drug-device classification system. The discussion remains important, however, because the classification of a nanoproduct as a drug or a device may have high stakes.

The actual FDA designation – "drug" or "device" – has significant consequences for both manufactures and the public. Quite simply, the FDA standards and approval process for the approval of drugs is significantly more burdensome, time-consuming and scrupulous than that for devices. Products designated as "drugs" undergo far more careful pre-approval review, delaying public access to the remedies. In contrast, the designation of "device" means that the product will reach the bedside for clinical use in a much shorter period with less scrupulous review.

Applicants for new "drugs" must demonstrate the product's safety and efficacy. The FDA's multi-step approval process requires drug manufacturers to submit studies to the agency to evaluate the safety and efficacy of the product and can take several years. These studies include preclinical research, synthesis and purification, animal testing, and clinical testing on humans in randomized clinical trials (RCTs). RCTs are reviewed and approved by institutional review boards (IRBs) that ensure that human participants are fully informed and protected. Phase I trials involve a small number of human subjects and evaluates the safety of the product. Phase II studies involve a relatively small number of humans and are the initial tests of the effective of the drug or device. If Phase II studies are successful, Phase III studies are initiated and usually test the product on several hundred subjects. Double-blind, controlled studies are required with the study drug ordinarily tested against the then current standard-of-care remedy. All data is collected and presented to an FDA team of experts and an Advisory Committee to determine if the benefits of the drug outweigh its risks. If so, final approval is granted and manufacturing and distribution can proceed.

In contrast, the FDA approval process for pre-market approval of a medical "device" is less stringent. For device approval, the FDA requires that the manufacturer supply "reasonable assurance that the device is safe and effective" for its intended use. In contrast to the time-consuming three-phased system of RCTs that is required for drug approval, the FDA will accept any "valid scientific evidence" regarding the safety of the device. Valid evidence might include partially or non-controlled studies or any other reports of its safety. Less emphasis is placed on the efficacy of the device and there is no requirement that it be compared to currently available alternatives. The less meticulous approval process required for devices is reflected in many third-party payers' refusal to compensate for new device use until additional evidence of safety and efficacy are produced. In contrast, third party payers typically reimburse for FDA approved drug remedies.

On one hand, the FDA approval process is expensive, time-consuming and delays availability of the drug or device on the market. On the other hand, consumers are protected by the staged oversight and review before drugs or devices are released on the market. This tension is no different for nano-sized medical products than it is for other medical drugs and devices. Indeed, at this point, the FDA does not appear to view nanotechnology as a unique regulatory challenge. As an FDA document explains:

> Existing requirements may be adequate for most nanotechnology products that we will regulate. These products are in the same size-range as the cells and molecules with which FDA reviewers and scientists associate every day. In particular, every degradable medical device or injectable pharmaceutical generates particles that pass through this size range during the processes of their absorption and elimination by the body.

According to FDA sources, there are currently no reports of adverse reactions of re-absorbable drugs similar to nano-sized particles. Therefore, "[e]xisting requirements may be adequate for most nanotechnology products we will regulate" (FDA, 2006). The FDA has already approved a new "nanoparticulate drug", aprepitant (Emend®), designed to prevent delayed nausea and vomiting that occurs more than 24 hours after receiving treatment for chemotherapy. Emend®, was a new chemical entity and received approval based on complete studies for toxicity and safety suggesting that the FDA system of drug investigation and approval will adapt seamlessly to nanomedicine (Till et al., 2005).

A final FDA approval option may prove uniquely relevant to production and distribution of nanomedical products. Drug manufactures are not required to submit new applications to the full range of FDA testing requirements, including animal studies and RCTs if the manufacturer can demonstrate that a new drug application is "bioequivalent" to one that has already been approved (C.F.R., 2005). As mentioned early, one of the anticipated usages of nanotechnology will be to reduce known medications in size in order to target affects, decrease toxicity, reduce overall dosages, accelerate onset of action of the conventional medication. Tricor®, a nanoparticulate drug was approved by the FDA in late 2004. Tricor® was based on a previously approved nanoparticulate version of fenofibrate also produced by Abbot labs. Fenofibrate had already been subjected to the standard series of submitted, tests, RCTs and analyses by the FDA. Therefore Abbott had only to present studies showing that Tricor® had a matching drug release profile, i.e. bioequivalence with fenofibrate (Till et al., 2005).

This accelerated process of approval through the bioequivalence option profoundly decreased the amount of time for approval and expedited public access to a purportedly superior medication. The bioequivalence option, however, may be a doubled-edged sword. The expedited process of approval, and foregoing of RCTs and other scrutiny applied to other new drugs undercuts the ability of the FDA system of review to identify unknown risks or consequences that might be associated with the size and operation of nanoparticles. Although they too are limited, RCTs at least ensure a close, prolonged look at the affect of the medication on real patients. Using the bioequivalence option to bypass the longer approval process applied to new medications means that the potential mechanical effects of the new mode of delivery receive a different and lower level of scrutiny than do new drugs.

These comments should not suggest that nanotechnologies never be candidates for the bioequivalence option. Instead, FDA officials and other policy observers should be sensitive to the tradeoffs implicitly accepted when bioequivalence option is used in nanotechnology. Finally, it is fairly clear that drug manufacturers who produce nanoparticle duplicates of previously approved lager particle original drugs reap another benefit of this approach. Because the resulting nano-version of the medication is technically a new drug has been developed the sponsor may receive as much as a 5 year exclusivity protection that in addition to the patent protection they already possess on the source drug. But that underlying patent, too, might be eligible for a patent extension equal to the time lost in securing regulatory approval for the nano-version of the original drug (Till et al., 2005, p. 167). This observation does not raise questions about patient safety *per se*, does suggest that the public good might be undermine by a remedy that extends that time that the public has access to economic generic equivalents to the affected drugs.

The bigger concern, however, is that the current drug and device approval system conducts very little effective oversight and investigation beyond the initial round of laboratory research, animal studies and RCTs. The FDA investigatory and regulatory paradigm, as well as other aspects of the FDA's current mandate, appears to adapt well to nanotechnology applications in medicine. Disturbingly, however, FDA oversight is weakest in the very areas in which concerns regarding nanotechnology are most high.

FDA pre-approval studies including RCTs are designed to measure risks and harms that manifest themselves during the pre-approval process and are clearly related to the study item, drug or device. RCTs and other pre-approval are by necessity limited in time, scope and number and composition of patient population. To expand this portion of the drug and device approval process would dramatically increase expense and time and delay society's access to presumably beneficial products. The unjustified delay and diminution of available remedies can represent as much harm to society as the premature release of a drug or device. It is, in short, impossible to escape the fundamental trade-off between more and better testing, and delayed access to innovative drugs and devices – between the benefits of avoiding a drug disaster and the costs of denying beneficial therapies to patients. Therefore, although the current pre-approval FDA model is limited, it is, on balance defensible and should not be fundamentally altered to meet the new challenges posed by nanomedicine.

Instead, additional attention needs be focused on the oversight of nanomedicine in the post-approval period. It is in the post-approval period that regulatory scrutiny is most lax. More significantly, it is also the time that the unanticipated and longer-term impact of nanomedicine innovations are most likely to surface. Confirming this view, a 2006 study by the Office of the Inspector General (OIG) concluded that the "FDA lacks clear and effective processes for making decisions about, and providing management oversight, of postmarketing safety issues." The Office of Drug Safety (within the FDA) "does not track information about ongoing postmarketing safety issues." In addition the:

> FDA faces data constraints in making post marketing safety decisions. There are weakness in the different types of data available to FDA, and FDA lacks authority to require certain studies and has resource limitations for obtaining data.
> (GAO, 2006)

The current formal FDA approval process provides some assurance that the nano-remedy has been evaluated for safety and efficacy before widespread human use. In addition, its categories and tests are flexible enough to apply to any nano-application available. Unfortunately, that assurance is severely undercut by our current knowledge of the life-cycle of innovative medical technologies and limitations of the FDA's postmarket decision-making and oversight process.

The Life Cycle of Medical Innovation and the Limits of FDA Review and Approval

A growing number of scholars now recognize that new medical technologies proceed through a fairly predictable life cycle from introduction-to proliferation-to maturation (Grady, 1988; DeVille, 1998; Jacobson, 2005; Mastroianni, 2006). This life cycle suggests that all new technologies have unintended consequences and limitations which are epistemologically impossible to recognize before their substantial proliferation and widespread use in the general patient population. After a technological innovation first becomes a conventional and widely used procedure, the medical profession does not fully realize: all the potential side-effects; the precise degree of caution or degree of skill required to employ it; the patient populations on which it is most and least effective and most and least dangerous. However, when a new drug, device or procedure is used on a larger and broader scale than a clinical trial, these limitations eventually surface. Without a systematic means of identifying, tracking, and analyzing these limitations as they occur in general usage, however, it frequently takes time for remedies to be developed and disseminated to the clinical population. During this period, the limitations and the peculiarities of the innovation surface as errors, injuries and less than perfect results for patients. Sometimes the profession itself through anecdote discovers the limitations of the modality and initiates the safeguards required to protect patients. Sometimes the previously unknown limitations of newer innovations are publicized through lawsuits or through follow-up studies conducted by scholars. Eventually, through clinical experience with larger and more diverse patient and provider populations, the medical profession as a whole progressively discovers the types of problems associated with a new procedure and institutes new precautions to deal with newly discovered dangers. Once identified, these second-generation safeguards are frequently adopted and enforced by professional societies, institutions or regulatory bodies to ensure that patients are protected against the newly discovered limitations of the technology. As these precautions are publicized and enacted, outcomes improve and adverse events associated with the innovation decline. The time required identifying the chief limitations and error traps associated with a particular innovation varies wildly depending on such factors as the frequency of the occurrence, the scrupulousness of the practitioner, publicity, the complexity of the remedy, the formality of the inquiry, etc. But the consequence of delayed identification is clear. Until the set of inevitable but hidden limitations is identified, patients will be harmed.

Uncovering the Unknowable

The efficacy of the current gold standard RCT, and accompanying FDA regulation, is undermined by built-in and unavoidable limitations related the number and profile of the physicians and patients in the trials and the chief harms and benefits that are the main focus and scrutiny of the study. FDA approval cannot uncover the kind of information that only surfaces over time, from unrestricted usage on the general patient population. RCTs, by their very nature, are controlled with strictly defined and monitored protocols and carefully delineated physician and patient inclusion criteria. Study pools are not large for most drug studies, and even smaller for device studies, leaving the possibility that lower frequency but still serious adverse events may not surface during the course of the RCT, either by chance or by the vicissitudes of particular study population. Consider the paradigmatic case of laparoscopic cholecystectomies as an alternative to open surgery for gall bladder disease. The laparoscopic procedure proliferated rapidly because the option was attractive to patients, physicians and payors on many fronts. After proliferation, however, the procedure unexpectedly generated adverse events at a rate much higher than open surgeries. After several years use, and not a few lawsuits, the profession and institutions uncovered a series of incipient error traps and limitations related to the procedure. As a result, the institutions and the medical profession instituted precautions, safeguards and additional training (including certification) to decrease them (Mastroianni, 2006; Jacobson, 2005; DeVille, 1998). While this process and unavoidable period of adjustment occurs with any medical innovation, it takes time, time during which patients can be injured. Lawsuits sometimes play an important role in identifying the seams in the new technology. But the litigation system is both under- and over-inclusive (and sometimes arbitrary) and cannot be trusted as the means to uncover all or even the most important limitations of the innovations in a timely and consistent fashion. Clinicians, in many ways, are the best parties to identify potential injuries associated with the technology. But they, too, will do so only on a case-by-case basis until the sheer number of adverse events associated with the procedure reaches a critical mass and comes to the attention of the entire profession. The key point is that discoveries made through litigation or routine clinical usage are ordinarily not collected, analyzed and transmitted to the rest of the profession and the manufacturer of the drug or device in a systematic way. As a result, discovery of treatment limitations and error seams is at the mercy of haphazard identification by clinicians and by plaintiff's attorneys and the tort litigation system.

Nanotechnology and the Life-Cycle of Medical Innovation

What is true for other medical innovations is especially poignant for medical treatments involving nanotechnology. In fact, for a number of reasons, the interim adjustment phase of discovering the limitations and risks of individual nanoremedies is likely to last far longer than it might for other drugs or devices. The currently and virtually undetectable size of nanosubstances will make it harder to identify when the substance is culpable for injuries suffered by patients. Because nanosubstances have the ability to move freely through the human system, injuries and side-effects might surface far distal from the site of the illness or injury and will be less obviously linked to the nano-treatment received by the patient. Moreover, scientists and developers remain uncertain how even common materials and elements will react at the submolecular world inhabited by nanoparticles. Little is know of impact on nanosubstances of longer-term exposure to fluids, heat, enzymes, bacteria or other nanosubstances. Particle size may have chemical or physiological specific effects on the human body not yet identified. For example, nanoparticles might gain access to cells that are normally bypassed by larger particles of the same or similar substance. There is no information on whether and how long nanoparticles might remain in tissue once it has been entered. There is no clear notion of what kinds of injuries or illness might accompany the introduction of nano-sized foreign particles into the human body. Therefore, the impact of nanoparticles may manifest itself in a way that disguises the causal connection between cause and symptom making it far is less likely that these relationships would be identified and confirmed through RCTs of the size required for FDA approval. Therefore it is likely to be difficult to discover, during initial RCTs or in the ordinary course of treatment, links between some induced injuries or illness and the nanosubstance or device that was used on the patient. Recall that the RCTs were designed to identify predictable risks and those that surfaced during the study on a limited patient population. Consider again the history of laparoscopic cholecystectomies. In such a surgical procedure, clinicians and others would quite naturally be more sensitive to injuries or harms that related to bleeding, shock, infection and damage done to adjacent organs. When such injuries occurred, and recurred a sufficient number of times, an inferential correlation between treatment-to-injury would be made relatively easily. This correlation could then be relatively easily tested to ensure there was a genuine causal connection between treatment and injury and precautions could be devised and instituted.

In contrast, it will be far more difficult to forecast the kinds of injuries that will occur with nanotechnology. In addition, the harms could be localized to the site of the disease, distal to the injury or illness, or take the form of a systemic reaction apparently unconnected with the disease. Lawsuits have served as an adjuvant, albeit limited, means of identifying heretofore unknown risks and limitation of innovative medical practice. Because many of the limitations and dangers will likely be of an unanticipated nature and in an unexpected location, patients and attorneys will be less likely to connect the patient's problems with the previous treatment with a nanosubstance or device in any individual case. Even if the close time proximity of treatment-to-injury or illness raised suspicions, attorneys would be faced with a nearly overwhelming burden of demonstrating "causation." The demonstration of causation, scientifically, medically and legally is likely to be profoundly burdened by the absence of any easy mechanical means of detecting, tracing, and analyzing the presence, path and impact of nanoparticles.

What is needed instead is the post-RCT identification of previously unknown risks and limitations of a new technology and the communication of that information to the manufactures, clinicians and institutions in an expeditious way. Currently, there is no such routine mechanism for new drugs and devices, including those developed through nanotechnology – the class of medical products for which it is most needed.

Mandatory "Registry Studies" and Medical Nano-Devices and -Substances

Given the foregoing discussion, it is essential that clinicians, institutions and regulators identify unexpected limitations of nanotechnology as soon as possible after the innovation is released to medical community for widespread usage. Past informal identification of problems from ordinary clinical experience and lawsuits and chance is insufficient. Although many limitations eventually surface in ordinary use of a drug or device, in nanoapplications the process will likely be long-delayed and ultimately incomplete. Anne Mastroianni has advocated the use of mandatory drug and device "registry studies" as the most useful means of detecting the concealed limitations of innovative technologies after they have been released for use in the general patient population. "Registry studies" are not treatment studies. They do not compare one modality with another and they do not have a hypothesis. Instead, they compile clinical data related to safety and efficacy of a particular drug or device. Registry studies, by formally collecting and analyzing data from all uses of the innovative measure, would be able to significantly cut the "gap in time" between the first performance of the innovation and the detection, assessment and reporting rate of recurring, but unexpected complications (Mastroianni, 2006). Such registries would identify limitations not identified after the RCTs. Registries are frequently kept by individual institutions, by private device or drug sponsors, and by governmental sponsors such as the Pediatric Oncology Group (POG). The American College of Surgeons has developed one as has the Veterans' Administration. In each of these settings, they have been useful in quality control initiatives and helped identify previously invisible problems that only surface with ongoing use of devices and drugs. Registries are likely to yield important data and insights on the usage of any drug or device.

In the case of medical nanotechnologies, the use of post-approval registries should be mandatory. Although there are costs associated with both the collection and the analysis of the data, those costs will are justified by the uncertainties surrounding the use of nanotechnologies in medicine and the likelihood that ordinarily usage of these innovations are unlikely to reveal all the limitations and dangers in the short or even medium term. Given the difficulties in recognizing potential correlations between an intervention and the range and type of side effects that might occur, a mandatory formal process for collecting data is justified. Costs might be borne and reasonably distributed across the system. Sponsors could be required to develop and provide the datasheets and form fields in which information would be collected. Institutions that establish nanotechnology treatment centers could be required to serve as unpaid data-collectors, during treatment and probably throughout a substantial follow-up period in which patients' ordinary medical care and subsequent medical history are tracked and analyzed. This cost would not be inconsequential because the medical informational net would probably have to be thrown fairly wide given the limited knowledge that currently exists regarding nanoparticles and their interaction with the human body. Finally, a new governmental agency or office within an existing agency would receive the collected data from the institutions and analyze the data as it accumulates. These costs, too, would not be inconsequential, but they are unavoidable. As the analyzing body identifies new limitations in an ongoing manner, this information will be forwarded to the sponsor and to the institutions in order to allow the development of clinical or technical remedies at the suitable production or patient-care level. Such a proposal would require the costly establishment of a new regulatory office, or preferably, an expansion of the duties and authority of the Office of New Drugs and the Drug Safety

Board within the FDA. Some registries for standard post-approval drugs and devices are currently maintained by professional societies, sponsors or institutions. In the case of nanotechnology, this voluntary option is insufficient. It is crucial that the analysis of all post-approval medical drugs and devices that use nanotechnology be centralized in one database and organization. Because different nanosubstances used for entirely different therapeutic purposes might generate similar side effects or problems, analysis across the various therapeutic remedies is essential. It is only by analyzing data related to all the various therapeutic uses of nanomedicine that the most common hidden limitations and pitfalls will be forced to surface in the shortest period of time. A centralized repository is the only way of achieving this manner of cross-technology analysis.

Tort Litigation

Although clinicians and institutions will soon have the opportunity to implement a least a small number of medical applications utilizing nanotechnology, there will likely be at least a brief period before the medical and legal professions need to face directly the legal questions related to the widespread clinical use of nanotechnology. Clearly, nanotechnology will produce a broad assortment of unfamiliar and complicated factual scenarios with which litigants, attorneys and courts will have to wrestle. But despite their factual novelty and complexity, most cases will likely be compatible with and suitably resolved by reference to the familiar doctrines of negligence (medical malpractice) and products liability. For the most part, when legal complications arise, they will stem from the new and unfamiliar factual scenarios and the difficulty of procuring evidence, rather than the inability of existing tort doctrine to adapt to new technologies. Still, it is important to recognize that the litigation of nanotechnology claims in medicine is likely to generate a number of interesting challenges to the traditional array of defendants and the claims made against them. These challenges, however, should be resolvable through careful forethought and only mild or moderate alteration of existing practice and common law rather than a fundamental overhaul of the tort system.

Ordinarily, individuals who are injured as a result of the use of a medical product might, depending on the circumstances, initiate legal action against the manufacturer of that product, the physician who prescribes and employs the product, or both. The same will be true in personal injury suits for injuries purportedly stemming from the medical use of nanotechnology. Suits for nanotechnology injuries will likely interweave the potential product liability of the manufacturer with the medical malpractice liability of the treating physician and her institution (Eaton, 1995; American Law Institute, 1965, sec. 402A).

Product liability is referred to as "strict liability" because it does not require the existence of fault or negligence to justify a finding on behalf of the plaintiff. The "strict liability" designation is significant as well it allows the imposition of "punitive damages" over and above the actual damages allowed in negligence claims.

Pharmacological companies that produce, manufacture and sell nanosubstances for medical purposes will be subject to "strict liability" standards in product liability claims involving injuries to patients. Consider the fictitious case of NanoCure®, a start-up company, that develops and secures FDA approval of an injectable nanosphere, NanOxy2®. NanOxy2® will provide temporary supplemental oxygen support for patients who are hematologicaly compromised in some way. Suppose a patient receives the fictional NanOxy2® and dies of a stroke 12 hours after injection. NanoCure® as manufacturer and seller of NanOxy2® may be held liable under a theory of strict liability if the product "caused" the patient's injury.

Ordinarily, manufacturers and sellers of products are liable under strict liability if their product is in a "defective condition" and "unreasonably dangerous (Douthwaite, 1992, sects. 10-7, 10-4)." But, prescription drugs, the category most fitting for NanOxy2®, are typically considered "unavoidably unsafe" products and are not subject to the standard rules of product liability. Instead, as manufactures and sellers of NanOxy2®, an "unavoidably unsafe product," NanoCure® may be held legally liable if the NanOxy2® received by the patient had: a "design defect", a "manufacturing defect," or if the warnings provided by the company were insufficient (American Law Institute, 1965, sec. 402A, comment k).

The fictional NanoCure® will be found strictly liable for any "design defects" in NanOxy2® that cause injuries. In fact, producers and sellers of products will be strictly liable for injuries caused by a design defect unless there is no other cost-effective way of producing the product without impairing its legitimacy. Given the inherently dangerous nature of many or most biomedical agents, there is frequently no safer way to produce the product without undermining its efficacy. In these instances, the required warnings discussed below play a vital role in resolving liability. At any rate, the determination of a design defect is a factual issue that will be resolved by the jury with the help of a wide range of expert witnesses including physicians, biochemists, engineers, etc. In contrast, "manufacturing defects" occur during production when the resulting product, in this case, NanOxy2®, departs from its intended, and otherwise legitimate, design. This type of claim, as well, will be a factual issue that is decided by the jury after the introduction of empirical and expert evidence by both plaintiff and defendant. The strict liability aspect of product liability means that defendants are always liable for design and manufacturing defects, regardless of the degree of care exercised to prevent them. Therefore, nanoproducts "rushed to market" will put the producers at significant legal risk if their enthusiasm undercuts their quality control measures. In short, if design defects and manufacturing defects reach consumers, no level of care with protect the producer.

Given the fact that most medical products are "inherently dangerous" the manufacturer's "duty to warn" takes on special significance. NanoCure® has the duty to warn the patient of the all the risk that reasonably foreseeable product users and consumers would consider material in their decision to use the product. As in ordinary product liability cases, the question of whether the warning was sufficient is a fact issue which will be determined by the jury. NanoCure® will also reap significant benefits if its warnings conform to those required by the FDA during the approval process. Moreover, if the content of NanoCure®'s warning is found adequate and is satisfactorily conveyed to the treating physician, it will probably be insulated from liability through the learned "intermediary rule." Under this doctrine, it is the treating physician's responsibility to disclose the manufacturer's warning to the patient (Douthwaite, 1992, sects. 10-8, 10-9; Preiser, 1989, p. 658). If the physician fails to do so, NanoCure® or any other producer will be immune from liability on failure to warn grounds.

In the above respects, currently existing product liability law can be applied without modification to the fictional NanOxy2® or any other real medical device or substance produced using nanotechnology. Similarly, the physician who prescribes the use of NanOxy2® or other nanoproducts for a patient who is ultimately injured, will for the most part face standard medical malpractice law. In contrast to manufactures, physicians who prescribe nanoproducts for their patients are *not* "producers" or "sellers" of the product and cannot be sued under product liability doctrines (Hursh & Bailey, 1965, pp. 78–79). Sometimes, treating physicians are not even named as parties in cases involving product liability. That is not always the case, however, and physicians may be added to product liability cases involving nano-produced products under standard professional negligence and informed consent claims (Lagnese, 1993). Or in some scenarios, they may be sued for professional negligence even in the absence of a parallel product liability claim.

In order to collect damages from a physician for medical malpractice, the patient/plaintiff has to demonstrate that: (1) the physician had a *duty* to the patient; (2) the physician *breached the duty*, i.e. did not perform according to the standard of care; (3) the breach *caused* the damage suffered by patient; (4) the patient suffered *damage*. In medical malpractice suits involving the use or failure to use nanotechnology, the first and fourth required elements will pose few if any novel issues. The determination of whether a physician has a *duty* to a patient is based on the existence or non-existence of a physician-patient relationship. The evidence proving or disproving a physician-patient relationship will be no different in nanotechnology lawsuits than in conventional medical malpractice cases. Similarly, the determination of whether the patient/plaintiff suffered *damage* such as disfigurement, disability, death, etc. is an empirical question which will be resolved in nanotechnology cases in the same way as it is conventional cases – with the introduction of physical evidence. Two required elements of medical malpractice – *standard of care* and *causation* – will be more complicated in nanotechnology medical malpractice suits in at least the short term.

In order to satisfy the *standard of care* a physician is required to possess and exercise the same degree of learning, skill, and care that "a reasonably prudent physician" would possess and exercise in the same or similar circumstances. Expert witnesses typically are required to demonstrate that a defendant-physician failed to act as a reasonably prudent physician would have acted in a comparable situation. Although it is not always definitive, "customary practice" plays an important role in determining what is reasonable medical practice and what constitutes the standard of care. Medical experts ordinarily rely heavily on customary practice when evaluating a defendant-physician's performance.

At this stage of development, it is highly unlikely that the use of any current medical option utilizing nanotechnology will be deemed the standard of care. That is, no nanosubstance or device is currently considered part of ordinary and required medical practice. Thus, in the short term, it is unlikely that a physician could or would be held liable for the failure to use any particular nanodevice or substance. Physicians, however, are required to keep abreast of changing standards of care and the extent to which the use of any technology is customary or the standard of care is a rapidly moving target. Some nano-applications are already in limited clinical use. There is no bright line rule and no way to predict precisely when that threshold point will be reached for any particular nanotechnology, but physicians must stay cognizant of the profession-wide use of technology and the fact that, historically speaking, the transition may occur within a relatively short period of time. A possible early candidate for introduction into physicians' standard armory is nanosubstance imbued wound dressings to discourage infection.

In contrast, physicians who choose to employ nanotechnology applications may be held liable for their negligence under a number of circumstances. Physicians will of course be liable if they recommend the use of a nanoremedy that is inappropriate for the patient's condition. Like other routine medical malpractice cases, this determination will be made by a jury after hearing testimony from expert witnesses providing guidance on the best treatment for the patient's illness or injury. If the nanoremedy selected by a physician is within the acceptable standard of care, the physician will not be held liable for flaws or limitation in the program *per se*. As discussed above, flaws in the nanoproducts themselves that lead to patient harm likely will be litigated as a species of product liability for which physician will usually not be liable. But physician users will be held legally responsible for failing to be aware of known flaws or limitations in a nanotechnology, and for failing to take appropriate precautions in order to limit the injuries that might be caused by those flaws. The test of whether a clinician has breached the standard of care in reliance on an imperfect product will turn on whether a reasonably prudent physician would have selected the product, should have known of those flaws, been sensitive to how those defects could impact patient care, and taken steps to protect patients from those shortcomings. These required steps might include: sufficient warnings to patients through informed consent; appropriate monitoring for known side effects; the use of safeguards to limit known potential harms; and the refusal to treat patients with the remedy when it is counter-indicated for that class of patients.

Consider the use of a nanosensor/nanosphere drug delivery system that automatically monitors blood glucose and releases the appropriate level of medication. When depleted of active medication, these nanospheres are removed via a filter placed in the patient's vein. If known dangers of the remedy include high-blood pressure, then the physician's duty will include not only warnings to the patient, but also a reasonable means of monitoring, or if appropriate, self-monitoring for the patient. There may be known difficulties in removing "spent" nanospheres from the patient's body after they have served their purpose. Suppose the long-term presence of these depleted nanospheres in the patient's circulatory system pose a risk of kidney damage. The treating physician must not only warn the patient of these eventualities but monitor the patient with appropriate means to ensure that those know side-effects to the least amount of damage possible. Again, as in other malpractice suits, the jury will decide with the aid of expert witness, when there has been a breach in the standard of care expected by the physician.

Although the physician's liability for the use of nanoremedies will be resolved doctrinally and procedurally just as other medical malpractice cases are, there will be one important short-term distinction – the factual difficulty of proving "causation," a required element in suits for medical malpractice. Guidance on "duty" and "standard of care" are provided by expert witnesses. "Damage" can be demonstrated by the simple reference to the patient's condition. But in many cases "causation" may prove a significant hurdle to medical malpractice plaintiffs. But although damage may occur temporally soon after the application of a nanoremedy, "causation" may be difficult prove to juries. The virtually undetectable size of nanosubstances will undermine attempts to link the substances to injuries. Although nanosubstance might migrate in the body and inflict damage in parts distant from injury, this path and the mechanism will be difficult to reconstruct in court. Therefore it is likely to be difficult to prove the essential "causal" element of medical negligence cases until more knowledge regarding the behavior of nanosubstances in the human body exists, and quite simple, until more sensitive detection and tracking technology exists in order to determine if the nanoremedy indeed caused the injury.

While the factual difficulty of proving of "causation" in both product liability and medical malpractice suits will probably handicap plaintiffs in the short term, the tort system will likely operate as efficiently or inefficiently as it does in other medical injury lawsuits. But one possible technological possibility may portend palpable challenges for courts, attorneys, physicians and institutions in nano-related lawsuits. There appear to be two paths that medically-related nanosubstances or devices might take in reaching patients in clinics and hospitals. The route chosen will likely depend upon the interests of manufacturer, but may have profound implications for all potential parties in medical injury cases.

In the first model, a developer/manufacturer, like NanoCure®, chooses to produce a nanosubstance or device, like NanOxy2®, "in-house" and sell and distribute it to medical centers and health care providers as one would an ordinary medical drug or device. As discussed, this scenario probably does not raise substantially new doctrinal issues for FDA approval, liability, or even patent law. But consider a second mode of producing and distributing medical nanosubstances and devices. In the second model, a developer and manufacturer, like NanoCure®, patents the design, software and manufacturing equipment that are sold as a kind of "turn-key nanofactory" to medical centers. The nanofactory is employed onsite by the institution's engineers and medical staff to a nanosubstance or device, like NanOxy2®, for medical use (Phoenix & Treder, 2003). Only centers with a sufficient interdisciplinary team would possess the resources to operate such nanofactories. The nanofactory would be designed and programmed to produce only the nanoproducts or nanodevices approved by the manufacturer, and presumably, approved by the FDA. One author has speculated that "a typical nanofactory might be the size of a microwave oven" (Phoenix & Treder, 2003). The purchaser of these nano-factories

– institutions or medical centers – could then produce as many or as much of a particular nanodevice or nanosubstance as they chose for use at the institution.

The prospect of institution-based "nanofactories" that produce and sell substances and devices for patient care significantly shifts the conventional array of institutional players and actors for legal and regulatory purposes. Will the FDA approve and regulate the "nanofactory" as a medical device? They will, after all, produce articles intended for the use in diagnosis, cure, mitigation, treatment, or prevention of disease. But, there mere approval and regulation of the "nanofactory" hardware itself seems somehow insufficient. The institutional team of physicians, scientists, physicists and engineers that operate the nano-production centers may ultimately have as much to do with the quality and safety of the resulting nanoproducts as does the design of the software and equipment. Would this realization require licensure? Or inspection of facilities in order to ensure that each institution conformed to Good Manufacturing Practices? Such conclusions seem almost unavoidable and would require significant additions to the FDA's regulatory corpus as well as a considerable expenditure of new resources.

Putting the means of production in the hands of the medical teams and institutions will affect who is sued under what theories when patients appeared to have been injured by nanosubstances manufactured on-site at a local medical center. The producer of the "nanofactory" will undoubtedly retain responsibility under traditional product liability theories. They produced and sold a device that may have injured an individual. But the medical center, physician, and nanoproduction team may also now be accountable under strict liability product liability theories as well. It is, after all, the medical center and team who produce the actual drug or device that purported harmed the patient. Under this scenario appropriate warning, correct application and sufficient monitoring of the patients may not be enough to immunize individual physicians and institutions from nano-produced harm. As producers and sellers of the nanosubtances, they might now held liable for design and manufacturing defects, as well as for negligent misuse of the manufacturing system, the recommendation to use a nanoremedy, the misapplication of nanosubstance itself and the subsequent management of the patient.

Conclusions

The implications of the bedside use of nanotechnology do not support a radically different formal regulatory scheme or a thoroughgoing overthrow of tort principles. But nanotechnology's appearance on the medical and legal scene will be dramatic and will challenge the thinking of those in industry, law and medical care. In some specified arenas, additional and new licensing and regulatory oversight provisions will have to supplement those that currently exist. Clearly the novelty of the production, distribution and use of medical nanoproducts will pose palpable challenges for courts and regulatory agencies attempting to apply current rules to new situations. All new technology is unavoidably accompanied by unintended consequences and limitations which are epistemologically impossible to recognize before broad proliferation of the modality. Nanotechnology is no different. Policymakers, regulatory bodies, institutions and individual clinicians who will approve and regulate usage of nanotechnology must recognize this unavoidable fact and work toward a coordinated reporting system that will identify the untoward effects, error traps, and limitations of this highly complex intervention as soon as possible after proliferation. Clearly, a tremendous amount of data will be necessary to explore fully the safety and efficacy of the various nano-applications looming on the horizon. The proliferation and use of these clinical tools will not, and should not stop while all the limits and dangers of nanotechnology have been fully and comprehensively uncovered and managed. Such a standard is impossible to meet and unwise. Not only would exhaustive pre-approval scrutiny through clinical trials delay patient access to many important remedies, but it would fail to meet its ultimate objective as well. It is true that more arduous pre-approval studies and analysis may uncover some previously unknown risks and complications. But the lion's share of the limitations and dangers will not and cannot be identified until the technology is used on the general population by a wide-range of physicians. Instead, on balance, society will most benefit by required post approval registries to track the experience of all patients who receive nano-products. A commitment to a systematic, ongoing, near real-time technology assessment through the use of mandatory registry studies after the FDA approval and release of the nano-drug or device will best balance providing access to important technology for patients in the short term, while still protecting them from what we do not and cannot yet know about those technologies in the long term.

References

American Law Institute (1965). *Restatement of Torts (Second)*. Philadelphia, PA: American Law Institute.
Bourne, M. (2005). Product of the Quarter. *The Bourne Report Medical Device Quarterly*, Fourth Quarter 2005, 1–10.
De Ville, K.A. (1998). Medical Malpractice in Twentieth Century United States: The Interaction of Technology, Law and Culture. *International Journal of Technology Assessment in Health Care* 14, 197–211.
De Ville, K.A. (2000). The Ethical and Legal Implications of Handheld Medical Computers. *The Journal of Legal Medicine* 22, 447–466.
Douthwaite, G. (1992). *Jury Instructions on Medical Issues*. 4th Edition. Charlottesville, VA: Michie.
Eaton, A.T. (1995). The Interface Between Medical Malpractice and Product Liability. *Medical Malpractice: Law & Strategy* 12, 1–3.
Engelhardt, H.T. Jr. & Jotterand, F. (2004). The Precautionary Principle: A Dialectic Reconsideration. *Journal of Medicine and Philosophy* 29(3), 301–312.
FDA News (March 2006). Bourne Releases Report on BIOMEMS and Nanotechnology. *FDA News Device Daily Bulletin*.
Fielder, F.A. & Reynolds, G.H. (1994). Legal Problems of Nanotechnology: An Overview. *Southern California Interdisciplinary Law Journal* 3, 593–629.

Flattum, J. (2006). Nanomedicine. *Human Evolution*, available at: http://www.human.evolution.org/nano_medicine.php [accessed on Nov 2, 2007].
Food and Drug Administration (2006). FDA and Nanotechnology. *Administration*, available at: http://www/fda/gov/nanotechnology/faqs.htm [accessed Nov 2, 2007].
General Accounting Office (GAO) (2006). *Drug Safety Improvement Needed in FDA's Postmarket Decision-Making and Oversight Process*. GAO-06-402. Washington, DC: GAO.
Grady, M.F. (1988). Why Are People Negligent? Technology, Nondurable Precautions, and the Medical Malpractice Explosion. *Northwestern Law Review* 82, 293–334.
Grimes, D.A. (1993). Technological Follies: The Uncritical Acceptance of Medical Innovation. *Journal of the American Medical Association* 269, 3030–3033.
Hursh, R.D. & Bailey, H.J. (1965). *American Law of Products Liability*. 2nd Edition. Vol. 2. Rochester, NY: Lawyers Co-operative Publishing.
Iglehart, J.K. (2001). America's Love Affair with Medical Innovation. *Health Affairs* 20(5), 6–7.
Jacobson, P.D. (2005). *Medical Liability and the Culture of Technology. The Pew Charitable Trust*, available at: www.medliabilitypa.org [accessed Nov 2, 2007].
Kacmar, D.E. (1997). The Impact of Computerized Medical Literature Databases on Medical Malpractice Litigation: A Time for Another *Helling v. Carey* Wake-Up Call?. *Ohio State Law Journal* 58, 716–654.
Kahn, J. (2006). Nano's Big Future. *National Geographic* 209(6), 98–119.
Lagnese, J. (1993). Applying Product Liability Theories to Physicians. *Medical Malpractice: Law & Strategy* 10(11), 1–3.
Masci, D. (2004). Nanotechnology. *CQ Researcher* 14(22), 517–540.
Mastroianni, A.C. (2006). Liability, Regulation and Policy in Surgical Innovation: The Cutting Edge of Research and Therapy. *Health Matrix* 16(2), 351–442.
Merkle, R.C. (2006). Nanotechnology and Medicine. *ALCOR*, available at: http://alcor.org/Library/html/NanotechnologyAndMedicine.html [accessed on Nov 2, 2007].
Miller, J. (2003). Beyond Biotechnology: FDA Regulation of Nanomedicine. *Columbia Science & Technology Law Review* 4, 5–42.
Pease, P. (2006). Brush up on your nanotechnology. *BBC NEWS* June 12, 2006, available at: http://newsvote.bbc.co.uk/mpapps/pagetools/print/news.bbc.co.uk/1/hi/sci/tech/4085214.stm [accessed on Nov 2, 2007].
Phoenix, C. & Treder, M. (2003). *Safe Utilization of Advanced Nanotechnology*. CRN: Center for Responsible Nanotechnology, December 2003, available at: http:www.crnano.org/safe.htm [last accessed on Nov 2, 2007].
Preiser, S.E., Wecht, C.H., & Preiser, M.L. (1989). *Preparing and Winning Medical Negligence Cases*. Vol. 1. Charlottesville, VA: Miche.
Rothman, D.J. (1997). *Beginnings Count: The Technological Imperative in American Health Care*. New York: Oxford University Press.
Sidawi, D. (2004). Nanotechnology Takes Medical Intervention to a New Level. *R & D 2004* 46(10), 28–29 [accessed on Nov 2, 2007].
Till, M.C., Simkin, M.M., & Maebius, S. (2005). Nanotech Meets the FDA: A Success Story About the First Nanoparticulate Drugs Approved by the FDA. *Nanotechnology Law & Business* 2(2), 163–167.

Statutory Law

21 U.S.C. 321 (g)(1) (1988a).
21 U.S.C. 321 (h)(1) (1988b).
21 C.F.R. 320 et seq. (2005).

Human Enhancement and (Bio)Nanotechnology

Stage Two Enhancements

George Khushf

Abstract In this essay I consider two kinds of human enhancement technologies. Stage 1 enhancements are discrete, medical enhancements; they involve a modest augmentation of some specific function or capacity, and have quantifiable harms and benefits that are amenable to conventional study designs. Examples of these enhancements include psychopharmaceuticals for enhanced cognitive function, cosmetic surgery, and sports doping. Stage 2 enhancements build upon stage 1, but involve more than a simple quantitative extension of stage 1 capacities. Stage 2 enhancement technologies are multifunctional; have an autocatalytic aspect that leads to accelerating development; and involve the convergence of multiple kinds of technology and technology platforms. These enhancements can involve radically new capacities that provide significant advantages to those who obtain them. I illustrate the features of stage 2 enhancements by discussing a nano-enabled brain-machine interface (BMI) that might arise from the research of Miguel Nicolelis and Rodolfo Llinas. Stage 2 enhancement technologies undermine the conditions of traditional, post hoc modes of ethical reflection. The needed upstream ethical engagement requires a fundamental change in our cultures of research and development, and for this change, we must bridge the gap between the research cultures of science and engineering, on one side, and the humanities, law, and policy, on the other.

Keywords Human enhancement, NBIC convergence, upstream ethics

Introduction

For several decades people have speculated about the prospects of a radical enhancement of human capacities. However, until recently, the scientifically viable enhancements have only involved tinkering at the margins. I call these "stage 1

Center for Bioethics and Department of Philosophy, University of South Carolina,
Columbia, SC 29208 USA;
email: khushfg@gwm.sc.edu

enhancements," and distinguish them from a new, more radical kind, which I call "stage 2 enhancements." I'll show that stage 2 calls for a different kind of ethical analysis.

Discussion of radical enhancement is, of course, not new. But in the first stage, we never quite got at the features that make a viable stream of science and technological radically disruptive. Perhaps the most discussed transformative option involved germline genetic engineering of humans. But even this, if it were to be developed, would only be modest. In fact, today we see how the genetics-based utopian and dystopian dreams were a non-starter. They involved a complete alteration in how humans would reproduce; required highly problematic experimentation on vulnerable populations; and could only have involved the simplest of traits. It would take generations before the benefits (or failures) would be seen. While germline genetic interventions might yet be significant over the long-term, they clearly do not involve socially destabilizing short-term risks.

Stage 1 enhancements involve medical interventions that offer, at best, a small, quantitative increase in some narrow, proper function. Examples include cosmetic surgery, psychopharmaceuticals, and sports doping. These enhancements are at the margins in two senses. First, they have only a marginal increase in capacity. Second, they have been at the margins of most peoples lives. While high performance athletes or entertainers might face issues raised by them, most people aren't aware of the issues and don't worry about them, either in work or play.

However, I think the marginal character of the enhancement issues will soon change. By soon, I mean within a few decades. And we are not accustomed to thinking of such a time horizon as soon. But the decisions made now in research labs around the world already establish path dependencies that condition how we can address these issues in the future. If we don't address the topic now, we will be forced to address it in a different, less appropriate manner in the near future. And the radical character of the enhancements that are now in formation are potentially destabilizing.

It is, of course, very tricky to characterize the features of these new enhancements and specify how they will be disruptive. Those who have addressed the topic often put the issues in grand, sweeping terms. Examples include the following:

- Self-aware evolution: Until now, the processes guiding the development of life have been blind, depending on chance happenings, which provide selective advantage in contexts of competition for limited resources. But now, through our science, we have developed the capacities to engineer directly the next stages of this process. Whether by genetically altering existing living systems or through direct creation of artificial life, we can (or will soon be able to) directly manufacture the next stage (Silver, 1998; Stock, 2002; Garreau, 2005).
- Cyborgs (human/machine hybrids): Technology has already radically enhanced human capacities, so we are stronger, faster, and more agile than any other living system. But the capacities were provided by external means: rockets, cars, cellphones, and computers. However, slowly, these technologies are getting smarter, and the human-machine interface is getting more seamless. Now we are at the cusp of the next stage: where the technology gets directly incorporated, whether

by implanted chips, neural interfaces, or simply by remote sensing capacities. With this shift, the very character of the human changes. We move from the natural organism, Humanity 1.0, to the natural/artificial hybrid, Humanity 2.0 (Vickers, 2003; Hughes, 2004; Clark, 2003 emphasizes continuity in understanding the human/machine interface, but also points to a more radical transformation when our external tools themselves become smart; the phrase "humanity 2.0" is from Elliott, 2003).
- Nanomedicine: Until now, medicine has largely focused upon therapy. It advanced tools that enabled the detection and mitigation of disease, or the restoration of function for those disabled. But the tools for doing this were rather crude, involving ad hoc, cookbook recipes, rather than a deep understanding of the underlying, molecular mechanisms of life. Now, thanks to developments in nanoscale science and technology and systems biology, we are moving to a different kind of interface with natural systems. Once we have this interface, we will have an unprecedented control. Instead of simple restoration of lost function, we can enhance normal function and introduce new capacities that humans have never had before (Kurzweil, 2004; Bayertz & Schmidt, 2004).

I think there is a kernel of truth in each of these formulations, and each provides valuable ways of framing issues that need further attention. But the grand and general tone also works against the kind of philosophical and ethical analysis that might inform specific, research initiatives advanced by scientists working on, e.g., human/machine interfaces or nanomedicine. These general accounts also don't enable us to discern how the more radical variants might be distinguished from the more modest enhancements that have been the focus of decades of ethical reflection.

In this essay, I want to sketch the features of stage 2 enhancements, so we can identify them. I'll also try clarifying how they are different from stage 1, and show how they call for a change in the way we approach the ethical issues. Put simply, stage 2 enhancements involve a double convergence: first, a convergence of multiple sciences and technologies; and second, a convergence of earlier kinds of enhancements. This form – as a convergence of processes and technology platforms – alters how we identify the enhancements and how we should reflect upon them. At preliminary stages, we simply have the sum of previous enhancement efforts, along with the initiation of some newer, innovative projects. But there comes a stage when a quantitative increase leads to a qualitative transition, and where discrete components interact in ways that involve an emergent dynamic. As the more radical, qualitative transformations take form, they lock us into pathways that become extremely difficult to alter by traditional, post-hoc modes of ethical reflection.

To illustrate the issues associated with stage 2 enhancements, I will consider an example from nanomedicine: a nano-enabled human/machine interface discussed as part of the US NBIC Convergence initiative. Although my primary concern is to elucidate the contrast between stage 1 and stage 2 enhancements, I also take this case example as an illustration of the kinds of ethical issues raised by nanoscale science and technology more generally. Nano is enabling. In many cases, the ethical issues raised by such research will not be unique. Instead, one must attend to diverse streams of research that converge, and the role of nanoscale science and technology might simply involve the enabling of a more seamless, integrated variant of some project already conceptualized in another domain of research. In these contexts one cannot make headway on the ethical issues by asking about any of the components or processes in isolation. Instead, we must attend to the visions and ends that currently organize the otherwise disconnected bits into coherent research programs, and we should proactively consider how these initiatives might be configured so downstream ethical and social problems don't arise or so more disruptive variants are at least mitigated. This, in turn, will involve a significant change in how we conduct our research in the first place. My goal here will be to sketch the kind of changes we need, if we are to appropriately address the nano-enabled enhancements that lie on our mid-term horizon.

Stage-One: The Old Enhancement Debate

A geographical image can be used to illustrate the development of *enhancement technologies*. First, stage 1: there are a few trickling streams, each rather sparse, arising in isolated, lonely mountains far removed from the hustle and bustle of ordinary life. Each trickle is separate from the other, spawns its own little trenches, with its own culture and life. But then, gradually, we arrive at stage 2: these streams come together, speed up, grow larger, and they come down from the isolated heights and enter into the hills and valleys of ordinary life.

In this initial section, I'll consider a few of the isolated streams, and show how the earlier stage of the enhancement debate has been framed. Then, in the next section, I'll consider the immediate horizon in front of us, where we start to see these streams merge, and where the enhancement debate takes on a new, more troubling and radical form.

Sports Doping

One of the best illustrations of the "old debate" (which is, of course, still a current debate) can be found in the area of sports doping. Most people are aware of some notorious athletes: Ben Johnson's revoked gold medal for the 100-yard dash in the 1988 Seoul Olympics, or perhaps Mark McGuire's steroid enhanced home run record in baseball. Perhaps some are even aware of the tragic side, of fatal blood clots associated with pharmacologically boosted red blood cell counts in bicycling, or of testosterone induced rage in weight lifters. Perhaps some are even aware of the pervasive pressure that high performance athletes face to use drugs for enhancement purposes. But all this just touches the tip of the problem in this one little tributary of the enhancement debate.

High performance athletes, even at college and high school levels, train in ways that combine diet, specialized exercise routines, and continued medical attention to mitigate the ever-present problems of injury and diminished function. In a sense, all their training involves a controlled tearing down and building up of their body, and the "dietary supplements" or "medical treatments" sometimes blur the lines between therapy and enhancement (Hoberman, 2002). In fact, the ability to draw a line and to determine when it is crossed involves a whole industry. There are national and international commissions that attend to the issues, and the rules of diet and medical treatment are now so complex that the average athlete often must depend on experts for making the appropriate distinctions. There are also complex strategies that have been developed for detecting violations of doping rules, and equally complex games played by athletes and their trainers for navigating these rules (Mendoza, 2002). In some areas of sport, organizers have capitulated to the evolving drug culture, and they make distinctions between doped and non-doped leagues (perhaps a new way of addressing the eroding distinction between amateur and professional athletes, if only people could find ways to determine who is really playing by the non-doping rule).

High performance athletics is still at the fringes of ordinary life. Although we actively track their play, the problems such athletes face seem distant and surreal to most of the population. But within this small world, the problem of doping has moved from a relatively small, peripheral role to a pervasive one, and all such athletes confront it and the complex rules spun by it on a daily basis. Later I shall suggest that this arena can be taken as a microcosm of larger developments that will soon influence all of society.

Cosmetic Surgery

Athletes dope for strength or speed. But entertainers pursue a different kind of enhancement: the face lift, boob job, or tummy tuck. At first, there was a similarly distant, surreal aura to such technological or surgical fixes. They were for the rich and famous who had nothing better to do with their vast cash holding, and who were so vain and image-focused that they were willing to submit their bodies to neurotoxins and the surgical knife. But in this area, the widening of the stream has touched the "ordinary" person, although we still find relatively few who take this path.

One of the more interesting developments in this area also concerns the kinds of enhancements that are emerging, and the ways they are integrated into social discourse. At first, there were few options. But now, in addition to the conventional nose job (rhinoplasty) or botulism-mediated tightening of a wrinkled face, we see enhancements of voice (for those who want a smooth sounding voice on the telephone) and even the prospect of a complete face transplant. We also find a multiplication and refinement of the tools used for the more conventional cosmetic interventions, and we see the technology integrated into alternate visions of human excellence. Thus, for example, there are popular television shows like "Extreme Makeover," where a somewhat homely, out-of-shape and poorly dressed housewife gets transformed into a vibrant, in-shape beauty by the appropriate combination of exercise, a new wardrobe, and the surgeon's knife.

The key thing to notice with this development is the shifting appraisal provided by society; namely, how we move from a fringe practice for the rich and famous to a readily available strategy for morphing the ordinary person into a vibrant, extraordinary one (Davis, 2000; Little, 2000; Alam, 2001).

Smart Drugs

Although each of these enhancement streams affect people of all kinds, our public images and stereotypes link them to specific groups. Thus the sports doping problems are linked to men, and the enhancements reinforce stereotypes of brute, mindless strength. Cosmetic surgery is linked to women, with stereotypes of an equally mindless concern for appearance and external beauty. These two strands are quite visible in the public eye. But there is another one, which again affects all, but which now takes unique form among youth, especially the smart, academically oriented kind. "Mindless" is not quite the right word for this kind of enhancement.

Among academically oriented youth in colleges, there is an extensive interest in pharmacological means for enhancing intelligence. This goes beyond an interest in new energy boosting drinks or the caffeine and glucose shot in a Starbucks latte. Students are interested in the prescription variety – whether Ritalin or newer memory enhancing drugs originally developed for age-related dementia or Alzheimer's. And here we don't simply have a new iteration on the older problems of recreational drug use among youth. Even those students who normally would not experiment with drugs like Marijuana now seriously struggle with questions about whether to use the "smart drugs." And the problems raised by such psychopharmaceuticals clearly go far beyond the youth culture (Gardner, 2000; Elliott, 2000; Gazzaniga, 2005a, 2005b).

Common Features of the First Stage of the Enhancement Debate

Although these three areas of enhancement are generally isolated from one another, and we usually don't see the same person pursuing more than one kind, they still share common features. I'll just focus on five of these shared attributes:

1. The enhancements are *medical* and require a physician to prescribe legally the treatment. Generally, the technologies are developed in a context of treatment, addressing the deformation or debilitation that attends illness or trauma. Since they are medical, they are regulated, and access depends on a gatekeeper, whether to prescribe the drug or provide the surgery. While this gate keeping can be, and increasingly is sidestepped on a black market, we find the regulatory mechanisms providing a social constraint on the way enhancements are accessed. Also, the background medical ethical norms provide guidance,

whether in sports medicine for drug use, pharmaceuticals for attention, memory or mood, or surgery for improving form or figure (Miller et al., 2000).
2. The enhancements are *discrete*. We can usually neatly isolate the specific intervention at issue, and then consider its effect. This discreteness simplifies analysis, and it enables us to distinguish the "medical enhancement" from other kinds of enhancements associated with diet, exercise, or study.
3. The enhancements usually *serve a narrow, specific purpose*. They might help mood or memory, or build muscle, or improve figure.
4. The enhancements have *harms that can be studied and quantified* in the same way, and by the same tools we use to study benefits and harms in other areas of medicine. Although there has been relatively little study of the overall utility of these discrete medical enhancements, it is not difficult to see how such study might be conducted. As a result, we can generally consider the weight of benefit against the harms.
5. While the enhancements do have clear, documented effects (although many are not well-studied), they are, in the end, relatively *modest effects*. An athlete using steroids might have a significant boost in strength or speed, but he will not gain radically new, superhuman powers. The student on Ritalin might gain 100 points on an SAT entrance exam, but he or she will not instantly become a genius or know things she would not have been able to know before.

On the basis of these shared attributes, broader ethical and social analysis of the enhancements can take the form of an assessment of the risks and benefits of such enhancements. Many people believe that the questions of usage should be left up to each individual, since they are in the best position to assess the risks and benefits. Since the public generally sees the benefits as modest and the risks as significant, few have felt threatened by those who take the riskier path.

The one exception has been in the area of sports, and this context is instructive. We need to ask: Why has the otherwise marginal practice become so significant in the area of sports? Here three factors are important. The first concerns the level of specialization in sport. For most people, it would be very difficult to isolate specific functions or abilities that could by directly enhanced by current medical technologies, and, once enhanced, could provide significant benefit in realizing life goals. Our goals are generally too diffuse, or when specific, we couldn't better attain them by some specific medical/technological enhancement of some designated ability.

The second factor concerns the level of commitment found in a high performing athlete. One of the things we value most about athletes is the strength of will, the courage, and the persistence in a demanding activity, even when many obstacles arise. In most sports, this includes a significant risk of injury and harm, which the athlete willingly accepts in the pursuit of excellence. This virtue – with the attendant risk-taking and singleness of purpose – makes the athlete willing to accept a gamble that others would normally not accept.

The third factor is, perhaps, the most significant, and it provides the key to the ethics of sports doping. Put simply, sports assume a competitive context, and they require a "fair" game. The meaning of such *fairness* is not easy to specify, and it ultimately depends on background notions of human flourishing and the kind of excellences manifest in sport. Consider one of the simplest and purest examples: runners in a race; a dash, like the 100-m dash. What do we value in the athlete? It is surely not just the final time. Many animals can easily best the fastest human, and a "human + machine" could easily best the fastest animal. The time doesn't matter. Instead, we consider the somewhat nebulous combination of natural ability, intense training, diet, and so on. Now, with this as our context, we need to ask: What would sports doping add to the sport? What excellence does it enhance? And here is the rub: While the drug might lead to a better final time, it doesn't offer anything else. It "enhances" the final time, but it does not enhance any excellence we value in the sport. Yet it brings risks, and most runners would prefer to avoid these.

When we couple the risks with the competitive context, we now see the full problematic. Most athletes would prefer not to use the drugs, since they bring risks, but don't add in any way to the excellence of the sport. But the margin in sport is small, and if some avail themselves of the pharmaceutical enhancement, then all must follow if they want to remain competitive. Thus, if some are allowed to use them, all will be compelled to use them. This would introduce an accelerating risk, but add nothing to the beauty of the sport itself. Thus doping is banned. And if an athlete secretly dopes, this is just cheating, and that is something antithetical to human excellence.

Now we need to ask: What would happen if enhancement technologies would provide the ordinary person advantages in ordinary life comparable to the kinds of advantages doping provides in sport? What if these move from the margin to the center of life? And most significantly, what would happen if the kinds of enhancements provided are more radical, perhaps even leading to abilities people never had before?

Stage-Two: Enhancements on the Horizon

We are just at the stage where the new kinds of enhancements are taking form, and it is still difficult to characterize them clearly. They seem to hover in a strange world between real and make-believe, science and fiction. But they are not just future and speculative. Real scientists are conducting the research, influential companies funding it, and national agencies promoting their development. These enhancements are of an unprecedented kind. To illustrate this new stage of the enhancement debate, I consider a representative initiative called "NBIC Convergence," and then address some of the ethical and policy issues that arise from this.

NBIC Convergence

"NBIC Convergence" grew out of a December 3–4, 2001 workshop jointly sponsored by the National Science Foundation (NSF) and the U.S. Department of Commerce (DOC). (For an overview of the initiative and of the ethical and philosophical issues, see the essays in Roco and Bainbridge, 2002; Roco and Montenagno, 2004; and

Khushf, 2007). It included leaders in government like former Congressman and Speaker of the House Newt Gingrich (2002), and the Undersecretary of Commerce, Philip Bond (2002, 2004). Many leaders in industry played an organizing role, including technological powerhouses like IBM and Hewlett-Packard. And it was lead by some of the most influential people in our national science agencies, including Mihail Roco, perhaps the most influential architect of U.S.-funding policy in nanotechnology, and William Sims Bainbridge, a current director of NSF initiatives in information technology. The initial report of this workshop, published as a government volume and later by Kluwer Academic Publishers, was edited by Roco and Bainbridge (2002), and has the title: *Converging Technologies for Improving Human Performance. Nanotechnology, Biotechnology, Information Technology and Cognitive Science* (Thus, the N, B, I and C).

The kinds of enhancements of human performance contemplated in the report are radical. They include advanced human/machine interfaces, significant extension of the human life span, genetic engineering, and the complete transformation of formal educational systems. The overview of the workshop states that with a concerted effort, any of a list of 20 enhancements could be realized within the next few years. To give a sense of the tone of the document, I'll list the first six of these 20:

- Fast, broadband interfaces directly between the human brain and machines will transform work in factories, control automobiles, ensure military superiority, and enable new sports, art forms and modes of interaction between people.
- Comfortable, wearable sensors and computers will enhance every person's awareness of his or her health condition, environment, chemical pollutants, potential hazards, and information of interest about local businesses, natural resources, and the like.
- Robots and software agents will be far more useful to human beings, because they will operate on principles compatible with human goals, awareness, and personality.
- People from all backgrounds and of all ranges of ability will learn valuable new knowledge and skills more reliably and quickly, whether in school, on the job, or at home.
- Individuals and teams will be able to communicate and cooperate profitably across traditional barriers of culture, language, distance, and professional specialization, thus greatly increasing the effectiveness of groups, organizations, and multinational partnerships.
- The human body will be more durable, healthy, energetic, easier to repair, and resistant to many kinds of stress, biological threats, and the aging process." (Roco & Bainbridge, 2002, pp. 4–5)

After providing the list, Roco and Bainbridge summarize:

> If we make the correct decisions and investments today, any of these visions could be achieved within twenty years' time. Moving forward simultaneously along many of these paths could achieve a golden age that would be a turning point for human productivity and quality of life. Technological convergence could become the framework for human convergence.... The twenty-first century could end in world peace, universal prosperity, and evolution to a higher level of compassion and accomplishment. It is hard to find the right metaphor to see a century into the future, but it may be that humanity would become like a single, distributed and interconnected 'brain' based in new core pathways of society. This will be an enhancement to the productivity and independence of individuals, giving them greater opportunities to achieve personal goals.
>
> (Roco & Bainbridge, 2002, pp. 5–6)

There are several things that are remarkable about this list, and about the optimistic conclusion. Perhaps most striking is the radical character of the enhancements that are contemplated. Consider the first item on their list. When they speak of "fast, broadband interfaces directly between the human brain and machines," they don't just mean a better Road Runner connection for your home computer. They are speaking of direct neural/digital interface. The kind of linkage they have in mind can be seen in the work of one of the conference speakers, Miguel Nicolelis (2002a, b, 2003), a researcher at Duke University Medical Center. He implanted electrodes into the sensor motor-cortex of monkeys, and utilized a computer algorithm to translate the signals, so a robotic arm could directly mimic the monkey's own arm. After the algorithm was worked out for translating the brain's signals, the monkey was rewarded with fruit juice when it activated the robotic arm without moving its own arm. Eventually, the robot arm was moved miles away to another university, and the monkey was trained to manipulate the arm simply by looking at a computer image on the screen. The monkey could use the robot arm to pick up items and move them around – all this, without moving its own arms, and by means of a direct interface between its own brain and the computer/robotic arm/remote communication complex.

At the NBIC Convergence workshop, Nicolelis reflected on the implications of his research:

> The full potential of the 'digital revolution' has been hindered by its reliance on low-bandwidth and relatively slow user-machine interfaces (e.g., keyboard, mice, etc.). Indeed, because these user-machine interfaces are far removed from the way one's brain normally interacts with the surrounding environment, the classical Von Neuman design of digital computers is destined to be perceived by the operator just as another external tool, one that needs to be manipulated as an independent extension of one's body in order to achieve the desired goal. In other words, the reach of such a tool is limited by its inherent inability to be assimilated by the brain's multiple internal representations as a continuous extension of our body appendices or sensory organs. This is a significant point, because in theory, if such devices could be incorporated into 'neural space' as extensions of our muscles or senses, they could lead to unprecedented (and currently unattainable) augmentation in human sensory, motor, and cognitive performance.
>
> (Nicolelis, 2002a, p. 223)

At a subsequent NBIC workshop in 2004, another researcher, Rodolfo Llinas, from the New York University School of Medicine, considered how the electrodes utilized in the brain machine interface might be introduced in a new manner. He began his talk in an almost joking manner, telling how his friends at the medical school informed him that it's usually not a good idea to drill holes in peoples head and then insert foreign objects. (A reference to the implanting of electrodes associated with Nicolelis-like research.) He thus came up with the idea of using "nanowires" – wires so small they could be inserted by a catheter into the blood lines that feed the appropriate region of the brain, enabling a remote brain/machine interface to accomplish the kinds of things Nicolelis did with his implanted electrodes. Llinas then showed research results on how, in a Petri dish, the wire can trigger a neuron, and the neuron, when fired, can trigger the wire. He provided a Functional Magnetic Resonance Image or fMRI showing how, in a mouse, the wires cross the blood/brain barrier and snuggle up next to the neurons. (The account of Llinas' presentation comes from my recollections as a participant at the conference; his research is summarized in Llinas et al., 2005; Jain, 2006 and Leary et al., 2006 show how this research is viewed in neurosurgery and nanomedicine.)

These are just two of the many examples that could be provided on current research related to brain/machine interfaces. Some of this research is already in the human testing stage, enabling paraplegics to control "smart rooms" or blind people to detect objects through video glasses that completely bypass the natural eyes with electrodes that feed directly into the brain. And while the initial stages of human subjects' research will largely focus on medical treatments, many of the most prominent researchers have broader interests, seeking what Nicolelis has called "a major paradigm shift in the way normal healthy subjects can interact with their environment" with "unprecedented ability to augment perception and performance in almost all human activities" (Nicolelis, 2002a, p. 224). Some researchers are also engaging in self-experiment; for example, Kevin Warwick at the University of Reading in the UK has implanted microchips within his body in order to advance cybernetic research on the interface with "smart environments," viz., his laboratory. (A nice review of Warwick's work is provided in Clark, 2003.)

This research – just a glimpse at the first item in their list of 20 – already illustrates the kinds of enhancements that now lay on the horizon. Such "second stage enhancements" have the following features:

1. The enhancements provide *radically new capacities*. While many of these enhancements are initially developed for medical applications, the researchers involved immediately see their radical enhancement potential, and they openly explore ways to augment human ability. Even when the focus is on medical applications, treatment no longer just concerns the fixing of a broken machine. The very line between cure and enhancement blurs. Once a paraplegic has an interface with a smart room or a robotic prosthetic, it is a simple matter to upgrade the hardware to give superhuman strength or a new environmental awareness. A blind person with neural/video interface can as easily see in the infrared or ultraviolet range, as he could light of the visible spectrum.
2. The enhancements are *multi-functional*, and alter how we approach disability. In older disabilities research, there was an interest in restoring typical human function, and mimicking normality by means of prosthetics. Recently, there has been a shift to maximizing function, even if this involves completely new ways of developing prostheses. Thus, for example, a lost leg might be replaced with a strange looking prosthetic that might actually improve speed relative to the normal person. (There is now debate about whether individuals with such prostheses should be allowed to compete in normal races, since they now seem to have an advantage.) It is also being discovered that the technological linkages might introduce additional functionalities that might complement the restored kinds. Each of the 20 enhancements contemplated in the NBIC report concern a broad cluster of abilities, which are jointly enhanced by the new technology. The brain/machine interface, for example, might initially focus on some specific disability (like lost sight), but the technology itself is like a cell phone: once it is introduced, many other functions are enabled. The technology thus can provide a broad range of opportunities for the person who receives it, and by means of the new enhancements, people can explore new modalities of human existence.

Thus research on a disability like paraplegia might be the avenue for exploring new forms of human life. The "treatment" becomes an early experiment in how to transform human existence. (Many of the researchers on the cutting edge view this research in such a way, and we find strange combinations; e.g., military funding and medical research.)
3. The lines between diverse enhancements blur, and they involve the *convergence* of multiple kinds of technology. In the earlier stages, we had isolated streams, and the enhancement applications were usually associated with a specific function. Problems were localized in specific enhancement cultures. Now we find the streams starting to converge. In fact, a careful reading of the 20 NBIC items (or even the six highlighted earlier in this chapter) will show that there is considerable overlap between them. Thus, for example, developments in robotics, smart environments, and information technology turn out to be of central importance for brain/machine interfaces, and, likewise, advancements in the brain/machine interface will significantly accelerate and alter research in these other areas. The very name selected for the NBIC initiative is telling: there is a concern with a "convergence" of multiple streams of research, with an attendant recognition that each area complements the other.
4. The enhancements develop at an *accelerating rate*. A central theme in the initial NBIC workshop concerned how best to catalyze the enhancement research. It was recognized that initially some investment of effort and resources is needed to bring together the diverse streams of research, but once this linkage is made, there is an explosion of development. The enhancements thus advance at an accelerating rate, and, as they enable new capacities, these feed back upon the research process, making possible forms of advancement that cannot even be imagined at this stage. An example of such exponential advancement in capacity can be found in the Human Genome Project. When the project was first initiated, available technology for sequencing the genome would have required a century for completion. Planners recognized that enabling technology emerges and accelerates development. In fact, such development took place even faster than originally anticipated, so that the project was completed ahead of schedule.
5. The enhancements will provide *significant advantages* to those who obtain them. In competitive contexts of education, business, and the military, the pressure to use these enhancements will grow, and the problems raised by this will become prominent and pervasive in the everyday of life of all people.

Ethical and Social Issues Integral to Stage-Two Enhancement Technologies

The more radical enhancements (the stage 2 kind) are still a decade or two in the future, and it is tempting to say that we should not worry too much about them until they take a more concrete and specific form. At this stage, can we do much more than speculate? Such a "wait and see" attitude would, in fact, represent the ways we normally address the ethics and policy of emergent technology. Initially, we have researchers who push the envelope in science and technology, and industries, which find ways to link these emergent developments with human needs. As new technology takes form, we have different levels of ethical and social analysis. Thus, for example, a new pharmaceutical slowly percolates upward from the initial science and lab work to animal tests and ultimately to diverse phases of human subjects research, with their drug trials. As the drug comes on the market it is then further regulated by the medical norms which govern their prescription and use. If downstream some start to use these drugs for enhancement purposes, we simply take this as a new topic, and work through the ethical and policy issues at that stage Shouldn't we do the same thing with stage-two enhancements?

For several reasons, I think this traditional two step model (first R&D, then ethics and policy) is no longer appropriate. In fact, such assumptions are deeply problematic, because they don't sufficiently account for the radical character, accelerating rate, or transformative potential of stage-two enhancements.

While we are still a decade or two away from the full introduction of stage 2 enhancements, the research that determines the character of these enhancements is already well underway, and vast resources by industry and government are being invested in their development. We are already witnessing the initial stages of human experimentation, not just in medical arenas, but also in industry and the military. When these technologies are mature, they will not just be like a new gadget, even a highly influential one like computers or cell phones. They make possible radically new forms of human interaction, and with this, they alter the rules of post-hoc ethical and policy reflection. This is the key component to recognize. Many of the enhancements will be of such a kind that those who control them may have capacities to directly manipulate the rules of social engagement in ways we now might consider unfair. I'm not saying this will happen, but it might, and these kinds of radical shifts in power and control should be explored in tandem with the development of the technologies. If we wait until a later stage, when the developments are apparent, we may be too late to alter the path of the technology and channel it in the most constructive directions.

We also should consider the implications of the accelerating rate of development. Convergence catalyzes diverse streams of research, and the resulting explosion leads to countless specific technological innovations. This tremendous advance in productivity is a central concern of the U.S. Department of Commerce, and a reason why they have fostered such technological convergence. But the benefit has a flip side. When so many new, significant technologies emerge together, they outstrip our capacity to reflect on the ethical and policy implications. We can already see this in the field of bioethics. When I first entered the field about two decades ago, there was a relatively small bioethics community, and most had a general sense that the core ethical issues were being addressed. They were difficult new issues, for example, associated with the technological extension of life or allocation of organs for transplant. But since that time, the health care transformations and technological developments have accelerated to such an extent that only the most visible of topics are addressed. The new issues raised by any topic, for example, something like "medicine and the Internet" (a topic hardly addressed by the bioethics community) could occupy the whole bioethics community for countless years. The older model where you have an ethics and policy community that considers the results of science and industry is no longer a viable model, because the expanding ethics community remains too small, and the technological developments are too rapid and increasingly radical.

Instead, we today need a deeper integration of ethical and policy reflection into the diverse streams of research and industrial development. Those who create the new stage-two enhancements need to be actively involved in an ethical and policy discussion that considers how these capacities should be best advanced, and the larger public should be initiated into the radical, transformative projects that are now integral to our cutting-edge science and industry. This bridging poses deep challenges, both to our academic cultures of research, where there is a divide between the sciences and humanities, and to our industry and science agencies, which are vigilant against any outside intrusion into their sacrosanct domains. If we cannot move beyond the current polarizing discourse, still largely on the fringes of human life, what will happen when we fully enter the brave new world that already opens up in front of us?

Conclusion

While I think there are some clear things we can do to initiate a more responsible integration of research and ethical/policy reflection, I won't even attempt to outline these here. My goal has been more limited: *to sketch the kinds of enhancements that now lie on our horizon, and make apparent the depth and scope of the challenges they will pose.* Until we see this, any concrete recommendations will not make sense, especially since they must involve a commitment of effort and resources to change long-standing traditions of science policy and these traditions aren't easily altered. But if we see the challenge, then the excitement and gravity of the questions might lead to greater commitment to proactively address the ethical and social aspects of enhancement research. After all, shouldn't we all be interested in a project that seeks to reengineer our basic human capacities? Though few name it or even see it as such, many are involved in the project, and we already see some stage 2 enhancements taking form.

Whether we like it or not, Humanity 2.0 has entered the stage of beta testing, and we all have a stake in the new product.

Acknowledgement This essay is based on work supported by the National Science Foundation awards 0304448-NIRT: From laboratory to society: Developing an informed approach to nanoscale science and technology, and EEC-0646332: Complexity, systems and control in nanobiotechnology: Developing a framework for understanding and managing uncertainty associated with radically disruptive technology. The views are, of course, my own and do not necessarily represent the views of the National Science Foundation. An earlier version of this essay was distributed in electronic form as a briefing paper to policy makers in South Carolina through *Public Policy and Practice* vol 4, no 2 (November 2005), and was used for a June, 2007 meeting of the Expert Working Group on Enhancement of the European Union sponsored initiative, NanoBioRAISE.

References

Adam, D. (2001). Gene Therapy May be up to Speed for Cheats at 2008 Olympics. *Nature* 414: 569–570.
Alam, M. (2001). On Beauty: Evolution, Psychosocial Considerations, and Surgical Enhancement. *Archives of Dermatology* 137: 795–807.
Bayertz, K. & Schmidt, K. (2004). Testing Genes and Constructing Humans – Ethics and Genetics. In Khushf, G. (ed.), *Handbook of Bioethics*. Dordrecht, The Netherlands/Holland: Kluwer, pp. 415–438.
Bond, P. (2002). Converging Technologies and Competitiveness. In Roco, M. & Bainbridge, W.S.(ed.), *Converging Technologies for Improving Human Performance: Nanotechnology, Biotechnology, Information Technology and Cognitive Science*. Arlington, VA: National Science Foundation and U.S. Department of commerce, pp. 28–30. Available in PDF format at www.wtec.org/ConvergingTechnologies/.
Bond, P. (2004). Vision for Converging Technologies and Future Society. In Roco, M. & Montemagno, C. (eds.), *The Co-evolution of Human Potential and Converging Technologies*. Annals of the New York Academy of Sciences 1013, pp. 17–24.
Clark, A. (2003). *Natural-born Cyborgs: Minds, Technologies, and the Future of Human Intelligence*. Oxford/New York: Oxford University Press.
Davis, K. (2000). The Rhetoric of Cosmetic Surgery: Luxury or Welfare? In Parens, E. (ed.), *Enhancing Human Traits: Ethical and Social Implications*. Washington, DC: Georgetown University Press, pp. 124–134.
Elliott, C. (2000). The Tyranny of Happiness: Ethics and Cosmetic Psychopharmacology. In Parens, E. (ed.), *Enhancing Human Traits: Ethical and Social Implications*. Washington, DC: Georgetown University Press, pp. 177–188.
Elliott, C. (2003). Humanity 2.0. *Wilson Quarterly* Autumn: 13–20.
Fukuyama, F. (2003). *Our Post Human Future: Consequences of the Biotechnology Revolution*. New York: Picador.
Gardner, H. (2000). *Intelligence Reframed: Multiple Intelligences for the 21st Century*. New York: Basic Books.
Garreau, J. (2005). *Radical Evolution: The Promise and Peril of Enhancing our Minds, our Bodies – and What it Means to be Human*. New York: Doubleday.
Gazzaniga, M. (2005a). Smarter on Drugs. *Scientific American Mind* 16(2): 32–37.
Gazzaniga, M. (2005b). *The Ethical Brain*. Chicago, IL: Dana Press.
Gingrich, N. (2002). Vision for the Converging Technologies. In Roco, M. & Bainbridge, W. (eds.), *Converging Technologies for Improving Human Performance: Nanotechnology, Biotechnology, Information Technology and Cognitive Science*. Arlington, VA: National Science Foundation and U.S. Department of Commerce, pp. 31–47. Available in PDF format at www.wtec.org/ConvergingTechnologies/.
Hoberman, J. (2002). Sport Physicians and the Doping Crisis in Elite Sport. *Clinical Journal of Sport Medicine* 12: 203–208.
Hughes, J. (2004). *Citizen Cyborg: Why Democratic Societies Must Respond to the Redesigned Human of the Future*. Boulder, CO: Westview Press.
Jain, K.K. (2006). Role of Nanotechnology in Developing New Therapies for Diseases of the Nervous System. *Nanomedicine* 1(1): 9–12.
Khushf, G. (2007). The Ethics of NBIC Convergence. *Journal of Medicine and Philosophy* 32: 185–196.
Kurzweil, R. (2004). *Fantastic Voyage: Live Long Enough to Live Forever*. London: Rodale Books.
Leary, S.P., Liu, C.Y., & Apuzzo, M. (2006). Toward the Emergence of Nanoneurosurgery: Part III – Nanomedicine: Targeted Nanotherapy, Nanosurgery, and Progress Toward the Realization of Nanoneurosurgery. *Neurosurgery* 58(6): 1009–1025.
Little, M. (2000). Cosmetic Surgery, Suspect Norms, and the Ethics of Complicity. In Parens, E. (ed.), *Enhancing Human Traits: Ethical and Social Implications*. Washington, DC: Georgetown University Press, pp. 162–176.

Llinas, R.R., Walton, K.D., Nakao, M., et al. (2005). Neuro-vascular Central Nervous Recording/stimulating System: Using Nanotechnology Probes. *Journal of Nanoparticle Research* 7: 111–127.
McGee, G. (n.d). Parenting in an era of Genetics. *Hastings Center Report* 27(2): 16–22.
McGuire, G. & McGee, E. (1999). Implantable Brain Chips? Time for Debate. *Hastings Center Report* 29(1): 7–13.
Mendoza, J. (2002). The War on Drugs: A Perspective from the Front-line. *Clinical Journal of Sports Medicine* 12: 254–258.
Miller, F., Brody, H., & Chung, K. (2000). Cosmetic Surgery and the Internal Morality of Medicine. *Cambridge Quarterly of Healthcare Ethics* 9: 353–364.
Naam, R. (2005). *More than Human: Embracing the Promise of Biological Enhancement*. Calgary, AB, Canada: Broadview.
Nicolelis, M. (2002a). Human-machine Interaction: Potential Impact of Nanotechnology on the Design of Neuroprosthetic Devices Aimed at Restoring or Augmenting Human Performance. In Roco, M. & Bainbridge, W. (eds.), *Converging Technologies for Improving Human Performance: Nanotechnology, Biotechnology, Information Technology and Cognitive Science*. Arlington, VA: National Science Foundation and U.S. Department of Commerce, pp. 223–227. Available in PDF format at www.wtec.org/ConvergingTechnologies/.
Nicolelis, M. (2002b). The Amazing Adventures of Robot Rat. *Trends in Cognitive Sciences* 6(11): 449–450.
Nicolelis, M. (2003). Brain-machine Interfaces to Restore Motor Function and Probe Neural Circuits. *Nature Reviews/Neuroscience* 4(May): 417–422.
Nicolelis, M. & Chapin, J. (2002). Controlling Robots with the Mind. *Scientific American* (October): 46–53.
Nordmann, A. (2004). *Converging Technologies – Shaping the Future of European societies*. Available at www.ntnu.no/2020/pdf/final_report_en.pdf.
Porod, W. et al. (2004). Bio-inspired Nano-sensor-enhanced CNN Visual Computer. In Roco, M. & Montemagno, C. (eds.), *The Coevolution of Human Potential and Converging Technologies*. Annals of the New York Academy of Sciences 1013, pp. 92–109.
Roco, M. & Bainbridge, W. (2002). Converging Technologies for Improving Human Performance: Integrating from the Nanoscale. *Journal of Nanoparticle Research* 4(4): 281–295.
Roco, M. & Montemagno, C. (eds.) (2004). *The Coevolution of Human Potential and Converging Technologies*. Annals of the New York Academy of Sciences 1013.
Sandro, M.I. (2000). Real Brains for Real Robots. *Nature* 408: 305–306.
Silver, L. (1998). *Remaking Eden*. London: Harper Perennial.
Silvers, A. (2000). A Fatal Attraction to Normalizing: Treating Disabilities as Deviations from 'Species-typical' Functioning. In Parens, E. (ed.), *Enhancing Human Traits: Ethical and Social Implications*. Washington, DC: Georgetown University Press, pp. 95–123.
Stock, G. (2002). *Redesigning Humans: Our Inevitable Genetic Future*. Boston, MA: Houghton Mifflin.
Talwar, S. et al. (2002). Behavioral Neuroscience: Rat Navigation Guided by Remote Control. *Nature* 417: 37–38.
Vickers, M. (2003). Cyberhumanity: the Blurring Boundaries Between People and Technology. *Employment Relations Today* 30(2): 1–13.

Nanotechnology, the Body and the Mind

M. Ellen Mitchell

Abstract Nanotechnology promises to deliver the next wave of transformative inventions and devices with the potential to affect quality of life, longevity, the body, the brain, and the mind itself. However, there is not an arena for the sharing of knowledge across disciplinary realms to enable a deeper consideration of, and appreciation for, the potential and real impact that these technological advances will likely have on individuals, relationships, behavior, and culture. Knowledge of psychological science of human mind and behavior should be interwoven with the discovery and invention process if biomedicine and nanotechnology are to bring forward an improved quality of life. Human behavior tendencies surely will influence the way these technologies are used. We need to resist our very human tendency to think that anything new is an improvement, or that so called enhancements are some how better, and take lessons from our experiences with other advances that have demonstrated that there consequences, both intended and unintended, as well as error with new ventures. Above all, we need to develop structures and mechanisms to assure that interdisciplinary discourse can occur to maximize the probability that these advances are consistent with human flourishing.

Keywords Enhancement; brain, human behavior, executive functioning, intelligence, relationship violence, genetic engineering

Overview

Biomedical science is enjoying a new wave of innovation and financial support coincident with the shift from mere medical science or biology, to the integration of multidisciplinary findings and the advent of an array of nanoscale advances. These developments include, but are not limited to, novel materials, substances, informatics, and devices. Born out of the application of knowledge from traditional academic spheres applied to unique problem domains, these advances have the associated promise of improved

---

Director, Institute of Psychology, Fellow Institute Biotechnology and the Human Future, Illinois Institute of Technology

quality of life, health, and longevity. Science now allows us the possibility of knowing our future with respect to genetically linked disorders and diseases, the option to grow new tissue and soon whole organs, and the possibility to enhance our fundamental intellectual capacities. While these advances will undoubtedly influence health and well being, there is little research on, or understanding of, what this particular facet of scientific progress might mean for quality of life, culture, and our sense of being human.

Methodologies for assessing these technological impacts require long term longitudinal research that is cross disciplinary and difficult to conduct. Further, there is little incentive for designers and inventors of technology to undertake this type of investigation, and it is not practical to expect that it be completed. This should not, however, stop us from exploring what we have learned and how the many lessons of the past may help us to establish a meaningful research agenda and in turn to integrate that knowledge into our unfolding technological work.

The past century has brought forward an incredible array of inventions that have had a fundamental impact on morbidity, mortality, quality of life, life style, culture and society. Virtually all industry sectors and all academic disciplines have contributed in one manner or another to a changed face of life. The obvious transformations include changes associated with utilities, transportation, medicine, computing, telecommunications, and materials, a list too long to even attempt to enumerate, but all of which have been cast as revolutionizing, as indeed they were.

Many of the changes connected with invention have been direct, as in the case of antibiotics for infection, and most also have had other far-reaching and some times indirect influences. For example, coincident with recovery from specific infections, longevity has been extended through enhanced immune function and the benefits of early intervention and halting diseases that would have become ravaging if untreated by modern medicine. Indeed, the top ten causes of death have changed dramatically since 1900 from pathognomonic diseases like pneumonia and tuberculosis (TB), to illnesses best characterized as life style related like heart disease and cancer.[1] Powerful treatments for some diseases like TB have spawned resistant strains that now elude effective intervention protocols. For example, the media has recently been reporting on what have been termed superbugs or drug resistant strains of bacteria.[2] These strains are one example of many unintended consequences of our use of biological and pharmaceutical advances.

The advent of antibiotics is but one of a host of markers within a period of medical discovery that has had far reaching effects and implications. It was not merely antibiotics, or electricity, or high speed air flight that changed the world, as it were, but the confluence of many technological inventions collectively that has had a profound impact on all individuals and society. We now consider disease to be not merely a function of pathogens invading the body but the product of immune function, resistance, and changing, ever mutating pathogens. Our technologies have changed how we think about events and the world, as well as how we behave and feel. These concrete changes are evident to most who pause to reflect on them. It is more difficult, however,

---

[1] CDC (2005).
[2] Society for General Microbiology (2007).

to identify the not so subtle but very insidious ways that technological advances have changed the fundamental ways in which we think about problems, our communities, and ourselves. The changes before us portend even greater change for the future. For this reason, it is important to briefly address advances and promises of technology along with human psychological phenomena. As Parens[3] noted, "It would be a mistake to think that new biotechnologies are just more of the same" and it is important to be mindful of how and what benefits we would like our technologies to deliver.

Progress and Promise

Biomedicine and Nanotechnology

The average American would probably be hard pressed to define nanotechnology, list any of the medical advances that are promised by this area of investigation, or explain how it is core to biomedical advances.[4] It would be instructive, for the purposes of this paper, to outline briefly this area and to identify some of the discoveries that are being pursued under the umbrella of nanoscience. Broadly, the interest in nanoscience arises from the recognition that materials on the nanoscale do not behave in a manner consistent with the same substance in bulk.

Nano-science sits somewhere between molecular chemistry and quantum physics and is not one thing but a descriptor of size or scale. Nanoscale refers to sizes at the atomic level wherein, for example, a red blood cell is roughly 10,000 nm and a vaccinia virus, related to smallpox, is 200 nm. Size is thought to be a causal variable in transforming substances, which some believe is related to the high level of exposed surface area. Simplistically, the transformation of activity from that which is evident in the substance in bulk, as compared to, and differentiated from the activity of the substance at the nanoscale, is posited to occur because most molecules are exposed or on the surface thus creating optimal conditions for changes in atomic composition.

At the nanoscale, materials demonstrate unique magnetic, optical, and electrical properties such that materials actually become devices. For example, gold, which is inert in mass quantities, becomes conductive on the nanoscale and demonstrates distinctive optical qualities that are different from what is reliably produced by larger quantities. On the nano scale, single molecules take on the capacity to function as switches, gears, or other devices. Presumably, it's just a matter of time until scientists will be able to assemble molecules to build or create virtually anything. In the case of biomedical science and nanotechnology, the anything we are targeting, is ourselves. As Garreau[5] noted, "...we have started a wholesale process of aiming

---

[3] Parens (1998).
[4] For essays on the social, ethical and legal issues associated with nanotechnologies see: Cameron and Mitchell (2007).
[5] Garreau (2005).

our technologies inward. They have started to alter our minds, our memories, our metabolisms, our personalities, our progeny, and perhaps our souls. The shift is so profound that serious people are calling it radical evolution".

This shift, and the associate hyperbole regarding nanotechnology, may be unparalleled by any other technology or advance yet to come. While the public has been told numerous times that medicine will, for example, cure cancer, extinguish infant mortality and wipe out other major diseases, no area before nanotechnology has promised to cure all the ails of the body, deliver immortality and solve problems of the planet. In an NSF/DOC sponsored report,[6] 20 ways that converging technologies could benefit humanity with in the next couple of decades were identified. These included but were not limited to the following:

> ...The human body will be more durable, healthier, more energetic, easier to repair, and more resistant to many kinds of stress, biological threats, and aging processes.... A combination of technologies and treatments will compensate for many physical and mental disabilities and will eradicate altogether some handicaps that have plagued the lives of millions of people.... Anywhere in the world an individual will have instantaneous access to needed information, whether practical or scientific in nature... Engineers, artists, architects, and designers will experience tremendously expanded creative abilities... The vast promise of outer space will finally be realized... Average persons as well as policy makers will have a vastly improved awareness of cognitive, social, and biological forces operating in their lives...
>
> The authors conclude in their overview that If we make the correct decisions.... The twenty first century could end in world peace, universal prosperity, and evolution to a higher level of compassion and accomplishment.

This is quite a tall order. Perhaps it would be useful to examine what has come forward thus far, and focus on what is real rather than projected.

Nanotechnologies have already penetrated the market place. Rajeski[7] noted in 2004 that there were, then, already more than 131 products on the market that contained nanoparticles. Among some of the nano-bio based products or inventions already complete or underway is an incredible array of items. There are a variety of every-day items that incorporate nanoparticles including sunscreens, cosmetics, baby products, dyes, paints, tennis balls, fibers, coatings, etc. All these have gone forward without fanfare, unregulated and undisclosed, because the FDA declared nanoparticles to be no different from the same substances in bulk.[8] Products with nanomaterials are currently being developed across many sectors including cosmetics, leisure goods, transportation products, the military, nano electrics, computing, pharmaceuticals and biological materials. It is the last two categories that are the subject of this paper, albeit, the categories are not discrete. Nano-materials and nano electrics will undoubtedly have implications for biomedical applications for use as sensors, therapeutics, genetic engineering, pharmacology, prosthetics and more.

Other more dramatic discoveries and inventions have also occurred. For example, researchers at the Max Planck laboratory controlled the movement of a leech

---

[6] National Science Foundation (2004).
[7] Rajeski (2004).
[8] Hood (2004).

via communication between neurons and electronically based neuron transistors.[9] In essence, a computer controlled a living organism. Researchers at the *University of California, Santa Barbara*, created a nanoscale hybrid named the "smart bio-nanotube".[10] It is thought that this creation might serve for precise drug or therapeutic gene delivery. The nanotubes are labeled "smart" because they can open or close at the ends, depending on how the researchers manipulate the electric charge of the two components. So in principle, a nanotube could encapsulate a drug or a gene, and then open on command so that the contents would be released in the precise location desired. Similarly, free tissue transfer is being accomplished with the use of polymers that function as cell delivery systems. In essence, these polymers are used as a vehicle to deliver cells to a specific site where tissue growth is needed, as in the case of liver, heart, or bone regeneration or reconstruction. In addition to delivery of cells to a site, the polymer provides a structure or three dimensional spaces for the new cell growth.[11] Moreover, research into tissue manufacturing is moving full bore with the hope of development of whole organs as well as replacement tissues ranging from skin to muscle to brain.

DARPA (the Pentagon's Defense Advanced Research Projects Agency) is concentrating its efforts to create better soldiers, not through equipment but through enhancement. "Soldiers having no physical, physiological, or cognitive limitations will be key to survival and operational dominance in the future".[12] As noted by Washington Post reporter Farhi,[13] "... researchers are starting to see a more mundane, but culturally significant, sideline to gene therapy: the potential to create nearly superhuman athletes. The same techniques that could repair diseased muscles may enable athletes to heft more weight, run faster or jump higher than ever thought possible". The projected uses of these inventions reach into realms usually accorded only to science fiction.

Yet another area, nanobiosensing for use in medical diagnostics, is also being vigorously explored. Nanoparticles with optical fiber or biomarker capacities are being investigated for use in the identification of infection or individual cell changes. It is hypothesized that information gleaned by these sensors will be transmittable through tissue. The nanobiosensor diagnostic industry is expected to reach the billion-dollar mark by 2007–2008.[14] Researchers at private labs, as well as our national laboratories, are incorporating mono-metals like iridium, ruthenium, and rhodium, into RNA[15] with the intention of developing biogenetically based therapeutics and determining

---

[9] Kurzweil (2005).
[10] Press release, UC Santa Barbara (2005).
[11] Terada et al. (2000).
[12] Goldblatt (cited in Garreau, 1997 p. 105, op. cit.).
[13] Farhi (2005).
[14] See for example, Chu (2005).
[15] See for example, RNA Could Form Building Blocks for Nanomachines. Posted 2004. Available at http://www.azonano.com/news_old.asp?newsID=274.

how to enhance the information capabilities of these strands. This work is intended for eventual human uses, perhaps by expanding memory and intelligence.

While biomedical engineering and nanoscience predict great future strides, there is much occurring in medicine now that casts a pall on the unbridled optimism of nanoscience hype. Transplantation is an area that promised fabulous outcomes but which is not without its problems. People who receive transplanted organs also frequently endure negative sequelae, some of which are linked to fundamental psychological processes and others to physiological covariations. For example, optimism accounts for 40% of the variance in outcomes for heart transplant patients, unique from the effects of preoperative physical health.[16] Heart transplant patients have also been shown to have guilt and concerns about their donors, many expressing the presumably magical belief that the donor heart might have an affect on their personality.[17] Heart recipients show high levels of stress even years post transplant surgery.[18] Cognitive deficits have been documented to occur after cardiac surgery.[19] Similarly, people with pancreatic transplants have a significant incidence of peripheral neuropathies. Side effects attendant with biomedical advances can contribute to disability that, while perhaps preferable to death, can be substantial. The promise of perfection has yet to be delivered.

For transplant recipients, there are many transplant side effects associated with the immunosuppressive drugs that are necessary to avoid organ rejection. These problems include but are not limited to depression, headache, and changes in physical appearance that are sometimes dramatic.[20] Kidney transplant patients have higher than average risks for some cancers, perhaps associated with anti-rejection drugs.[21] One to two percent of people die within a year of pancreatic transplantation; survival rates for people with functioning kidneys who receive a pancreatic transplant are lower than those who manage their diabetes with conventional drugs and diet.[22] People who receive organs also are more likely to be susceptible to diseases from animals and other sources.[23] These negative side effects and outcomes, and particularly the psychological aspects, are seldom the focus of public discourse, nor do the scientific designers of the technology generally anticipate them. The public tends to leap in their thinking from new to great. The idea of organ replacement in the abstract is more appealing than the reality that is faced by many if not most transplant patients.

Biomedical work in the area of fertility is another source of lessons for consideration as we move forward with developing technologies. Research on assisted reproduction has gone beyond the study of life outcomes for quintuplets to more

---

[16] Leedham et al. (1995).
[17] Kaba et al. (2005).
[18] Inspector et al. (2005).
[19] Di Carlo et al. (2000).
[20] Rodrigue et al. (2005).
[21] Kyllonen et al. (2000).
[22] Venstrom et al. (2003).
[23] CDC (2000).

careful research of the incidence of a whole host of problems correlated with the various techniques used to help couples bear children. Ovulation induction, a procedure used to stimulate ovulation and one of the least invasive of the assisted reproduction technologies (ART), has been shown to be associated with statistically significantly higher incidences of placental abruptions, fetal loss after 24 weeks, and gestational diabetes, albeit the causal mechanisms are unknown. In vitro fertilization is associated with a higher frequency of preeclampsia, hypertension, placental abruption and placenta previa.[24] A sample of infants conceived with ART was shown to have breast development and public hair present at birth.[25] Several studies have shown an increased incidence of birth defects among babies conceived with ART.[26]

In addition to medical interventions for organ failure, people presently exercise many options for enhancement. These include but are not limited to breast enhancement through implants, cosmetic enhancement though surgery, and mood and performance enhancement through drugs.

Medical science has allowed us to use surgery to enhance appearance, change body shape, or restore failed organs but, enhancement's past record has not always been correlated with improvement. One can, for example, elect to have gastric bypass surgery to diminish obesity. Ironically, significant proportions of persons who have this surgery regain their original weight[27] and develop loss of control of their eating.[28] While surgery results in initial weight loss, it does not result in improvements in attendant depression and low self regard.[29] Dermabrasion and botox enhance appearance by decreasing wrinkles and blemishes, but these procedures need to be repeated after relatively short periods of time. The human body has a tendency to return to its former status despite heroic interventions and inventions. This migration to status quo, or regression to the mean, and return to prior states is ubiquitous and calls into question the value and wisdom of the pursuit of enhancement, particularly when our resources might be better used for other purposes.

The history of human use of substances and devices to maintain functioning, recover functioning, augment functioning, enhance functioning, or change functioning, spans across time. The use of chemical substances for enhancement has had a history that parallels the historical use of devices insofar as most drugs, both legal and illegal, have been ingested for centuries for many purposes. Historical writings are replete with references to drug ingestion for more than simple medicinal uses[30] including laudanum, various so called elixirs, and opiates. Certainly the use of opiates in Asian countries dates back thousands of years. Recreational use of

---

[24] Shevell et al. (2005).
[25] Rojas-Marcos et al. (2005).
[26] Hansen et al. (2005).
[27] Powers et al. (1997).
[28] Saunders (2004).
[29] Mathus-Vliegen et al. (2004).
[30] See e.g., Whitebread (1995).

substances is as old as time, indeed mammals, birds, and humans have all exhibited a propensity to ingest intoxicating substances, frequently fermenting fruit. For the most part, these uses impair rather than enhance functioning.

Human nature being what it is, people often undertake actions for purposes that are at variance with the planned uses. Prozac for depression is not the same as Prozac for escaping the human condition of occasional anxiety and dysphoria. Kramer[31] noted that there is an increasing usage of drugs for the purpose of shaping personality to make us more attractive personally, which he refers to as cosmetic pharmacology. This of course presupposes the idea that one can define an ideal personality. The use of steroids to reduce swelling, as in the case of demylenating polyneuropathies, is qualitatively different than the use of steroids for sport and performance enhancement. All of us know of examples wherein individuals ingest drugs in search of personal gain, escape, mind expansion, or other transformation.

Genetic engineering or the isolation, manipulation and reintroduction of DNA into an organism to introduce new beneficial characteristics, is another area of research and discovery offering the hope of curing a number of serious and debilitating conditions including Alzheimer's, Huntington's, and Parkinson's diseases. By manipulating the human genetic code and by adding and subtracting genes to replace defective or missing ones, researchers hope to repair problematic cells and sequences.

Work in the past two decades in the area of genetics and behavior has brought widespread acceptance of the view of genetics as a causal factor in individual differences, personality, and behavior. Indeed there is a risk of assigning genetic attributions that overlook the fact that complex psychological traits are subject to environmental influences.[32] Genetic research is changing the conceptualization of behavior and individual differences so that the public has an increasing sense of genetic determinism while researchers are finding that there are multiple genetic and environmental influences that exist in a complex interplay in which single gene effects are less important than quantitative trait loci or gene systems.[33] The Human Genome Project raised the hope and prospect of change at the most fundamental level, promising to bring forward changes that would endure, thus solving the other problem of incomplete and impermanent treatments.

The recognition that genetic defects can give rise to problems ranging from disease through disorder overshadows the fact that the average person has between 5 and 50 genetic mutations and it supports the belief of the value and possibility to be genetically perfect; verily defect free.[34] The idea that gene defects can be fixed in some process of genetic engineering fails to take into account the complex interaction affects that scientists are documenting.[35] Genetic science is forging ahead, driven by the idea that if people had perfect genes, then they would be problem free. This notion is wishful thinking; an empirical question at least and not a foregone conclusion.

---

[31] Kramer cited Parens. op. cit.
[32] Plomin and College (2001).
[33] Plomin and Crabbe (2000).
[34] Andrews (2001).
[35] See e.g., Wong et al. (2005).

Application of knowledge from biogenetic research is also associated with new approaches to the diagnosis and treatment of disease. It was reported, for example,[36] that scientists in Israel used DNA strands to detect specific RNA that signaled the presence of active diseases as in the case of prostate cancer. Presumably, information gleaned from genetic testing or genetically based medical testing should provide impetus for people to be more conscientious about their health and to be observant about early signs of the disease. However, research suggests that the stress generated by such findings may diminish the propensity for monitoring and negate the benefits that the knowledge was supposed to provide.[37]

So called enhancement often does not persist and it may contribute to ongoing chronic adversity and challenge. The iatrogenic effects of medical procedures are a well documented consequence of modern medicine. Moreover, the area of enhancement is not without controversy because the line between therapy and enhancement has become hazy with increasing acceptance of enhancements. Gains in acceptance have shifted the topic of debate from one of deciding if an intervention is medically necessary and therapeutic, versus elective or cosmetic, to the question of feasibility and gross risk. The distinction between health related and non-health related enhancement is not always clear.[38]

Use of devices for personal enhancement is not new. One might even assert that human kind, with its problem solving capacities and delight of novelty, has an innate tendency to use a variety of methods and materials for all of these purposes. For example, when George Washington obtained wooden teeth, presumably it was to recover and then maintain functioning, in that case mastication. Some might characterize his false teeth as medical/therapeutic because they better enabled him to eat; others might speculate they were cosmetic and elective. Most human actions are multi-determined, the cleverness and ingenuity of people will undoubtedly continue to contribute to the boom of advances and unique unexpected usages of our discoveries. Herein lies a predicament: if people can benefit from enhancement technologies for medical and health reasons, then why not psychological reasons like well being, for no reason other than desire and sheer ability to do so? If technology allows us to improve on ourselves then why not improve on our nature? If technology will allow us to craft a vision of beauty, no matter how idiosyncratically, culturally or historically imbedded it might be, then why not? What we know about human nature is that if we can, we will. The dilemma is not that the line is blurry but that there is no longer a line to hold. Biomedical intervention and treatment decisions become individual, posing difficulty for the development of regulation. Some options are available only to those with the necessary monetary resources. Even when we agree that medical choices may not be sensible or advisable, we know, as Parens ludically commented, that there will always be schmocters[39]

---

[36] Pollack (2004).
[37] See Andrews op. cit. for a discussion of the impact of genetic testing on prevention activities, p. 42.
[38] Meilaender (2001).
[39] Parens, op. cit.

(doctors in the business of making money) willing to provide treatments that other physicians may eschew. He also noted that, ironically, enhancement never works because, if everyone achieves enhancement, then no one has a competitive or distinctive advantage.

Nanotechnology and genetic technologies are promising to take enhancement to the next level, offering repair for defective genes, organ regeneration, tissue replacement, cognitive enhancement, memory enhancement, aging cessation, and augmentation of physical strength. It is unclear if many of the predicted advances will occur, or whether we should even label or think of them as advances a priori rather than faddish opportunities for personal expansion or decoration, or something else. It is important to question whether or not so called advances should, de facto, be considered advances or enhancements[40] or if we should suspend our judgment and enthusiasm, casting them as items that are simply interesting, to be approached with curiosity and patience until we observe and learn more. It seems, consistent with our human tendencies that we jump to conclusions before we even have the discoveries and we become embroiled in the mere prospect of something new, which then becomes imbued with possibility, subsequently resulting in disappointment and fueling more vigorous improvement seeking. Perhaps we should intervene on this vicious cycle rather than on our bodies.[41]

What is evident is that there is no aspect of human life or the human body that is exempt from being targeted for and by these advances. A question that has not been asked, however, is what these interventions might mean to the recipients of such materials and those who live around them? What will be the ripple effects of these advances? For example, if we achieve the goal of making people stronger, what will this mean for problems of violence? Violence is epidemic in America and relationship violence, between people who know and care about each other, accounts for the largest proportion of that violence.[42]

The proclivity to use drugs and devices to improve one's self is not a function of the technology per se, but is more an attribute of human nature. Hence, problem solving ingenuity and curiosity have contributed to great discoveries and transformation. These same qualities and attributes contribute to novel uses of that which we create. Thus, it is important to consider what happens when technology allows us to take the penchant to pursue beauty, novelty, and the desire to better ourselves to another level so that instead of, for example, using necklaces to elongate the neck, or alcohol to dull pain, we imbed devices in our bodies to eavesdrop, achieve infrared vision, copy ourselves, or transform our offspring into some sort of superhero. What is the line between cosmetic enhancements, extreme makeovers (to quote Hollywood), and re-crafting components of the human person? Is the pursuit of perfection life enhancing or, a distraction from more important and fundamental life issues? The intersection of human tendencies and technology is the nexus from which thorny problems are arising.

---

[40] For a discussion of enhancement see President's Council on Bioethics (2003).
[41] See also, Mitchell (2007).
[42] See e.g., Langhinrichsen-Rohling (2005).

A profound difference between the historical uses of devices and drugs and modern day uses pertains to the intention of the use and the invasive capability of the inventions. Use of drugs for individual palliative or recreational purposes is quite different from large, mass scale use of drugs for enhancement. It is not the use or misuse of a substance per se, but its intended purposes, the pervasiveness, and the invasiveness, that are key here. What we have before us is qualitatively, and quantitatively, different: the possibility of mass use of technologies that have the potential to change not an individual, but whole populations at the level of the brain and bodily functions. Such activities will have far reaching implications for culture; transhumanists predict that this will give rise to a new species.[43]

Human Tendencies

The problem with nanotechnology, biomedical science, and psychology is that they are all encompassing and much too large for easy integration of findings.[44] Changes growing out of science and technology do not occur in one fell swoop, nor are they uniform in distribution within or across people. Consequently, promises to render people more intelligent, for example, even if they could occur, would not occur in a manner akin to turning on a light bulb. Human behavior and psychological phenomena will inevitably exert important influences on change processes. It is therefore useful to note some of the essential areas in which people struggle because, while medical advances may result in many changes, our fundamental nature and the struggles that arise from who we are as people will probably not change, and if and when it does, it certainly will not be instantaneous. If biomedicine and nanotechnology brings forward a human transformation, our human qualities will influence what we do with invention and it will exert pressures that affect the uses of biomedical advances and nanotechnology.

There are several arenas around which human problems coalesce that will likely be a source variance. While there are many aspects and conceptualizations of human behavior that are not the subject of review or discussion for this paper, there are several domains of behavior that seem most relevant. These will be glossed over; each is a field unto itself that fills many books, so this chapter will of necessity, present a simplistic and overly brief description of a few of the most relevant areas.

The majority of psychological constructs that are pertinent to emerging technologies fall under the rubric of executive functioning. A second relevant area is self-regulation, which in turn bears a relationship to emotional intelligence. A third area of human behavior that will influence our uses of biomedical technologies includes our capacity

---

[43] See e.g., Bostrom (2005).
[44] The areas discussed in this paper are too big and in many respects, I have not done them justice. It is important for readers to seek out much more material on all the diverse topics touched on herein.

for delay and overall strength of negative capability. Finally, it is important to bear in mind that our pursuit of novelty, and our fundamental capacities for ingenuity and creativity, guarantee that this work will go forward and so we should face the possibilities before us that emerge from technology in combination with people.

In the psychological literature, the definition and measurement of executive functioning is not wholly agreed upon.[45] That debate not withstanding, the construct refers to one of our most important sets of cognitive functions. Broadly speaking executive functioning refers to higher order brain activities including planning, organization, attention, concentration, judgment, and the like. Most of these functions are critical to effective navigation on a day-to-day basis but they are difficult to define and measure. Most of us have a sense about what constitutes good judgment, and understand that this is not a unitary construct. Good judgment depends on weighing many factors including context variables, as in the case of deciding when to be assertive and when to adopt more passive stance. Planning is temporally grounded and shifts depending on the time phase adopted. Short and long term plans can yield very different perspectives on what constitutes the best practice, decision, or action. Executive functioning is an area of complex skill that requires coordination of sensory information, emotion, intellect, behavior, reason, the environment, and interpersonal interactions.

Executive functioning skills develop across time with practice, many on the basis of trial and error. Some individuals however, struggle to attain these skills, even with many trials across the entire life span. Such individuals face myriad challenges, and present challenges to others as well, because of the many ways in which these skills intersect with interpersonal interactions and life transactions. There are a range of behaviors and situations that reveal deficits in executive functioning. The most critical include problems with self-regulation, clarity of thought, attention, management of emotion, difficulties with capacity for delay of action and limitations associated with judgment and decision-making. Of significance, our technologies are making demands on our attention that are taxing. In large measure, various technological devices claim our attention rendering us a nation of partial attention. Consequently, regulatory schemes for limiting cell phone use while driving, for example, have become increasingly popular. These policies and laws fail, however, to address the real problem: it is not the absence of one's hands on the wheel of a car that is the central problem but rather it is the absence of one's brain power – cognitive capacities and attention – that places us at risk. Moreover, the frequent functioning with partial attention, on individuals and culture across time, is an unknown.

While neuropsychological research has revealed that many executive functions are controlled by the frontal lobes of the brain, it has also shown that brain functioning is both localized and diffuse. Improving frontal lobes of the brain is not parallel to breast augmentation because all brain parts are interconnected with feedback occurring between sections of the brain, organs, senses, and the body generally. The notion that one could go to a single place in the brain or alter a particular

---

[45] See for example, Barkely et al. (2001).

neurotransmitter, to bring about change in something as general as intelligence or executive functioning, for example, fails to take into account the complexities of the brain.

Self regulation[46] is a significant domain of psychological functioning encompassing many different facets that connect to an array of areas in which most people are challenged, to a greater or lesser extent. These challenges can range from the most ubiquitous, for example, regulation of diet or spending, to the more esoteric as in the regulation of expression of thought. In point of fact, most people struggle in some area(s) of regulation, whether related to bodily function (e.g., insomnia, dysregulated eating), mood (e.g., dysphoria, irritability, anger), behavior (e.g., spending, gambling, completion of work), thought (e.g. irrationality, rumination), feeling (e.g., chronic pain or lack of joy), interaction with others (e.g., promiscuity, withdrawal), or violence (e.g. partner violence, war crimes), to name some but not all of the most common domains that are expressions of this struggle.[47]

People tend to operate with excesses and deficits, some times overdoing it, and other times failing to act. Omissions and commissions both contribute to success and error. Potentially, regulation problems may be manifest in the usage of any item; it is neither unique nor confined to nanotechnological advances. Thus, why should there be any cause to even comment on this particular quality in relation to biomedical advances, whether on the nanoscale or some other scale? If in fact nanotechnologies evolve in the directions now predicted, they will have intense far reaching implications for how we are fundamentally, one to another, and as individuals. Our best and our worst qualities may well be amplified. If we develop devices that, for example, correct DNA defects, will we not want to correct all of our defects? If we undertake to perfect ourselves, to what end and what vision do we select as our goal? How do we, as a pluralistic society, decide how much of a good thing is enough, when as human beings we have a basic challenge in our capacity for regulation thus blocking us from ever reaching the point of enough?

In addition to struggles with executive functions and self regulation, most individuals experience uncertainty, confusion and indecision in the face of ambiguity. The strength of our negative capabilities, or our capacity to manage negative states like anxiety, suspicion, ambiguity, sadness and loss are a source of human skill or, conversely, limitation. We are, of necessity, sometimes suspicious because we sense our own inability to be sure of what and who is before us. We can act with great conviction only to discover later that we were mistaken. This is illustrated, for example, by the fact that many states have banned capital punishment because, despite our certainty of correctness, mistakes have been made.

The period of the 1960s, and the use of hallucinogenic drugs in religious ritual across the millennia, demonstrated to us some of the things that occur when experiences are confusing and hallucinogenic; nightmares teach us the responses we can have to wholly fictional experiences while harboring conviction that the experience

---

[46] Kanfer and Schefft (1988).
[47] Ibid.

is real. Moreover, empirical research demonstrates that rational decision-making can be fully compromised by mere arousal and emotional perceptions.[48] Nanotechnologies promise to be fully immersive wherein we will be able to enjoin in a virtual world. How will we distinguish between the real and the virtual when all our senses are subject to input from realms, the real and the invented? How will people manage heightened levels of arousal from the virtual world and greater frequency of experiences that tax our negative capabilities and executive functioning?

The challenges associated with human incapacities for uncertainty will be compounded by our growing reactivity or inability for postponement. In this information age, people have an increased intolerance for delay. Who among us has not had the experience of waiting all of 90 seconds for a web page to load and perceived it to feel interminable? The efficiency and speed of our technologies have infused us with a sense that everything should occur instantaneously. Computers have spawned a belief that if we enter things correctly then anything can be done rapidly and accurately every time.[49] Perhaps emerging nanotechnologies will usher in a shift of expectations to the even smaller metric of nanoseconds.

Our tendency to want things now, right this moment also has an attendant corollary of action in which we act and react in the moment, occasionally without sufficient latency for considered thought. It is sometimes more important to sleep on matters, as it were, than to bang out a response in the moment. There are multiple indicators of growing reactivity in the culture as exemplified by new words and pithy labels like road rage, hooking up, hyperdating, flaming, and just-in-time life style.[50] Our media fuel this by delivering news in real time in which unfolding events take on snap shot meaning so that conclusions are some times reached prematurely and incorrectly. Actions taken on the basis of ill conceived conclusions can be problematic. As a culture, we have become less and less inclined to take the time to find things out, than to be proactive, forward looking, get a leg up, engage in scenario planning, invent the future, hit the ground running, be cost effective, time efficient, and take calls during dinner, on the train, or in the wash room.

Our capacity to wait and postpone action is diminishing; so too is our sense of time being truncated, although the gravity of this may not be evident. The ability to understand cause and effect, to grasp the meaning of events and intention, is not an easy matter. Human interaction is rich and complex, wherein fluidity and perception are dominant characteristics of daily life. Our inferences and grasp of causes or, more personally, our assignment of blame, are antecedents to action. If we are always acting and reacting, but seldom reflecting, it is difficult to achieve integrated understanding. Perspectives can change in the fullness of time when they become disconnected from the heated moment of the event. In a diverse global world, taking time to consider events, our actions, others, and ourselves becomes even more important if we are to live peaceably with celebration of differences. Conflict resolution

---

[48] Shafir and LeBoeuf (2002).
[49] Turkel (2004).
[50] Wallace (1999).

depends, in large measure, on the ability to take the perspective of another. This capacity to integrate perceptions, knowledge, and skill is, in major portion, the essence of emotional intelligence.[51]

A great deal has been written about emotional intelligence, both what it is and about its relevance and importance.[52] There is recognition that raw intellectual intelligence of the type that is necessary to achieve superlative grades in school is not the totality of intelligence in the broader sense.[53] The ability to interact productively with others, to function well in a dynamic world, to engage fully in the activities of daily life across the multiple demands of work, home, school, and family, requires a deeper set of skills and abilities. Nanoscience is promising to be the vehicle for that delivery by perhaps enhancing memory, problem solving, and capacity for computation, logic, reasoning and the like. However, altering one of these skills, or raw ingredients as it were, may tax other human capabilities. Even highly intelligent people can make very poor decisions if self-interest, distorted perspectives, erroneous conclusions, intense emotion or, biased and unclear judgment co-exist. Indeed research on intelligence and school performance[54] suggests that intelligence is not a stable trait that a person has but a complex and fragile state that can fluctuate dramatically.

The drive to enhance human capability, presumably, is to make a better world. A vision for a better world however, requires an extraordinary perceptiveness, deep understanding and wisdom. The risks of enhanced raw intelligences in the face of unchanged emotional intelligence, executive functioning and other skill areas, in combination with human foibles, may not result in the outcomes that inventors and scientists seek.

The attitude that people should avail themselves of technologies that promise improvements gives rise to a creeping sense that there is something wrong with the status quo that requires change. Bostrom[55] has written directly on status quo bias and suggests that reluctance to embrace new technologies arises from this cognitive predisposition. Briefly, his view is that since we do not know the future, we can use only intuitive judgments in forecasting, and he presents data that these forecasts are prone to status quo bias. This presupposes, however, that there are not relevant data from past events or scientific findings, to be factored into those judgments. There is a critical difference between making decisions based on intuition and making decisions based on experience and/or other relevant empirical scientific data.

There is great pressure on individual patients, and guardians of minors, to make medical decisions quickly and early with little time to learn about the long term implications of those procedures. Medical information, both positive and negative, is typically provided by physicians and staff in terms that are intended to be understandable, and which are spelled out on consent forms, but which do not necessarily give full meaning to all the words. Actual data are not typically available, and not

---

[51] Goleman (1997).
[52] Emmerling and Goleman (2003).
[53] See, e.g., Sternberg (2004).
[54] See e.g., Aronson and Steele (2005).
[55] Bostrom, op. cit.

all parents would be able to interpret the findings if they were available. For example, what parent(s) would have the wherewithal to inquire about what it would actually mean for a diabetic child receiving a pancreatic transplant to have peripheral neuropathies? Would the average parent be able to connect the fact that most brittle diabetics have substantial vision loss, that can be managed through learning Braille and realize that peripheral neuropathies might have implications for being able to read Braille? The absence of data is a problem; likewise a flood of data is also problematical. In this information age, we have so much information that it is often difficult to discern the meaning from the data and make connections between and across diverse sources and domains of information.

The Road Ahead

Most medical and psychological research uses methods that emphasize similarities and differences from a statistical perspective based in norms. Thus, research on medical improvements frequently analyzes group differences, as in the case of comparing one group of people receiving a treatment to another group not receiving the treatment of interest. This approach usually involves statistical means comparisons to determine if the groups are sufficiently different to think about them as coming from different populations. Individual difference approaches on the other hand target variances and differences, not means or averages and similarities. Our whole approach to evaluating the impact of technologies thus rests on a concept of a normal population and an average response. In reality, the average response may not objectively exist and may only be a statistical artifact. Further much of the research that is undertaken in biomedicine is aimed at factors that might increase acceptance of the technology, rather than on the impact of the technology in relation to life style or quality of life.

An enormous proportion of biomedical research on human subjects emphasizes topics like adherence to medical regimens, which begin from the view that, if people can be convinced to just follow physician orders, then outcomes will be better. These approaches are useful, and in reality many health outcomes are better when people follow directions. However, these approaches do not necessarily question the wisdom of the goal and do not factor individual differences. It is inevitable that there will be outliers; it is inevitable that there will be differences that distribute in some uneven fashion because there are simply too many human variables to line up. Medical science contributes to the perspective that failures to achieve positive outcomes are a result of the patients' failure to conform and follow directions, or that possibly some component of the treatment is inadequate. Seldom do we consider the view that the absence of positive outcomes might be an expression of the inadequacy of our whole approach to developing improvements or that individual variation should perhaps be encouraged, if not nurtured. Medical science does not know or, forgets, that failure to follow medical regimens is a great example of a problem with self regulation.

Perhaps, as Parens[56] suggested, there are natural differences that medicine should not be used to erase: "The great challenge is to find ways to relieve suffering that do not perpetuate harmful conceptions of normality".[57] Hintermair[58] similarly noted, for example, that our approach to persons with hearing impairments is to develop technologies that solve the problem of being deaf, rather than make room in the culture for diversity. The populace is stuck on the notion that differences are deficiencies that should be fixed and that technology should provide the cures. Technology, with its assembly line approach to creation is aimed at reliable production of accurate or correct products and yet, it is human diversity that enables us to adapt, prosper, and transform. Human diversity accounts for the fact that responses to, and uses of, a single device or substance are not identical.

While the view of the human body as a machine that can benefit from replacement parts may be compelling, the body and mind are unique so that one part does not fit all. Interventions that are effective for some are simply not viable for everyone. The idea of a unified approach to the body overlooking the mind and the complex variations of individuals is doomed to failure. Perhaps our inability to regulate ourselves is now being expressed in this unending drive for the silver nano bullet that will deliver improvement and achievement of a unitary image of perfection. The image that we appear to be pursuing may be distracting us from other more important tasks like figuring out how to live well, and work together in the face of ambiguity about the future with all of our various foibles and differences.

The assessment of advances, the evaluation of their added value to determine if they are in actuality life enhancing, is extremely difficult to undertake. Serious, systematic and interdisciplinary discourse and research are necessary to integrate important findings already available so that we can avoid the next wave of disappointment and negative sequelae that will inevitably accompany this most recent burgeoning area of discovery. Integrated, systematic interdisciplinary research of type that would be necessary is uncommon, if not unheard of. The mechanisms for discourse between people from various disciplines and sectors also are limited. Such research and discussion are extremely difficult, and few scientists want to be distracted from their primary work or engage in broad inquiry but instead prefer to focus and produce findings that are extensive with respect to depth and limited with regard to breadth. This is not a fault of scientists; it is a strength that enables critical brilliant work but it is an impediment to other activities. This brings up a final essential point about human tendencies: That which is a strength in one arena, may be a limitation in another. Thus, how can we conceive of enhancement or perfection when many qualities have more than one value or face?

It has been suggested[59] that machines help us manage our lives and compensate for human frailty, that, "We make our technologies and our technologies make and shape us".[60] The recognition that the environment in which we live exerts a bi-directional

---

[56] Parens, op. cit.
[57] Parens, ibid., p. 19.
[58] Hintermair (2005).
[59] Turkel, op. cit.
[60] Turkel, ibid., p. 28.

influence is important and should compel us to examine the course we are on. We need to learn from our selves, and the past, and perhaps relinquish our efforts to achieve perfection, immortality, and sameness.

It is incumbent on all of us to remember that experience has taught us that steep discovery trajectories, such as those we are experiencing with biomedical advances, have attendant error trajectories. It is imperative to recall that when scientists conceive of discovery for the benefit of, and from the perspective adopted by, the average person that in many respects, scientists are not themselves, average. This renders them particularly ill suited to conceive of the ways that the average person might benefit from or use any particular invention and it exemplifies the need for inclusion of multiple other perspectives and disciplines.

Perhaps we should step back from our work and tendency for action to reflect on how we might use medical technology to better be ourselves, embracing the perspective that our lives are less problems to be solved but individual journeys to be regarded from the unique and vulnerable places that we each represent and occupy. Above all, we need considered and extensive interaction and discourse between the disciplines that is coincident with the developments in biomedicine because the impacts are likely to be vast.

References

Andrews, L.B. (2001). *Future Perfect: Confronting Decisions About Genetics*, New York: Columbia University Press.

Aronson, J., & Steele, C.M. (2005). Stereotypes and the Fragility of Human Competence, Motivation, and Self-concept. In C. Dweck & E. Elliot (Eds.), *Handbook of Competence & Motivation*, New York: Guilford.

Barkely, R., Edwards, G., Laneri, M., Fletcher, K., & Metyevia, E. (2001). Executive Functioning, Temporal Discounting, and Sense of Time in Adolescents with Attention Deficit Hyperactivity Disorder (ADHD) and Oppositional Defiant Disorder (ODD). *Journal of Abnormal Child Psychology*, 29(6), 541–556.

Beyond Therapy: Biotechnology and the Pursuit of Happiness. A Report of the President's Council on Bioethics. 2003. Washington, DC: U.S. Government Printing Office.

Bostrom, N. (2005). "Transhumanist Values". *Review of Contemporary Philosophy*, 4(1–2), 87–101.

Cameron, N., & Mitchell, M.E. (2007). *Nanoscale: Issues and Perspectives for the Nano Century*. New York: Wiley.

CDC. Guidelines for Preventing Opportunistic Infections among Hematopoietic Stem Cell Transplant Recipients. *Morbidity and Mortality Weekly Report*, October 20, 2000, 49(RR10), 1–128.

CDC. *National Vital Statistics Report*, 2005, 53(17). Available at www.cdc.gov, last visited 6/19/05.

Chu, W.L. (2005). Trend Sees Boom in Pharmaceutical Analytical Tools, November. Available at http://www.labtechnologist.com/news/ng.asp?n=63923-separation-technology-chromatography-biosensors.

Di Carlo, A., Baldereschi, M., Amaducci, L., Maggi, S., Grigoletto, F., Scarlato, G., & Inzitari, D. (2000). Cognitive Impairment without Dementia in Older People: Prevalence, Vascular Risk Factors, Impact on Disability. The Italian Longitudinal Study on Aging. *Journal of American Geriatric Society*, 48(7), 775–782.

Emmerling, & Goleman. *Emotional Intelligence: Issues and Common Misunderstandings. The Consortium for Research on Emotional Intelligence in organizations*, 2003. Available at www.eiconsortium.org, last visited 9/15/2005.

Farhi, P. (2005). Beyond Steroids: Designer Genes; for Unscrupulous Athletes, Better Bodies Are a Tweaked Chromosome Away, March 17. Washington, DC: *Washington Post*, p. C-01.
Garreau, J. (2005) Perfecting the Human (p.103). *Fortune, 151*(11), 101–108.
Goleman, E. (1997) *Emotional Intelligence*. New York: Bantam Books.
Hansen, M., Bower, C., Milne, E., deKlerk, N., & Kurinczuk, J. (2005) Assisted Reproductive Technologies and the Risk of Birth Defects – A Systematic Review. *Human Reproduction, 20*(2), 328–338.
Hintermair, M. (2005) Ethics, Deafness, and New Medical Technologies. *Journal of Deaf Studies and Deaf Education, 10*(2), 184.
Hood, E. (2004) Nanotechnology: Looking as we Leap. *Environmental Health Perspectives, 112*(13). Available at http://ehp.niehs.nih.gov/members/2004/112-13/focus.html.
Inspector, Y., Kutz, I., & David, D. (2005) Another Person's Heart: Magical and Rational Thinking in Psychological Adaptation to Heart Transplant. *Israeli Journal of Psychiatry and Relational Science, 41*(3), 161–173.
Kaba, E., Thompson, D.R., Burnard, P., Edwards, P., & Theodosopoulou, E. (2005). Somebody Else's Heart Inside Me: A Descriptive Study of Psychological Problems after a Heart Transplantation. *Issues in Mental Health Nursing*, July, 26(6), 611–625.
Kanfer, F., & Schefft, B. (1988) *Guiding the Process of Therapeutic Change*. Champaign, IL: Research Press.
Kurzweil, R. Foreward to Virtual Humans. Available at http://www.kurzweilai.net/meme/frame.html?main=/articles/art0600.html?, last visited 9/15/05.
Kyllonen, L., Salmela, & Pukkala, E. (2000) Cancer Incidence in a Kidney-transplanted Population. *Transplant International, 13*, 394–398.
Langhinrichsen-Rohling, J. (2005) Top 10 Greatest "Hits": Important Findings and Future Directions for Intimate Partner Violence Research. *Journal of Interpersonal Violence, 20*(1), 108–118.
Leedham, B., Meyerowitz, B., Muirhead, J., & Frist, W. (1995) Positive Expectations Predict Health After Heart Transplantation. *Health Psychology, 14*(1), 74–79.
Mathus-Vliegen, E.M., deWeerd, S., & deWit, L.T. (2004) Health Related Quality-of-life in Patients With Morbid Obesity After Gastric Banding for Surgically Induced Weight Loss. *Surgery, 135*(5), 489–497.
Meilaender, G. (2001) Designing our Descendants. *First Things, 109*, 25–28.
Mitchell, M.E. (2007) Scientific Promise: Reflections on Nano-hype. In N. Cameron & M.E. Mitchell (Eds.), *Nanoscale: Issues and Perspectives for the Nanocentury*. New York: Wiley.
National Science Foundation. *Large Benefits from a Small World*. Available at http://www.nsf.gov/od/lpa/nsf50/discov/nanoadver.htm, last visited 07/19/2004.
Parens, E. (1998) Is Better Always Good? The Enhancement Project. *Hastings Center Report, 28*(1), 1–22.
Plomin, R., & College, E. (2001) Genetics and Psychology. *European Psychologist, 6*(4), 229–240.
Plomin, R., & Crabbe, J.C. (2000) DNA. *Psychological Bulletin*, 126, 806–828.
Pollack, A. (2004) A Glimpse at the Future of DNA: MD's Inside the body. *New York Times*, April 29, 19.
Powers, P.S. Rosemurgy, A., Boyd, F., & Perez, A. (1997) Outcome of Gastric Restriction Procedures: Weight, Psychiatric Diagnosis and Satisfaction. *Obesity Surgery, 7*(6), 471–477.
Press release, UC Santa Barbara "Smart" Bio-Nanotubes Developed; May Help in Drug Delivery, August 2, 2005. Available at http://www.ia.ucsb.edu/pa/display.aspx?pkey=1325, last visited 9/15/05.
Rajeski, D. (2004) The Next Small Thing. *The Environmental Forum*, March/April, 42–49.
RNA Could Form Building Blocks for Nanomachines. Posted 2004. Available at http://www.azonano.com/news_old.asp?newsID=274.
Rodrigue, J.R., Baz, M.A., Kanasky, W.F., & MacNaughton, K.L. (2005) Does Lung Transplantation improve Health Related Quality of Life? *Journal of Heart Lung Transplant, 24*(6), 755–63.
Rojas-Marcos, P.M., David, R., & Kohn, B. (2005) Hormonal Effects In Infants Conceived by Assisted Reproductive Technologies. *Pediatrics, 116*, 190–194.
Saunders, R. (2004) "Grazing": A High Risk Behavior. *Obesity Surgery, 14*(1), 98–102.

Shafir, E., & LeBoeuf, R. (2002). Rationality. *Annual Review of Psychology, 53*, 491–517.
Shevell, T., Malone, F., Vidaver, M.A., Porter, F., et al. (2005). Assisted Reproductive Technologies and Pregnancy Outcome. *Obstetrics and Gynecology, 106*, 1039–1045.
Society for General Microbiology. MRSA in the Community: A New Threat to Children's Health?. *Science Daily.* Available at http://www.sciencedaily.com /releases/2007/11/071127212208.htm, last visited 11/29/2007.
Sternberg, R. (2004) Culture and Intelligence. *American Psychologist, 59*, 491–517.
Terada, S., Sato, M., Sevy, A., & Vacanti, J.P. (2000) Tissue Engineering in the Twenty-first Century. *Yonsei Medical Journal, 41*(6), 685–691.
Turkel, S. (2004) Whither Psychoanalysis in Computer Culture?. *Psycholanalytic Psychology, 2*(1), 16–30.
Venstrom, J.M., McBride, M.A., Rother, K.I., Hirshberg, B., Orchard, T., & Harlan, D.M. (2003) Survival after Pancreas Transplantation in Patients With Diabetes and Preserved Kidney Function. *Journal of the American Medical Association, 290*, 2817–2823.
Wallace, P.M. (1999) *The Psychology of Internet.* Cambridge: Cambridge University Press.
Whitebread, C. (1995) *History of Non-medical Use of Drugs in the United States.* Available at http://www.druglibrary.org/schaffer/History/whiteb1.htm.
Wong, A.H., Gottesman, I.I., & Petronis, A. (2005) Phenotypic Differences in Genetically Identical Organisms: The Epigenetic Perspective. *Human Molecular Genetics, 14*(1), 11–18.

Nanotechnology and Human Flourishing: Toward a Framework for Assessing Radical Human Enhancements

Ronald Sandler

Abstract Any robust human enhancement needs to be assessed according to the risks and costs involved in both pursuing and realizing the enhancement, how it will affect those who are enhanced, how it will affect those who are not, and how it will shape our form of life. These sorts of assessments are often underwritten by beliefs about what constitutes human nature, the human condition, and human flourishing. In this chapter I identify and review these beliefs and the way that they are employed in discussions regarding human enhancements. I then propose a framework for assessing human enhancements that is oriented around human flourishing, rather than human nature.

Keywords Human Nature, Human Flourishing, Human Enhancement, Ethical Evaluation

> What [existentialists] have in common is simply the fact that they believe that existence comes before essence… What do [atheistic existentialists] mean by existence precedes essence? We mean that man first of all exists, encounters himself, surges up in the world – and defines himself afterwards…He will not be anything until later, and then he will be what he makes himself…; man simply is.
> (Sartre, 1948, pp. 338–339)

> Now we say that the function of a [kind of thing]…is the same in kind as the function of an excellent individual of the kind… And the same is true without qualification in every case, if we add to the function the superior achievement in accord with the virtue; for the function of a harpist is to play the harp, and the function of a good harpist is to play it well. Moreover, we take the human function to be a certain kind of life, and this life to be activity and actions of the soul that involve reason; hence the function of the excellent man is to do this well and finely.
> (Aristotle, 1999, pp. 1098a8–1098a15)

Ronald Sandler
Department of Philosophy and Religion
Nanotechnology and Society Research Group
Northeastern University

> If the *Cognitive Scientists* can think it,
> the *Nano* people can build it,
> the *Bio* people can implement it, and
> the *IT* people can monitor and control it.
>
> (Roco & Bainbridge, 2003, p. 13)

Introduction

The social and ethical issues associated with emerging nanotechnologies are diverse and legion. Many of the issues are familiar – e.g., privacy and information security, environmental impacts, work place and consumer safety, distributive justice, public engagement, intellectual property, and informed consent. While certainly significant, these issues are nevertheless in direct continuity with already well recognized and often well understood aspects of responsible technology development and implementation, though they may exhibit novel aspects due to the characteristic features, rate of innovation, or potential volume of nanoscale technologies, or else the distinctive features of the social contexts into which they are emerging (Sandler, 2007b). However, not all social and ethical issues associated with nanoscale science and technology are of this type. Nanoscale science and technology has the potential to enable or otherwise contribute substantially to technological innovations, such as robust artificial intelligences, synthetic biology, artificial life-forms, virtual worlds, and direct brain-machine interfaces, which raise social and ethical issues that do not have any direct precedent. These issues, though perhaps not as immediate or urgent as the familiar issues, are as much within the domain of nanoethics, since they are issues associated with innovations that nanoscale science and technology help to realize, and addressing them is as much a part of responsible development of nanotechnology, since they are relevant to the extent to which nanotechnologies contribute to human flourishing in socially just and environmentally sustainable ways, within appropriate moral boundaries (Sandler, in press). In this chapter I consider one such set of technologies, human enhancement technologies, contextualizing them within ongoing (and in some respect perennial) discussions regarding human nature and human flourishing.

Human enhancement through technology is ubiquitous. Pharmaceuticals enhance our immune system. Pacemakers enhance our circulatory system. Hearing aids enhance our perceptual system. Fertility treatments enhance our reproductive system. Nutritional supplements enhance our physical performance. Most people embrace a wide range of areas of enhancement, methods of enhancement, and purposes of enhancement, including and beyond therapeutics. If so many forms of human enhancement are already generally considered permissible, if not desirable, then there should not be anything problematic with more extensive forms of enhancement. After all, expanding human cognitive, perceptual, and physical capabilities by means of brain-machine interfaces, exoskeletons, genetic engineering, and biochemical reprogramming, for example, differ only in degree, not in kind, from what we now regard as mundane and ethically benign enhancements.

This common line of reasoning is fallacious. The fact that two things differ only in degree does not imply that there is no ethical distinction between them. Sending a child to his room alone for 5 minutes can be appropriate punishment, sending him there for two days is child abuse. The difference is "only" a matter of degree, length of time, but it makes an ethical difference. Good humored joking differs "only" in degree from ridicule. Strict dieting differs "only" in degree from anorexia. Torture differs "only" in degree from interrogation. To take a different sort of example, Clovis spear points differed "only" in degree from their technological predecessor, but they helped enable a human population explosion and the extinction of North American mega-fauna. Identifying a property of something, even a crucial one, and showing that it differs only in degree from some ethically (or socially or environmentally) unproblematic object or practice does not establish that it is also unproblematic. The difference in degree may be ethically significant or there may be other properties that are relevant to ethical evaluation. If particular forms of extreme human enhancement are ethically acceptable, it is not because they are in some respects "only" an extension of the sorts of enhancement commonly performed and accepted today.

This is not to claim that extreme human enhancement is ethically problematic. It is only to point out that the ethical status of human enhancements is not determined by the acceptability of currently widely employed enhancement technologies. Any human enhancement needs to be assessed on its own merits, not on the extent to which it is like what has come before. It needs to be assessed according to the risks and costs involved in both pursuing and realizing the enhancement, how it will affect those who are enhanced, how it will affect those who are not, and how it will shape our form of life. These sorts of assessments are often underwritten by beliefs about what constitutes human nature, the human condition, and human flourishing. Ethical evaluation of extreme human enhancements therefore requires identifying and reviewing these beliefs and the way that they are employed in discussions regarding human enhancements.

Human Nature Defined and Human Nature Discovered

Some philosophers have argued that what is distinctive about human beings is that our existence precedes our essence. We have no function, no purpose, prior to defining it for ourselves. Darwinism would seem to support this. We are just one species among many, the product of non-intentional, non-rational, non-orchestrated physical processes. We are self-conscious and self-reflective, but this only enables us to recognize and wonder about this situation in which we find ourselves; it does not alter its parameters. To some this is disturbing; to others it is liberating. We are not constrained by external designs, expectations, or requirements. We place the limits on our possibilities. Human nature is what we make it. Human flourishing is what we want it to be. We create ourselves.

Others have argued that what is distinctive about human beings is the particular sort of essence that we have. We differ innately from individuals of all other species.

We have psychological and emotional capacities that they lack. We are reflective and deliberative in ways that they are not. These innate features both open and circumscribe the possibilities for our life-form. Darwinism does not belie this, it describes the mechanism by which it was brought about. Human nature is not constructed by us. It is has been "constructed" through the process of evolution and is discovered by us. It is not in our choices. It is in our biology, our genes. Human flourishing is not wide open. It is determined by our innate biology.

The truth is in between. The science of human nature has flourished. Neuroscience, behavioral genetics, cognitive science, and evolutionary psychology have been successful in establishing an innate, evolved human nature that substantially explains our form of life. Our nature is in the structure of our brains, our modes of learning and cognition, our behavioral and emotional dispositions, and our intuitive tendencies. We are social because of our biology; we are rational because of our biology; we have culture because of our biology. We have the faculties, dispositions and tendencies that we do because they were adaptive for our ancestors in the environments in which they lived (Pinker, 2002; Richerson & Boyd, 2005; Ehrlich, 2000).

But biology is not destiny, and genes are not identity. First, environment is relevant to how a particular genotype is expressed or affects phenotype. What a particular gene or gene sequence "does" within a person often is dependent upon internal and external environment. Second, the variation among human traits and behaviors is only partially explained by genetic difference. Environmental factors also explain a significant amount of the variance for most psychological, cognitive, behavioral, and personality traits (Pinker, 2002). Third, although the broad outlines of our form of life are established by our innate human nature, the details are not. Forms of sociability, methods of parenting, ways of taking pleasure, and modes of production, for example, have differed widely among cultures, as well as among individuals within cultures. There are limits on this variation. Some norms, such as prohibition of murder and care for infants, appear in all cultures and are necessary for a social group to persist or function at all. Nevertheless, the human form of life admits considerable variation in the details of its realization (Ehrlich, 2000; Brown, 1991). Fourth, culture is a cause of human behavior that is not reducible to genes and physical environment. Culture is only possible for us because we are capable of certain forms of knowledge and technology creation, transmission, and accumulation; and we are only capable of these because of our innate evolved human nature. However, once culture emerges it exists independently of our genes and is partially explanatory of many human behaviors and practices (Richerson & Boyd, 2005). Fifth, our innate human nature is not prescriptive. "Human nature" refers to how our brains are structurally and functionally organized, the innate tendencies and dispositions we have, and our range of physical, cognitive, and psychological possibilities. It does not indicate which tendencies ought to be encouraged or discouraged, which possibilities ought to be pursued or avoided, or by what means these ought to be done. Characterizing innate human nature is a project in description, while deciding among social, ethical, and political norms and institutions is a project in prescription. Those norms and institutions must be sensitive to our dispositions and possibilities.

The prescriptive project needs to take into account the best information about ourselves and our world, since a scientifically informed understanding of human nature may provide support for some prescriptions or undermine support for others (Rachels, 1990). However, the science of human nature does not determine what ends we ought to promote or how we ought to promote them.

Human Nature and Human Flourishing

The general point defended above – that a proper understanding of our innate, biological human nature must inform, but does not fully determine, social and ethical ends and norms – is applicable to accounts of human flourishing. We are living, sentient, social, rational animals. Each of these aspects is essential to us. Each contributes to shaping our form of life and thereby informs what is constitutive of human flourishing. As living beings our good involves survival, biological health, and reproduction. As sentient beings our good involves enjoyment and the absence of pain. As social beings our good involves being part of well-functioning social groups and healthy relationships. As rational beings our good involves autonomy, meaningfulness, and the accumulation of knowledge. These are not controversial claims. Most people would agree that, in general and under most circumstances, longevity, health, knowledge, pleasure, well-functioning social groups, and so on, are goods. These two things are related. It is because they are rooted in our biology that shapes our form of life that they are widely regarded as constituents of human flourishing (Sandler, 2005, 2007a).

One need not realize each of these to some maximal or ideal degree in order to live well or flourish. Nor is there a hierarchical ordering of their importance. Certain goods figure more prominently in the lives of some people than in others. A person might flourish or live well in this or that way, or flourish or live well over all, even while some of these goods are not substantially realized in her life. Furthermore, there is not one correct way of realizing these ends. All human societies care for their young, have forms of recreation, artistic expression and medicine, and innovate and transmit knowledge. Different cultures, sub-cultures, and individuals simply do so in different ways. This cultural and individual variation is possible because of human commonalities that substantially differentiate us from individuals of other species (Ehrlich, 2000; Richerson & Boyd, 2005). For individuals of other species, because they are non-rational or significantly less or differently rational, both what constitutes their flourishing and how they realize it are strongly determined by their biological natures (Foot, 2001). There is much less (and so much less variable) technological innovation, complex social structuring, and accumulation of these over generations. So while social learning and social structure play a significant role in the form of life of some species, and innovation and technology has been observed in some species, these still fall far short of that which is widely considered distinctive of human culture: the accumulation and transmission of knowledge, technology, and social practices over generations.

Although individuals and cultures vary in how they aim and strive to realize the ends constitutive of human flourishing, not all realizations of them are as good as any other. Realizations of one end that are detrimental to accomplishing others, or the same end in the future, are less good than those with greater fecundity. For example, pleasure from recreational methamphetamine use is pleasure, but it undermines realizing most of the other ends, and is therefore unendorsable. Similarly, social groups that stymie the intellectual and cognitive development of their members, such as some college fraternities, are unendorsable for that reason, even if they function well. Some realizations of the ends may be unendorsable because they are detrimental to the flourishing of others or are premised on false beliefs. For example, the Ku Klux Klan and other racially motivated hate groups are detrimental to those outside of the group, are socially disruptive, and are premised on false beliefs about human biology and the moral significance of skin color and ethnicity. So even (or, more to the point, particularly) when these types of social groups are well-functioning, they are not endorsable. These standards of evaluating possible realizations of the ends constitutive of human flourishing do not follow strictly from the science of human nature or the natural sciences more broadly. They involve, for example, non-empirical, normative accounts of what constitutes a morally relevant difference. Because we are rational, moral agents, not all realizations of the ends constitutive of human flourishing are equally endorsable. They cannot all be seen as equally good (Hursthouse, 1999).

Human nature defines the broad outlines of our form of life, and in so doing provides the basis for a general account of the constituents of human flourishing. However, it does not settle the details of how those ends are to be realized or pursued, or their role in the context of different people's lives. Nor does it fully determine which variations of realization and pursuit are endorsable and which are not. What constitutes human flourishing is biologically informed in the general, but it is not biologically determined in the particular.

Human Flourishing and Nanotechnology: The Radical and the Routine

When Richard Sylvan asked in 1973, "Is There a Need for a New, an Environmental, Ethic?" he answered, correctly, that there was (Sylvan, 1973). The prevailing social and ethical norms were no longer informed by or consistent with our best understanding of ourselves, our world, and the relationships between the two. Over the course of only a few generations, scientific discovery and technological innovation, implementation, and dissemination had altered radically our population size, our capacity for consuming and depleting environmental resources, our understanding of the links between our health and the environment, our geographical mobility, and accounts of our origins and uniqueness. As a result, many thinkers are working to develop a new ethic for our new, scientifically and technologically transformed, situation. They are reconsidering human chauvinism, exploring the possibility of human-independent values in nature, and developing ethics capable of handling non-local, non-immediate, impersonal, and collective action problems. Of course, science and technology have continued to advance, and there is now a new technological revolution underway. Nanotechnology promises to improve everything from consumer electronics to military weaponry, clothing to medicine, tennis balls to energy production, sunscreen to building materials, and environmental remediation to beer containers. Predictions are that it will be transformative, encompassing, and inevitable.

Just as we ought to try to anticipate challenges posed by nanotechnology to current regulatory systems before they materialize, we should try to anticipate challenges to social and ethical norms grounded in obtaining conceptions of the human person and situation. This is not merely because a technological onslaught may be coming and we should begin preparing ourselves. It is also because appreciating the possibilities for how some nanotechnologies might alter us and our relationships would seem a crucial part of making informed and discriminating judgments regarding whether to support particular types of nanotechnology or lines of nanoscience research. Although a nanotechnological revolution may be inevitable, its particular shape and trajectory are not. Still less defined are any social or ethical transformations that might result. Even if it is true, as many claim, that the rate of technological progress is exponential, it does not follow that social change also proceeds at that pace. Indeed, there is no way to measure social change in a way that makes such a claim intelligible. Moreover, not all social change engendered by technological progress is properly considered social progress. For these reasons, social progress is neither fully determined nor well measured by technological progress.

One of the implications of previous revolutions in science, particularly the Darwinian revolution, is that human flourishing must be understood naturalistically. To flourish as a human being is to flourish as a particular kind of living, sentient, social, rational animal. For human beings, the constituents of flourishing are (something like) longevity, health, reproduction, pleasure and the avoidance of pain, well-functioning social groups and healthy relationships, autonomy, knowledge, and meaningfulness, realized in an endorsable form by endorsable means. Are any of the facts about us that inform what counts as a flourishing human life so understood going to be altered by nanotechnology? Probably. Nanomedicine, for example, promises to allow us to avoid and eliminate some diseases, recover from some previously terminal injuries, and slow bodily and cognitive deterioration. If it delivers on this promise, then what can reasonably be considered a long, healthy life might be different in 30 years than it is now. Nanotechnology also has the potential to enable novel forms of realizing several of the other constituents of human flourishing. Nanotechnologies might make possible new forms of sociability; they might enable new ways of reproducing ourselves; and they might make possible new sources of pleasure and new ways of avoiding pain. Therefore, advances in nanotechnology might require that we rethink what counts as a long, healthy, socially rich, pleasurable, meaningful human life.

This sort of rethinking is familiar. Current conceptions of what counts as health and longevity in industrialized nations differ from what could have been reasonably maintained 200 years ago, and in just the last decade information technologies and the internet have enabled novel forms of sociability that have implications for what can reasonably be considered well-functioning social groups and relationships. So the impacts of nanotechnology on human flourishing described above, while potentially profound in many ways, are nevertheless similar to those of previous technologies. Most nanotechnologies are *routine nanotechnologies* in this respect: They are intended to improve on what we already have in historically familiar sorts of ways, although their cumulative effect might be to significantly alter what can reasonably be maintained as realizing the constituents of human flourishing.

But could nanotechnology go further? Is it possible that nanotechnology will enable altering the constituents of human flourishing, not just what can be considered endorsable realizations of them? Many nanotechnologists believe so. For example, there is a research program underway, funded in part by the United States National Science Foundation and drawing participation from members of the mainstream scientific community, called nanotechnology, biotechnology, information technology, and cognitive science (NBIC) convergence. Its aim is to explore the possibilities for humans and society at the intersection of these technologies, and the expectation is that "converging technologies integrated from the nanoscale would achieve tremendous improvements in human abilities, and enhance social achievement" (Roco & Montemagno, 2004, p. vii). According to Mihail Roco, chair of the National Science and Technology Council's subcommittee on Nanoscale Science, Engineering and Technology and Senior Advisor for Nanotechnology at the National Science Foundation, "Accelerated improvement of human performance has become possible at the individual and collective levels. We have arrived at the moment when we can measure signals from and interact with human cells and the nervous system, begin to replace and regenerate body parts, and build machines and other products with finesse suitable for direct interaction with human tissue and the nervous system" (Roco, 2004, p. 3). Among the "key visionary ideas" of the NBIC program are "expanding human cognition and communication" and "improving human health and physical capabilities" with technologies located both inside and outside the human body (Roco & Bainbridge, 2003, p. 17).

If the NBIC research program, or other programs like it, such as the Defense Advanced Research Project Agency (DARPA) programs to radically enhance soldiers' cognitive and physical capabilities (Garreau, 2005, chap. 2), are successful, a description of the sort of beings that we are might not be exhausted by a strictly biological description, and an account of our flourishing might not be exhausted by a strictly naturalistic account. It remains to be seen whether the development of technologies that would challenge obtaining conceptions of human flourishing are science fiction or science-in-progress. Roco puts the timeline for "Converging technology products for improving human physical and mental performance (brain connectivity, sensory abilities, etc.)" at one generation and "evolution transcending human cell, body, and brain" at (a cautious) n generations (Roco, 2004, p. 6). Ray Kurzweil, a well known technologist and futurist, projects 2045 as the year when the merging of human and machine intelligence will have "a profound and disruptive transformation in human capabilities" (Kurzweil, 2005, p. 136). Even Bill Gates has said that he expects a future, still some generations off, which includes direct brain-machine interfaces (Sullivan, 2005). But while the timeline is vague and contested, the language is telling. Those who are promoting technologies of this sort speak of human-technology co-evolution, post-human evolution, human-cyborg evolution, and transhuman evolution. They seek to make possible extending human and social development beyond the "constraints" of our biology. This is *radical nanotechnology*. It does not merely aim at changing the ways in which people can realize human flourishing or offer methods of "optimizing" human nature and the pursuit of human happiness. It aims at altering the nature of human beings and thereby what is constitutive of human flourishing.

The Limits of Human Nature: How Not to Assess Human Enhancement

Since at least Aristotle, the questions that have dominated the ethics of human flourishing have been: What is human nature? What is constitutive of human flourishing, given human nature? How can we, individually and collectively, achieve human flourishing, given human nature? But now it seems, on the basis of what many prominent nanotechnologists consider to be real possibilities, that human nature no longer needs to be considered entirely given. We finally may be approaching the technological sophistication that will enable us to meaningfully ask: What do we want our nature and flourishing to be? This is a resurgence of Jean-Paul Sartre's existentialist question, since our existence would precede our essence, although in this case human nature would not be created by our actions but by our science and technology. The possibilities are not wide open, but constrained by human nature as it is given and the limits of our technological capabilities. Nevertheless, the science of human nature may soon pass from a period of discovery to a period of creation.

Let us set aside the technological, economic, political, and sociological questions regarding this possibility and focus instead on an ethical question: Is this a future that we ought to want or encourage? Those who answer affirmatively tend to see the central human challenge as pushing boundaries and defying limits. Representative of the *transcendence* view are Roco, William Bainbridge, Ramez Naam, and Kurzweil.

> Despite moments of insight and even genius, the human mind often seems to fall far below its full potential. The level of human thought varies greatly in awareness, efficiency, reactivity, and accuracy. Our physical and sensory capabilities are limited and susceptible to rapid deterioration in accidents or disease and gradual degradation through aging…All too often we communicate poorly with each other, and groups fail to achieve their desired goals. Our tools are difficult to handle, rather than being natural

extensions of our capabilities. In the coming decades, however, converging technologies promise to increase significantly our level of understanding, transform human sensory and physical capabilities, and improve interactions between mind and tool, individual and team...

Each scientific and engineering field has much to contribute to enhancing human abilities, to solving the pressing problems faced by our society in the twenty-first century, and to expanding human knowledge about our species and the world we inhabit; but combined, their potential contribution is vast.

(Roco & Bainbridge, 2003, p. 4)

[F]ar from being unnatural, the drive to alter and improve on ourselves is a fundamental part of who we humans are. As a species we've always looked for ways to be faster, stronger, and smarter and to live longer...

In the end, this search for ways to enhance ourselves is a natural part of being human...It's wired deep in our genes – a natural outgrowth of our human intelligence, curiosity, and drive. To turn our backs on this power would be to turn our backs on our true nature. Embracing our quest to understand and improve on ourselves doesn't call into question our humanity – it reaffirms it.

(Naam, 2005, pp. 9–10)

Although impressive in many respects, the brain suffers from severe limitations. We use its massive parallelism (100 trillion interneuronal connections operating simultaneously) to quickly recognize subtle patterns. But our thinking is extremely slow:

the basic neural transactions are several million times slower than contemporary electronic circuits. That makes our physiological bandwidth for processing new information extremely limited compared to the exponential growth of the overall human knowledge base.

Our version 1.0 biological bodies are likewise frail and subject to a myriad of failure modes, not to mention the cumbersome maintenance rituals they require. While human intelligence is sometimes capable of soaring in its creativity and expressiveness, much human thought is derivative, petty, and circumscribed...

Humans are already replacing parts of their bodies and brains with nonbiological replacements that work better at performing their "human" functions...To me, the essence of being human is not our limitations – although we do have many – it's our ability to reach beyond our limitations. We didn't stay on the ground. We didn't even stay on the planet. And we are already not settling for the limitations of our biology.

(Kurzweil, 2005, pp. 8–9)

The diagnosis of the human condition offered by advocates of the transcendence view is that we are restrained by our innate biological limits. Their prescription is to overcome these limits by means of technology. Their justification for this prescription is that part of the "essence" or what is "natural" for human beings is striving to bush boundaries and overcome limits, including our own.

Not everyone agrees with this diagnosis and prescription. Advocates of the *reconciliation* view believe that it is our limits that define us and make possible the richness, value, and meaning of human life. On this view, the prescription is not to try to overcome the limits embedded in our evolved human nature, but to recognize their centrality to the goods in human life and to seek accomplishment given them. Representative of this view are Leon Kass, former chair of the President's Council on Bioethics, Francis Fukuyama, and Bill McKibben.

A flourishing human life is not a life lived with an ageless body or untroubled soul, but rather a life lived in rhythmed time, mindful of time's limits, appreciative of each season and filled first of all with those intimate human relations that are ours only because we are born, age, replace ourselves, decline, and die—and know it. It is a life of aspiration, made possible by and born of experienced lack, of the disproportion between the transcendent longings of the soul and the limited capacities of our bodies and minds. It is a life that stretches towards some fulfillment to which our natural human soul has been oriented, and, unless we extirpate the source, will always be oriented. It is a life not of better genes and enhancing chemicals but of love and friendship, song and dance, speech and deed, working and learning, revering and worshipping. The pursuit of an ageless body is finally a distraction and a deformation. The pursuit of an untroubled and self-satisfied soul is deadly to desire. Finitude recognized spurs aspiration. Fine aspiration acted upon *is itself* the core of happiness. Not the agelessness of the body, nor the contentment of the soul, nor even the list of external achievement and accomplishments of life, but the engaged and energetic being-at-work of what nature uniquely gave to us is what we need to treasure and defend.

(Kass, 2003, pp. 27–28)

[T]he most significant threat posed by contemporary biotechnology is the possibility that it will alter human nature and thereby move us into a "posthuman" stage of history. This is important…because human nature exists, is a meaningful concept, and has provided a stable continuity to our experience as a species. It is, conjointly with religion, what defines our most basic values. Human nature shapes and constrains the possible kinds of political regimes, so a technology powerful enough to reshape what we are will have possibly malign consequences for liberal democracy and the nature of politics itself.

(Fukuyama, 2002, p. 7)

[A]ll those grander questions, especially the one concerning "the meaning of conscious existence," can only be usefully answered by people whose bodies eventually start to sag, by people who live and who grieve and who celebrate, by people who mourn and who know that they will someday die. There isn't a *right* answer to them, which we will find if only we summon enough brainpower. There are only the sweet answers worked out over time by real humans in real life.

(McKibben, 2003, pp. 226–227)

Advocates of both transcendence and reconciliation believe that a proper understanding of human "nature" (or "soul" or "essence") supports their diagnosis of the human condition and the prescriptions that they make on that basis, both in general and regarding radical human enhancement technologies in particular. They are mistaken.

For one thing, human history and culture are replete with *both* transcending limits and adhering to limits. Both define the texture of human experience. Both are characteristic of human beings. In science and technology are expressed the human impulse to reject existing limits on our knowledge and capabilities and to push the boundaries of what we can accomplish both individually and collectively. This impulse is ubiquitous, reaching into every area of human experience that science and technology touches, including our bodies and minds. The capacity for cumulative innovation over generations, which is enabled by our cognitive, imaginative, psychological, and social capacities, is among our most striking and distinctive characteristics. We are the *technological animal*. In ethics, religion, custom, and political institutions are expressed the human impulse to set and respect limits on what ends we seek, how we pursue acceptable ends, and how we interact with other people and non-humans. This impulse is also ubiquitous, reaching into the personal, social, political, and environmental aspects of human experience. The capacity to determine, identify and adhere to limits is also enabled by our cognitive, imaginative, psychological and social capacities, and it is also among our most striking and distinctive characteristics. We are the *moral animal*. These two aspects of human nature and the human condition *taken together* are what constitute our being the *cultural animal*. Human nature favors neither the transcendence nor the reconciliation view, since respecting and surpassing limits are equally in us, and are equally crucial to our form of life.

Moreover, what is remarkable about these new technologies, what makes them both technologically and ethically radical, is that they may provide the ability to alter our life-form and thereby our form of life. If we implement them, we will not merely be extending out our existing nature, expanding on it, or realizing it more fully. We will be changing it. So in making decisions about whether to pursue these technologies we must consider both what our form of life is now and what our form of life could be. *We must decide what we want our form of life to be.* This is the ethically complicating dimension that these technologies introduce. It is also what undermines appeals to the "authority" of our existing nature. In this context, to claim we should not alter human nature because it is against our nature to do so begs the question. Whether we should privilege and protect the "authority" of our existing nature is part of what is at issue.

Furthermore, human nature is not prescriptive. It consists of biological, psychological, and cognitive structures, dispositions, tendencies, and possibilities. Social and ethical norms must be sensitive to and informed by these. We need an ethic for beings like us in a world like ours. But our tendencies, dispositions and possibilities do not determine which ones we ought to encourage or attempt to realize, or the means by which we ought to do so. We have both transcendent and reconciliation tendencies, within individuals and distributed throughout the species, and we have both transcendent and reconciliation possibilities before us. Appeals to our existing human nature will not adjudicate here. We can go either way while being "true" to human nature. Do we want our existing limits to continue to define us? Do we want to push existing boundaries? Which ones and in what ways? What human nature consists of now does not answer those questions. It is prescriptively inert.

Beyond Human Nature: Toward a Framework for Assessing Radical Human Enhancements

Human nature cannot be used to guide decision-making about radical human enhancement. Rather than being distracted by questions about what is or is not consistent with human nature, we should focus on assessing the desirability of particular forms of enhancement: What goods are gained if the enhancement is realized? What goods are lost or altered if the enhancement is realized? What are the risks and costs associated with pursuing and realizing the enhancement? What are the risks and costs of not pursuing the enhancement? Once the prescriptive ruminations about human nature are cleared away, it becomes clear that these are the issues of primary significance and that each of the advocates of transcendence and reconciliation presented above has views on them. Advocates of transcendence generally believe that there are great goods to be gained with radical human enhancement, there will not be significant losses of existing goods, and the associated risks can be sufficiently managed. Advocates of reconciliation generally believe that there are not substantial goods to be gained through radical human enhancement, there will be significant losses of existing goods, and the risks involved are substantial. These competing evaluations, when made regarding a particular form of enhancement, can be assessed without falling back on assertions about the "soul," "essence," "dignity," or "nature" of human beings. We commonly identify, assess, and make decisions about risks under conditions of uncertainty, and reflecting on the constituents of human flourishing can help identify what human goods may be enhanced, supplemented, undermined, altered or lost, either in the pursuit of some enhancement or in its realization.

Take, for example, Kurzweil's vision for the human future,

> The Singularity will allow us to transcend these limitations of our biological bodies and brains. We will gain power over our fates. Our mortality will be in our own hands. We will be able to live as long as we want (a subtly different statement from saying we will live forever). We will fully understand human thinking and will vastly extend and expand its reach. By the end of this century, the non-biological portion of our intelligence will be trillions of trillions of times more powerful than unaided human intelligence...
>
> The Singularity will represent the culmination of the merger of our biological thinking and existing with our technology, resulting in a world that is still human but transcends our biological roots. There will be no distinction, post-Singularity, between human and machine or between physical and virtual reality.
>
> (Kurzweil, 2005, p. 9)

The questions highlighted above provide a productive framework for critically evaluating Kurzweil's view. It is clear that Kurzweil advocates an alternative form of life to the one we have now. He advocates radical expansion of human knowledge, which he believes will enable a concomitant expansion of pleasure, sociability, and other goods constitutive of human flourishing. Each of these other goods, he claims, supervenes or is constituted by recognition, appreciation, and construction of patterns. As our knowledge increases, we will be able to recognize, experience, appreciate, and construct increasingly more complex and compelling patterns (Kurzweil, 2005, p. 388). Therefore, he believes that the merging of human and machine intelligences will result in a fuller realization of existing human goods than currently can be accomplished. But is it really plausible that the goods currently constitutive of human flourishing would persist if we choose to become networked intelligences? Knowledge and longevity become primary goods, but what of the others? Is Kurzweil's account of them as forms of pattern recognition, construction, and appreciation accurate? Will they have the same prominence and play the same role in this other form of life as they do now? At a minimum, those realizations of them that are dependent upon features of our current form of life will not be retained. We cannot have the same family relationships that we have in this life, framed as they are by lifespan, dependencies, developmental trajectories, and forms of intimacy. This other form of life may have "family relationships", and they may be pleasurable, rewarding, and meaningful, but they will not be the familiar relationships, the known goods, which we have right now.

Furthermore, our psychological and cognitive finitude largely defines our individual subjectivities and perspectives. It is not obvious what it would be like to be a merged, networked intelligence. What will challenge, authenticity, meaningfulness, and even personal identity amount to in this form of life? At best we can know what it would be like for us to imagine what it would be like. But our limitations when it comes to taking the perspective of other life-forms, or even other humans, suggests this will not be particularly informative. We cannot have clear ideas of what achieving this "enhancement" would be like, so we can make only speculative judgments about the qualities of that form of life. Do we trade known realizations of the known goods of our known form of life for unknown goods of an unknown form of life? Kurzweil's radical enhancement, when considered from the perspective of human flourishing, involves radical risks to many of the constituents of human flourishing.

This is not to pass final judgment on Kurzweil's proposed enhancements. The questions and concerns raised above are preliminary and incomplete. They do not, for example, consider the implication for environmental values or distributive justice. Moreover, even if pursuing the singularity is imprudent, selfish, and hubris, as it has been characterized, it might not be ethically impermissible; and even if it is ethically impermissible, it might not be something that should be legally prohibited. There is thus a long way to go from these reflections to a prescription for social or political response. However, they do illustrate that a focus on the potential risks, losses, and gains of a form of radical enhancement, informed by a comprehensive account of human flourishing, provides a productive framework for assessing the enhancement. It is an effective approach to identifying and clarifying considerations salient to whether the enhancement's alteration to our current form of life is in fact desirable, something that can be endorsed upon informed reflection and discourse. This, not adhering to prohibitions or imperatives embedded in human nature, is ultimately what is at issue.

The Inevitable Conclusion

Nanoscale science and technology have been promoted as the platform for the next revolution in science, technology, and industry. The substance and rate of accomplishments in labs around the world increasingly indicate that this is more reality than rhetoric. Human culture may well be on the cusp of a nanotechnological makeover. There is no reason to believe that as nanotechnology, computer science, information technology, robotics, biotechnology, and neuroscience converge this makeover cannot reach our bodies and minds. Indeed, many advocates of transcendence are also advocates of the inevitability of transcendence (Stock, 2003; Kurzweil, 2005; Naam, 2005). They emphasize that radical enhancement technologies are often just therapeutic technologies electively implemented in healthy humans. (This is not the case with all radical human enhancement technologies. Some research programs, such as DARPA's soldier enhancement program and NSF's NBIC program have such technologies as their goal.) We cannot end Alzheimer's without making cognitive enhancement possible. We cannot cure muscular dystrophy without making physical enhancement possible. We cannot cure blindness without making visual enhancement possible. They believe that once these sorts of enhancements are possible we will not be able to indefinitely forestall people from using them to gain advantage for themselves and their children. Thus, expected nano-bio-info-cogno-robo-neuro advancements in therapeutics will lead to a human enhancement race, which will lead to radical human enhancement. From this point of view, social and ethical reflections on human enhancement are likely to have about as much an effect on the trajectory of human enhancement as shouting does on the trajectory of a tornado.

This line of reasoning is seductive. But we must be precise about what is and is not inevitable, as well as about what reflection and discourse on radical human enhancement can be expected to accomplish. While attempts to implement radical human enhancement technologies are likely to occur, the particular forms of enhancement, timeline for enhancements, patterns of implementation, success rates, and societal responses are certainly not settled. These are influenced by public policy, regulation, economics, public opinion, and ethical and religious commitments, among other interrelated factors. Unless technology drives history in an implausibly detailed way, the rate, shape, and extent of human technological enhancement is underdetermined by the technological possibilities and rate of technological progress. There is, therefore, important social, ethical, and politic work to do, and we ought to be developing effective frameworks for doing it. Focusing on what is constitutive of human nature is not a useful framework. It is conceptually misguided, leads to intractable disagreements, and distracts from the central issue: which deviations from our form of life are desirable and are the risks involved in pursuing them acceptable. Reflecting on the constituents of human flourishing and how proposed enhancements might alter them must be a prominent component of any framework for ethical evaluation of radical human enhancements.

Acknowledgments This chapter is based on work supported by the National Science Foundation under Grant No. NSE-0425826 and Grant No. SES-0609078. The author thanks Christopher Bosso, Jeffrey Sandler, Gabriel Plotkin, and the participants in his Spring 2006 'Ethics After Darwin' seminar for their helpful comments and discussions on earlier versions of material appearing in this chapter, and John Basl, Emily Volkert and Thomas Lodwick for their research assistance.

References

Aristotle (1999). *Nicomachean Ethics* (T. Irwin, Trans). Indianapolis, IN: Hackett.
Brown, D. (1991). *Human Universals*. New York: McGraw-Hill.
Ehrlich, P. (2000). *Human Natures: Genes, Cultures, and the Human Prospect*. New York: Penguin.
Foot, P. (2001). *Natural Goodness*. Oxford: Oxford University Press.
Fukuyama, F. (2002). *Our Posthuman Future: Consequences of the Biotechnology Revolution*. New York: Picador.
Garreau, J. (2005). *Radical Evolution: The Promise and Peril of Enhancing Our Minds, Our Bodies – And What It Means to be Human*. New York: Doubleday.
Hursthouse, R. (1999). *On Virtue Ethics*. Oxford: Oxford University Press.
Kass, L. R. (2003). Ageless Bodies, Happy Souls. *The New Atlantis*, 1, 27–28.
Kurzweil, R. (2005). *The Singularity Is Near: When Humans Transcend Biology*. New York: Viking.
McKibben, B. (2003). *Enough: Staying Human in an Engineered Age*. New York: Times Books.
Naam, R. (2005). *More Than Human: Embracing the Promise of Biological Enhancement*. New York: Broadway Books.
Pinker, S. (2002). *The Blank Slate: The Modern Denial of Human Nature*. New York: Penguin.
Rachels, J. (1990). *Created from Animals: The Moral Implications of Darwinism*. Oxford: Oxford University Press.
Richerson, P. J., & Boyd, R. (2005). *Not by Genes Alone: How Culture Transformed Human Evolution*. Chicago, IL: University of Chicago Press.
Roco, M. (2004). Integrating Science and Technology. In M. Roco & C. Montemagno (eds.), *The Coevolution of Human Potential and Converging Technologies. Annals of the New York Academy of Sciences*, 1013, 1–16.
Roco, M., & Bainbridge, W. (2003). Overview. In M. Roco & W. Bainbridge (eds.), *Converging Technologies for Improving Human Performance: Nanotechnology, Biotechnology, Information Technology and Cognitive Science (NBIC)* (pp. 1–27). Boston, MA: Kluwer.
Roco, M., & Montemagno, C. (2004). Preface. In M. Roco & C. Montemagno (eds.), *The Coevolution of Human Potential and Converging Technologies. Annals of the New York Academy of Sciences*, 1013, vii–viii.
Sandler, R. (2005). What Makes a Character Trait a Virtue? *Journal of Value Inquiry*, 39(3–4), 383–397.
Sandler, R. (2007a). *Character and Environment: A Virtue-Oriented Approach to Environmental Ethics*. New York: Columbia University Press.
Sandler, R. (2007b). Nanokchnology and Social Context. *Bulletin of Science, Technology, and Society*, 27(6), 446–454.
Sandler, R. (2008). Nanotechnology and Social Context. *Bulletin of Science, Technology, and Society* (in press).
Sandler, R. (in press). *Nanotechnology and the Other Ethical Issues*. Washington, DC: Woodrow Wilson International Center for Scholars, Project on Emerging Nanotechnologies.
Sartre, J. P. (1948). *Existentialism and Humanism* (P. Mairet, Trans.). London: Methuen and Co. In A. Frazier (ed.) (1975), *Issues in Religion* (2nd ed.) (388–394). Belmont, CA: Wadsworth.
Stock, G. (2003). *Redesigning Humans: Our Inevitable Genetic Future*. Boston, MA: Houghton Mifflin.
Sullivan, R. (2005, July 1). Gates Says Technology Will One Day Allow Computer Implants—but Hardwiring's Not for Him. *Associated Press*.
Sylvan, R. (1973). Is There a Need for a New, an Environmental, Ethic? *Proceedings from the XVth World Congress of Philosophy*. Varna, Bulgaria: Sofia.

Author Index

[*n* denotes the note numbers in text]

A

Ach, J. S., 20*n*34, 71, 72, 76, 77
Agar, N., 80
Akula, S. M., 117–127
Alam, M., 208
Aldred, M., 150
Al-Fandi, M., 120
Alivisatos, A. P., 119
Altmann, J., 147
Anders, G., 44*n*5, 49
Andrews, L. B., 226*n*34, 227*n*37
Arendt, H., 49
Aristotle, 20*n*35, 21*n*38, 22, 239, 247
Arnall, A. H., 70, 76
Aronson, J., 233*n*54
Ashammakhi, N., 122

B

Baba, Y., 120, 121
Bacon, F., 17*n*17
Bailey, H. J., 195
Bainbridge, W., 210, 211, 240, 246–248
Bainbridge, W. S., 4, 15*n*8, 20, 43, 69, 72, 80, 138, 211
Baird, D., 70*n*1, 125
Ball, P., 35, 36, 89
Barkely, R., 230*n*45
Barron, A., 176, 177
Bath, J., 19*n*30
Bauer, M., 135, 143*n*29, 148
Bauman, Z., 145
Baum, R., 165
Baumgartner, C., 7, 67–82
Bayertz, K., 205
Beck, U., 168
Bender, W., 49*n*21
Benford, R., 133, 135, 136, 142
Bensaude-Vincent, B., 5, 6, 19*n*30, 27–40
Berube, D., 73
Birch, K., 142, 144
Bloch, E., 49, 50*n*26
Blumenberg, H., 53*n*35
Boddington, P., 148, 150, 152
Böhme, G., 15*n*12, 73
Bostrom, N., 229*n*43, 233
Bourne, M., 182
Bowring, F., 147, 149
Boyd, R., 242, 243
Brown, D., 242
Brown, N., 46*n*9, 150, 152
Bruggen, B., 123
Bucciarelli, L. L., 105
Buchanan, A., 75
Bueno, O., 33

C

Callon, M., 168
Cameron, N., 221*n*4
Cao, Y., 121
Chen, C. J., 13*n*1
Chesters, G., 139*n*18
Chiaravalloti, F., 30*n*2
Chu, W. L., 223*n*14
Clark, A., 205, 213
Coffrin, T., 86
College, E., 226*n*32
Collins, H. M., 148*n*41, 171
Colvin, V., 86, 160, 167, 175
Cote, G. L., 119

Court, E., 77
Crabbe, J. C., 226n33
Croquette, V., 119

D

Davis, K., 208
Dawkins, R., 34
Decker, M., 92, 100
Descartes, R., 17n18, 19n30
DeVille, K. A., 8, 181–199
Diani, M., 139n18
Di Carlo, A., 224n19
Doherty, B., 139n18, 148
Doolittle, E., 114
Douthwaite, G., 194, 195
Drexler, K. E., 5, 6, 19n30, 20, 27, 28, 32–38, 40, 58, 69, 73, 77, 78, 90, 112, 165
Dupuy, J.-P., 6, 23n46–n47, 31, 32, 43, 44n5, 46n12–n13, 47n15, 48n16, 48n19, 49, 51n27, 52, 54n37, 57, 59n55, 72, 79n8, 165
Dyson, O. F., 117–140

E

Earley, J. E., 15n9
Eaton, A. T., 193
Edlinger, K., 15n11
Edwards, S. A., 19n31
Ehrlich, P., 242, 243
Eisenberg, R. S., 76n5
Elliott, C., 205, 208
Emerich, D. F., 121, 122
Emmerling, R. J., 233n52
Engelhardt, H. T. Jr., 184
Etzkowitz, H., 168
Evans, R. J., 135, 140–141, 148, 151

F

Fan, Z., 118
Farhi, P., 223n13
Featherstone, D. J., 125
Ferguson, J., 159
Ferrari, M., 120, 121
Feyerabend, P., 54n38
Feynman, R., 3, 58n52, 118
Fielder, F. A., 184, 185
Fischer, F., 136, 143n29, 148
Fischer, M., 159
FitzPatrick, W., 105
Flattum, J., 183
Fleck, L., 171

Fogelberg, H., 133–138, 142, 151, 152
Fohler, S., 14n4, 22n40
Foot, P., 243
Friedrich, H. E., 118
Fukuyama, F., 149, 248, 249

G

Galison, P., 55n41, 56n45–n46, 57–59
Gamm, G., 43, 55, 57–59
Gannon, F., 89
Gardner, H., 208
Garreau, J., 204, 221n5, 223n12, 246
Gazzaniga, M., 208
Gee, D., 86, 89, 99
Gibbons, P., 168
Giddens, A., 145
Gilette, S. L., 123
Gjerris, M., 44n4
Glasgow, J. N., 119
Glasner, P., 136
Glimell, H., 133–138, 142, 151, 152
Gloy, K., 15n7
Goleman, E., 233n51–n52
Goose, C., 119
Gordijn, B., 123
Grady, M. F., 188
Granovetter, M. S., 139n18–n19
Greenberg, M., 86, 89, 99
Grill, L., 30n1
Grimes, D. A., 183
Grinbaum, A., 31, 32
Grossman, T., 44n5
Grove-White, R., 135, 148
Grunwald, A., 6, 7, 45n6–n8, 46, 47n14, 51, 52n30, 54n37, 72, 85–100
Guchet, X., 5, 6, 27–40
Gupta, A., 159
Gutmann, W. F., 15n11
Guzzoni, U., 21n37

H

Habermas, J., 75, 145, 149
Hansen, M., 225n26
Hanson, V., 52n31, 60n58
Haraway, D., 14n2
Hardin, G., 76n5
Hardman, R., 118, 124
Harremoes, P., 88, 92
Hartmann, U., 13n1
Haum, R., 95, 96, 97n7, 98, 99
Hegel, G. W. F., 17n19–n20

Heidegger, M., 14
Heller, M. A., 76n5
Hetzel, A., 43
Hintermair, M., 235n58
Hla, S.-W., 19n30
Hoberman, J., 207
Holmes, D., 171, 172
Hood, E., 118, 222n8
Horlick-Jones, T., 148
Hösle, V., 17n16
Hubig, C., 43, 53n35
Hübner, K., 18n26
Hughes, J., 205
Hume, D., 107, 108n6
Hunt, G., 124
Hursh, R. D., 195
Hursthouse, R., 244

I

Igarashi, R., 121
Iglehart, J. K., 183
Inspector, Y., 224n18
Iqbal, S. S., 122
Irwin, A., 136, 148

J

Jacobson, P. D., 182, 188, 190
Jain, K. K., 212
James, W., 104n2
Jasanoff, S., 169
Jömann, N., 20n34, 71, 72, 76, 77
Jonas, H., 23n45, 49n21, 60n57, 92, 94, 95
Jones, M., 144
Jones, R. A. L., 19, 21n39, 35–38
Jotterand, F., 3–10, 23n48, 184
Joy, B., 77, 88, 90
Julliard, Y., 72

K

Kaba, E., 224n17
Kacmar, D. E., 183
Kahn, J., 185
Kaminski, A., 52n33, 54n37
Kanfer, F., 231n46
Kanitscheider, B., 21n39
Kant, 49n24
Kasili, P. M., 119
Kass, L., 74, 75, 248
Kass, L. R., 249
Kataoka, K., 120
Kather, R., 15n6

Kelty, C., 8, 157–178
Kerr, A., 148–150
Keulartz, J., 104n3
Khushf, G., 6, 9, 19n32, 49n25, 50, 51n27–n28, 52, 54n37, 60, 165, 203–216
Köchy, K., 15n12
Koselleck, R., 53n36, 60
Kramer, 226n31
Krug, H., 86n3
Kuekes, P. J., 69
Kurzweil, R., 20, 44n5, 205, 223n9, 246–248, 251–253
Kyllonen, L., 224n21

L

Ladikas, M., 100
Lafontaine, C., 31
Lagnese, J., 195
Lambert, P. J., 117–127
Lam, L., 121
Langer, R., 120
Langhinrichsen-Rohling, J., 228n42
Lanza, G. M., 121
Lash, S., 168
Latour, B., 14n3, 22n41–n42, 55n42, 56n43, 56n45–n46, 57–59, 168, 171
Law, J., 168
Leary, S. P., 212
LeBoeuf, R., 232n48
Leedham, B., 224n16
Lee, J. H., 119
Lehn, J. M., 21n39
Leydesdorff, L., 168
Li, J., 119, 121
Lin, P., 125
Little, M., 208
Litton, P., 76, 78
Llinas, R. R., 212
Lösch, A., 51n29, 53n34
Löw, R., 15n12
Lu, J. G., 118

M

MacDonald, C., 86
Mae-Wan Ho, 147n37
Mainzer, K., 21n39
Mannheim, K., 171
Marcus, G., 159, 171–172
Marshall, K., 53n34, 61n59
Masci, D., 181
Mason, D. S., 122
Mastroianni, A. C., 188, 190, 192

Mathus-Vliegen, E. M., 225n29
Matsudai, M., 124
Mayer, S., 142–143, 145–148
Mayne, A. J., 30n3
Mazzola, L., 119
McAdam, D., 138
McKibben, B., 249
McNeil, S. E., 117, 120
Meaney, M., 61n59
Meilaender, G., 227n38
Meininger, G. A., 119
Melucci, A., 135, 140, 151
Mendoza, J., 207
Merkle, R. C., 183
Merleau-Ponty, M., 28
Merton, R., 54n38
Michael, M., 136, 148, 150, 152
Miller, F., 209
Miller, J., 185
Mitchell, M. E., 9, 219–236
Mnyusiwalla, A., 70, 80, 119, 123
Moghimi, S. M., 119
Moltmann, J., 16n14
Montemagno, C., 121, 246
Moor, J., 71–72, 74n4, 77
Moore, G. E., 105
Moriarty, Ph., 19n30
Mukhopadhyay, R., 119
Müller, A., 15n9, 21n39
Murosaki, T., 122

N
Nicolelis, M., 212–213
Nietzsche, F., 17n21
Nordmann, A., 6, 14n5, 15n7, 15n10, 19n32, 23n48, 43–61, 69, 134, 137

O
Orland, B., 54n37

P
Peirce, C. S., 54, 104, 110, 111
Pitt, J. C., 7, 31, 103–115
Plows, A., 4, 8, 133–152
Popper, K., 54
Poser, H., 16n13

R
Rawls, J., 77n6
Reichle, I., 43

Reinsborough, M., 4, 8, 133–152
Rip, A., 45n8, 46n10, 72n3
Robinson, W. L., 72n2
Roco, M. C., 4, 246

S
Sandler, R., 239–253
Santayana, G., 3
Schummer, J., 17n22, 18n23–n24, 20n33, 20n35, 21n38, 23n44, 70n1, 134
Selin, C., 46n9, 46n11
Sellars, W., 106
Simondon, G., 37, 38, 40
Swierstra, T., 72n3

T
Tarrow, S., 138–139, 151
Thrall, J. H., 121, 123
Till, M. C., 186–187
Toensmeir, P. A., 124
Tolles, W. M., 122
Toumey, C., 165
Trache, A., 119
Traweek, S., 171
Treder, M., 90, 197
Turberfield, A. J., 19n30
Turkel, S., 232n49, 235n59–n60

U
Uesaka, M., 119
Urry, J., 168

V
Vasir, J. K., 121
Venstrom, J. M., 224n22
Vetter, R. J., 122
Vickers, M., 205
Vogt, T., 125
Von Neumann, J., 31
Vonnegut, K., 109
Von Schomberg, R., 93, 94

W
Wabuyele, M. B., 119
Wallace, P. M., 232n50
Wall, D., 139–140
Wardak, A., 91, 96, 98
Warren, C., 23n48

Weber, J., 43
Weber, M., 38, 54, 173
Webster, T. J., 122
Weckert, J., 70n1, 71, 72, 74n4, 77
Weil, V., 75
Weinberg, R. A., 121
Welsh, I., 135–136, 139n18, 143, 148, 151
Whitebread, C., 225n30
Whitehead, R., 15n9
Whitesides, G., 35
Whitesides, G. M., 35–36, 54n37
Whitman, A. G., 7, 117–127
Whittier, N., 139n18
Wickline, S. A., 121
Wieland, W., 21n36
Wiesner, M., 167, 175, 177
Williams, D., 123–124
Williams, R. S., 69
Willis, R., 151
Wilsdon, J., 151
Wittgenstein, 54n37
Wolbring, G., 150
Wong, A. H., 226n35
Wood, S., 86, 135, 137n8, 137n10, 138, 141, 142, 144n31, 146–147
Woolgar, S., 171
Woyke, A., 5, 6, 13–23
Wullweber, S., 133, 135
Wynne, B., 136, 143n29, 148

X
Xi, J., 119

Y
Yamaguchi, Y., 121
Yamato, M., 122
Yih, T. C., 120
Yoshikawa, T., 120
Yuan, G., 123

Z
Zhang, Y., 118, 119
Zoloth, L., 79n9, 80

Subject Index

[*n* denotes the note numbers in text]

A
Allotropes, 157–160, 178
Alzheimer, 208, 226, 253
The American College of Surgeons, 192
Anthropology, 4, 5, 8, 23*n*47, 79, 157–160, 169, 171–174, 178
Anthropotechnology, 80
Asbestos, 86, 88, 89, 98*n*8, 99
Assembler, 5, 6, 28, 33–36, 40
Assisted reproduction technologies (ART), 225

B
Biochips, 69, 71
Bio-ethics, 82
(Bio) nanotechnology, 3, 4, 6–9
Biosensors, 119, 122, 182, 223
Biotechnology, 7, 32, 33, 40, 55, 57, 68, 76, 78, 81, 88*n*5, 118, 134, 138*n*13, 140, 151, 211, 246, 249, 253
Bottom-up molecular manufacturing, 165
Brain, 9, 44, 69, 73, 81, 121, 203, 211–214, 219, 223, 229–231, 240, 246–248
Brain-to-machine-networks, 81
Brownian motion, 6, 35–38
Buckminsterfullerenes, 160
Buckyballs, 159, 169

C
California, 114, 223
Carbon nanotubes, 56, 58, 89, 118, 160
Cartesian automata, 5, 28
Cartesian automaton, 31
Cartesian model, 27, 31
Catastrophisme éclairé, 46
Catenanes, 29
CEMES, 30
Center for Biological and Environmental Nanotechnology (CBEN), 157, 158, 160, 161, 165–178
CEST, 70, 75
Chemical abstract service (CAS), 99
Chips, 33, 57, 58, 120, 122, 184, 205
Christian middle ages, 17
Classical, 5–7, 15, 16*n*13, 19, 27–30, 32, 54, 57, 76, 89, 91–95, 99, 105, 159, 212
Classical risk management, 89, 92–95
Co-evolution, 49, 247
Cold War, 14
Common sense pragmatism (CSP), 109–111
Complex machines, 5, 6, 28, 32, 33, 35, 37
Conditio humana, 79
Conquest of space, 43, 44, 53*n*34, 54, 57
Converging technologies, 8, 23*n*46, 51, 133, 134*n*4, 136, 142, 211, 222, 246, 248
Converging technologies for improving human performance. Nanotechnology, Biotechnology, Information Technology and Cognitive Science, 211
Corrective medicine, 121
Cosmetic surgery, 203, 204, 207–208
Cyborg age, 9
Cyborgs, 60, 88, 204

D
DARPA, 223, 246, 253
Darwinism, 241, 242
Delocalization, 55, 56, 58–60

261

Deus absolutus, 17
Device, 3, 8, 10, 34, 35, 39, 68, 69, 71, 72n2, 76, 77, 80–82, 112, 115, 117, 119, 130, 161, 181–193, 195–199, 212, 219, 221, 225, 227–231, 235
Diagnostics, 69, 72, 77, 120, 223
Direct action network, 141
Drug, 3, 8, 10, 38, 69, 71, 117, 120–122, 138, 140, 147n39, 150, 152, 182–199, 206–210, 215, 220, 223–226, 228, 229, 231
Drug delivery, 3, 38, 69, 120, 121, 138, 140, 152, 183, 196
Drug-targeting systems, 69
Duke University Medical Center, 212
Dutch Health Council, 70, 71, 74

E
Earth first, 139, 141
Earth summit, 92
EGE, 69–71, 75
Engines of creation, 27, 28, 32, 33
Enhancement, 158, 203–211, 213–216, 223, 225–228n40, 229, 235, 239–241, 247, 249–253
Environmental-ethics, 82
ETC-Group, 57n49, 70, 75
European Commission, 41, 61, 155
European Group on Ethics and New Technologies, 70
European Union, 14, 92, 93
Evolution, 16n13, 36, 49, 204, 211, 222, 242, 246, 247

F
Fat fingers, 113
Featherstone, 125
Food and drug administration (FDA), 8, 182, 184–189, 191, 193, 194, 197–199, 222
The Foresight Institute, 165
France, 30, 38, 56n43, 140
Frei Universität of Berlin, Germany, 30
Friends of the Earth, 91, 97
Fullerene, 118, 130, 160

G
Gene-ethics, 82
Genetically modified organisms (GMO), 3, 4, 43, 93, 98
Genetic engineering, 19, 39, 80, 97, 133, 139, 141, 142, 204, 211, 222, 226, 240

Genetic engineering network, 139, 141
Genetics, Robotics, Artificial Intelligence and Nanotech (GRAIN), 134
Genetic tests, 71
GeneWatch, 135, 139, 141
Germany, 91
The Good life, 7, 50, 75, 103, 104n1, 105, 106, 108–115
Greenpeace, 70, 141
Grenoble, 140
"Grey goo", 28, 32, 77, 78, 88, 90, 167
"Grey goo" problem, 77, 78
Guimbal turbine, 38

H
The Heavenly Righteous Opposed to Nanotech Greed (THRONG), 139, 140n20, 141, 145
Heuristics of fear, 49n21, 92
Human-cyborg evolution, 247
Human enhancement, 4, 9, 10, 44, 49, 60n56, 112, 158, 203, 239–241, 247, 249–251, 253
Human flourishing, 5, 10, 209, 219, 239–241, 243–247, 253
Human identity, 149
Human nature, 3, 7, 10, 60, 75, 77, 143, 149, 226–228, 239–244, 247–253
Human-technology, 28, 40, 125, 247

I
Implanted biosensors, 122
Individualized medicine, 69, 71, 72
Industrial Revolution, 69, 71, 72
International Council on Nanotechnology (ICON), 167, 168, 170, 172, 173
In vitro fertilization, 22, 225
Israel, 227

J
Justice, 15, 74, 76, 77, 79, 81, 82, 136, 138, 141, 143, 144, 149, 150, 229n44, 240, 252

L
Laboratoire de Photophysique Moléculaire, 30
Law of Unintended Consequences, 108
Leviathan and the Air Pump, 168

M

Machines, 5, 6, 9, 19n30, 27–29, 31–38, 40, 41, 60, 69, 74, 81, 112, 137n10, 212–214, 235, 246, 247, 251
Medical malpractice, 9, 181, 193, 195–197
Molecular assemblers, 5, 6, 73, 77, 78
Molecular machines, 27–29, 33–36, 40, 81
Molecular nanotechnology, 90

N

Nanates, 118, 120, 121, 124, 125
Nanites, 118, 121–125
Nano, Bio, Info, Cogno (NBIC), 32, 49, 51, 57n49, 68–75, 77–82, 134, 149n40, 205, 210, 212–214, 246, 253
Nanobiosensor diagnostic, 223
Nanobiotechnology, 5, 38–40, 68, 76, 79, 133, 134n2, 135–137, 140, 142–152, 164
Nanobot, 3, 6, 28, 88, 113, 119, 130, 138, 184
Nanodevices, 68, 182, 183, 196–198
Nano divide, 76, 77
Nanoethics, 7, 44n3–4, 70n1, 78, 240
Nanoforum, 86, 97n7
Nano-hype, 87
Nanomachines, 6, 19n30, 33–40, 69, 112, 118, 130, 223n15
Nanomedicine, 71, 77, 80, 182, 186, 188, 193, 205, 212, 245
Nanoremedy, 188, 196–198
Nanoscience, 3, 14, 19, 20, 22, 30, 33, 38, 39, 45n7, 145, 166, 168, 178, 221, 224, 233, 245
Nanoscience Group of the Laboratoire d'Architecture et d'Analyse des Systèmes (LAAS, France), 38–40
Nanoscience Group of the LLPM, 30
Nanosensors, 118, 183
Nanosphere-drug delivery systems, 183
Nanosurgery, 121
Nanotechnology, 4–10, 13–16, 18, 19n30, 20n34, 21, 22, 23n46, 27, 28, 32, 33, 35–37, 40, 43n1, 44n5, 45n7, 46n9,11, 47, 48n18, 49, 50, 51n27, 52n31–32, 53n34, 54n37
National nanotechnology initiative, 73, 118, 123, 160
National Science and Technology Council (NSTC), 14n4, 15, 19, 246
National Science Foundation (NSF), 4, 160, 210, 222n6, 246
NBIC Convergence, 49, 51, 57n49, 68, 69, 71–73, 78–81, 205, 210–214, 246

Neal, 160
Network, 31, 56n45, 57, 58, 81, 120, 125, 135–138, 139n18, 140n21, 141n23, 142, 144, 151, 152, 161, 163–165, 168, 251, 252
Neural/digital interface, 212
New Chartists, 135, 139, 141, 146
New Luddites, 135, 139–141
Newt Gingrich, 211
New York University School of Medicine, 212
Non-therapeutic enhancement, 74, 80
Nuclear technology, 88, 97

O

The Office of Drug Safety, 188
Office of the Inspector General (OIG), 188
Oppenheimer, 123
Orsay, 30

P

Para-ethnography, 171
Parkinson's diseases, 226
The patient, 79, 122, 124, 138, 189–191, 194–198, 234
Pentagon's Defense Advanced Research Projects Agency, 223
Pharmaceutical enhancement, 210
Pharmacogenomics, 69, 78, 81, 134, 152
Photodynamic therapy, 121
Physician, 9, 69, 71, 72, 79, 120, 124, 126, 183, 189, 190, 193–199, 208, 228, 233, 234
Physician-patient relationship, 195
Post-human evolution, 247
Pragmatism, 7, 103–105, 109–112, 114
Precautionary principle, 7, 51, 85–87, 89, 91–100, 146, 147
Prey, 167
Proteomics, 134, 145
Prozac, 226

Q

Quantum dots, 118–120, 130, 169
Quantum mechanics, 15, 19
Quantum physics, 19, 30, 57, 221

R

Radical human enhancement, 249–251, 253
Regenerative medicine, 121, 122

Registry studies, 191, 192, 199
Rice University, 157, 158, 160
Richard Dawkins, 34
Rio de Janeiro, 92
Rotaxanes, 29
The Royal Academy of Engineering, 78, 90
Royal Society, 70, 74, 78, 90, 95, 97n7, 99, 134n1, 137, 139n17, 141, 145, 146n34, 147, 152, 168

S
Schmocters, 227
Self-aware evolution, 204
Shaping the World Atom by Atom, 14, 15
Single Walled Carbon Nanotubes, 160
Singularity, 48, 251, 252
Smart drugs, 208
Soft machines, 6, 27, 36–38, 40
Sports doping, 203, 204, 206–210
Stage 1 enhancements, 203, 204
Stage 2 enhancements, 203–205, 215, 216
The Standard of care and, 9, 186, 195–197
Sticky fingers, 113

T
Technological imperative, 183
Tort Litigation, 9, 190, 193–198
Toxic Substances Control Act (TSCA), 91, 96, 98
Transhuman evolution, 247
Transhumanists, 20, 229
Transplantation, 122, 224

U
UK Earth First!, 141
UNESCO, 70
United Nations Conference on Environment and Development (UNCED), 92
United States, 4, 8, 14, 73, 81, 90, 91, 114, 118, 246
University of Reading, 213
U.S. Department of Commerce (DOC), 210, 215, 222
USSR's, 14

V
Virtual world, 232, 240
Von Neumann's model, 27

Printed in the United States
125274LV00002B/19-75/P